MAINTENANCE, MODELING AND OPTIMIZATION

MAINTENANCE, MODELING AND OPTIMIZATION

Edited by

MOHAMED BEN-DAYA
SALIH O. DUFFUAA
ABDUL RAOUF
Department of Systems Engineering
College of Computer Sciences & Engineering
King Fahd University of Petroleum & Minerals

Kluwer Academic Publishers
Boston/Dordrecht/London

Distributors for North, Central and South America:
Kluwer Academic Publishers
101 Philip Drive
Assinippi Park
Norwell, Massachusetts 02061 USA
Telephone (781) 871-6600
Fax (781) 871-6528
E-Mail <kluwer@wkap.com>

Distributors for all other countries:
Kluwer Academic Publishers Group
Distribution Centre
Post Office Box 322
3300 AH Dordrecht, THE NETHERLANDS
Telephone 31 78 6392 392
Fax 31 78 6546 474
E-Mail <services@wkap.nl>

 Electronic Services <http://www.wkap.nl>

Library of Congress Cataloging-in-Publication Data

Maintenance, modeling, and optimization / edited by Mohamed Ben-Daya, Salih O. Duffuaa, Abdul Raouf.
      p. cm.
   Includes bibliographical references and index.
   ISBN 0-7923-7928-4 (alk. paper)
    1. Plant maintenance. 2. Factory management. I. Ben-Daya, M. (Mohamed) II. Duffuaa, S. (Salih) III. Raouf, A. (Abdul), 1929-

TS192 .M345 2000
658.2'02--dc21   00-062493

Contents

Part I INTRODUCTION

1
Maintenance Modeling Areas   3
M. Ben-Daya and S. O. Duffuaa

Part II MAINTENANACE PLANNING & SCHEDULING

2
Mathematical Models in Maintenance Planning and Scheduling   39
Salih O. Duffuaa

Part IV CONDITION BASED MAINTENANCE

13

On-Line Surveillance and Monitoring   309

Hai S. Jeong Elsayed A. Elsayed

14

Maintenance Scheduling Using Monitored Parameter Values   345

Dhananjay Kumar

Part V INTEGRATED MAINTENANCE, PRODUCTION AND QUALITY MODELS

15

A General EMQ Model with Machine Breakdowns   377

V. Makis, X. Jiang and E. Tse

List of Figures

List of Tables

Preface

Production costs have been coming down over time due to automation, robotics, computer integrated manufacturing, cost reduction studies and more. On the other hand, these new technologies are expensive to buy, repair and maintain. So the demand on maintenance is growing and maintenance costs are escalating. This new environment is compelling industrial maintenance organizations to make the transition from being repair departments for fixing broken machines to that of high level business units for securing production capacity.

On the academic front, research in the area of maintenance management and engineering is getting tremendous interest from researchers. Many papers appeared in the literature dealing with the modeling and solution of maintenance problems using operations research (OR) and management science (MS) techniques. This area represents an opportunity for making significant contributions by the OR and MS communities.

This book provides in one volume the latest developments in the area of maintenance modeling. Prominent scholars contributed chapters covering a wide range of topics. We hope that this humble contribution will serve as a useful informative introduction to this field that may permit additional developments and useful directions for more research in this fast growing area.

This volume is divided into six parts and contains seventeen chapters. Each chapter has been subject to review by at least two expert. in the area of maintenance modeling and optimization.

In the first chapter which forms Part I, M. Ben-Daya and S.O. Duffuaa provide an introduction to major maintenance modeling areas illustrated with some basic models. Areas covered include maintenance planning and scheduling, spare parts provisioning, preventive and condition based maintenance, and integrated maintenance, production, and quality models. Directions for future research are outlined.

Part II contains five chapters (Chapters 2 to 6) dealing with maintenance planning and scheduling. In Chapter 2, S.O. Duffuaa presents an overview of maintenance planning and scheduling models. Models presented are mathematical programming based models. I. El-Amin focuses on electric generators maintenance scheduling in Chapter 3. He presented a general model and elaborated on various solution techniques. Chapter 4 by D.N.P. Murthy is about maintenance service contracts. This chapter provides an overview of the literature on maintenance service contracts, the issues involved and the models used to study these issues. In Chapter 5, A. Sheikh, M. Younes, and A. Raouf address issues related to spare parts procurement strategies based on reliability engineering. Simulation meta-modeling of the maintenance float system by C. Madu is the object of Chapter 6.

Part III deals with preventive maintenance in six chapters (Chapters 7-12). Preventive maintenance policies and their variations are discussed in Chapter 7 by T. Dohi, N. Kaio, and S. Osaki. Important preventive maintenance optimization models, which involve age replacement and block replacement, are reviewed in the framework of the well known renewal reward argument. A general framework for analyzing maintenance policies is discussed in Chapter 8 by V. Makis, X. Jiang, A.K.S. Jardine, and K. Cheng. A wide class of maintenance decision problems are analyzed using the optimal stopping framework. The necessary mathematical background from martingale dynamics and the optimal stopping theory are summarized and the approach is illustrated by analyzing a general repair/replacement model. Chapter 9 by T. Nakagawa deals with imperfect preventive maintenance. Three maintenance policies are analyzed. In Chapter 10, a single machine inspection problem is discussed by Hariga and Al-Fawzan. The inspection problem is formulated using discounted cash flow under different assumptions regarding the length an nature of the planning horizon. A general approach for the coordination of maintenance frequencies in cases with a single set-up is presented in Chapter 11 by R. Dekker. An optimal approach for solving the multi-component maintenance model is discussed. The approach is also extended to deal with more general maintenance models like minimal repair and inspection, and block replacement. The last chapter of Part III deals with maintenance grouping in multi-setup multi-component production systems. A general mathematical modeling framework is presented for this class of problems.

Part IV focuses on condition based maintenance and contains two chapters. The first one, Chapter 13, deals with on-line surveillance and monitoring. The second one, Chapter 14, presents an approach for maintenance scheduling using monitored parameter values (MPV). One of the advantages of the discussed approach is that the importance of the MPV in explaining the failure characteristics can be identified without making assumptions about the distribution of the hazard rate. The approach can be used for both repairable and irreparable systems.

Part V deals with integrated production and maintenance models and contains two chapters. Chapter 15 by V. Makis, X. Jiang, and E. Tse considers a general economic manufacturing quantity problem with machine breakdowns. The structure of the optimal policy is obtained using semi-Markov decision processes. In Chapter 16, E.K. Boukas and Q. Zhang deal with production and maintenance control for manufacturing systems using continuous flow models. Various types of models that incorporate production, corrective maintenance, and preventive maintenance are described.

Part VI addresses issues related to maintenance and new technologies and contains one chapter by G. Waeyenbergh, L. Pintelon, and L. Gelders dealing with Just-In-Time (JIT) and Maintenance. The chapter describes the consequences of the JIT philosophy on maintenance.

THE EDITORS

Acknowledgment

The Editors would like first to acknowledge the authors for their valuable contributions. This book would not have been in this form without their participation and cooperation throughout the stages of this project. We are also grateful to all the referees who reviewed the material submitted for this book. The critique and suggestions they made have definitely lead to substantial improvements.

It takes a lot of skill and patience to do all the word processing, text conversion, and type-setting, and our thanks go to Shahid Pervez and Arifusalam for their support. We are also grateful to Mr. John R. Macleod for his assistance with English editing.

We are indebted to Gary Folven, Publishing Editor OR/MS areas at Kluwer for his interest in this book, full cooperation and assistance since the initiation of this project. Carolyn Ford and other Kluwer staff members have been very helpful.

We would like also to express our gratitude to our families for their patience and encouragements. Preparation of this book has been sometimes at the expense of their time.

Finally, the Editors would like to acknowledge King Fahd University of Petroleum and Minerals for excellent facilities and for funding this project under the number SE/MODEL/199.

Contributing Authors

M. Ben-Daya & S. O. Duffuaa
Dept. of Systems Engineering,
King Fahd Univerity of Petroleum
and Minerals, Saudi Arabia

A. K. Sheikh & M. Younas
Dept. of Mechanical Engineering
King Fahd Univerity of Petroleum
and Minerals, Saudi Arabia

**V. Makis, X. Jiang,
E. Tse & A.K.S. Jardine**
Department of Mechanical
and Industrial Engineering,
University of Toronto, Canada

Abdul Raouf
Adjunct Professor
King Fahd Univerity of Petroleum
and Minerals, Saudi Arabia

T. Dohi & S.Osaki
Dept. of Industrial and Systems
Engineering, Hiroshima University.
Japan

Dhananjay Kumar
Nokia Svenska AB,
Sweden

Christian N. Madu
Lubin School of Business,
Pace University, New York, USA.

N. Kaio
Dept. of Economic Informatics,
Hiroshima Shudo University, Japan

D. N. P. Murthy
The University of Queensland
Australia

K. Cheng
Institute of Applied Mathematics,
Academia Sinica, China.

Ibrahim El-Amin
Dept. of Electrical Engineering
King Fahd Univerity of Petroleum
and Minerals, Saudi Arabia

Toshio Nakagawa
Dept. of Industrial Engineering,
Aichi Institute of Technology
Japan

Moncer Hariga
Industrial Engineering program
King Saud University
Saudi Arabia.

Hai S. Jeong
Dept. of Applied Statistics
Seowon University,
Korea

Mohammad A. Al-Fawzan
King Abdulaziz City for
Science and Technology
Saudi Arabia.

E. K. Boukas
Mechanical Engineering Dept,
École Polytechnique de Montréal,
Montréal, Québec, Canada

R. Dekker, R. Wildeman,
J. Frenk, and R. Egmond
Econometric Institute,
Erasmus University Rotterdam,
The Netherlands.

Q. Zhang
Department of Mathematics,
University of Georgia,
Georgia, USA

G. Waeyenbergh, L. Pintelon,
& L. Gelders
Center for Industrial Management
Belgium.

G. van Dijkhuizen
University of Twente
The Netherlands.

I

INTRODUCTION

Chapter 1

OVERVIEW OF MAINTENANCE MODELING AREAS

M. Ben-Daya and S. O. Duffuaa
Systems Engnieering Department
King Fahd University of Petroleum and Minerals
Dhahran-31261, Saudi Arabia.

Abstract   This chapter provides a brief overview of the research areas of maintenance modeling. It is not meant to be a complete review of maintenance models but rather as an informative introduction to important maintenance modeling areas. Areas covered include maintenace planning and scheduling, spare parts provisioning, preventive and condition based maintenance, and integrated maintenance, production, and quality models. Directions for future research are outlined.

Keywords: Preventive maintenance, imperfect preventive maintenance, condition based maintenance, inspection, replacement, maintenance planning and scheduling, spare parts provisioning, integrated maintenance models, delay time model.

1.   INTRODUCTION

Production and manufacturing problems received tremendous interest from the operations research (OR) and management science (MS) researchers. Many books and textbooks have been written and several journals are dedicated to these subject. These topics are part of the curriculum in various industrial, mechanical, manufacturing, or management programs.

In the past, maintenance problems received little attention and research in this area did not have much impact. Today, this is changing because of the increasing importance of the role of maintenance in the new industrial environment. Maintenance, if optimized, can be used as a key factor in organizations efficiency and effectiveness. It also enhances the organization's ability to be competitive and meets its stated objectives.

The research in the area of maintenance management and engineering is on the rise. Over the past few decades, there has been tremendous interest and a great deal of research in the area of maintenance modeling and optimization. Models have been developed for a wide variety of maintenance problems. Although the subject of maintenance modeling is a late developer compared to other area like production systems, the interest in this area is growing at an unprecedented rate.

The purpose of this chapter is to give a brief overview of the research areas of maintenance modeling. It is intended as a first informative introduction to the important areas of maintenance modeling and as a foundation to suggest useful directions for future developments. The area of maintenance modeling and optimization represents an opportunity for making significant contributions by the OR and MS communities. It should be made clear at the outset that this chapter is not meant to be a complete review of maintenance models. It is intended to simply give some ideas about the general areas of maintenance modeling illustrated using basic models.

This chapter is organized as follows. Preventive maintenance related models are discussed in Section 2. These include inspection , replacement, preventive and condition-based maintenance. Issues related to maintenance planning such as maintenance capacity planning , planning and scheduling, and spare parts provisioning are presented in Section 3. Section 4 deals with integrated models of maintenance, production, and quality. Future research directions are outlined in Section 5.

2.   MAINTENANCE POLICIES

Maintenance strategies can be classified into three categories:

1. Breakdown maintenance. Replacement or repair is performed only at the time of failure. This may be the appropriate strategy in some cases, such as when the hazard rate is constant and/or when the failure has no serious cost or safety consequence or it is low on the priority list.

2. Preventive maintenance (PM) where maintenance is performed on a scheduled basis with scheduled intervals often based on manufacturers' recommendations and past experience with the equipment. This may involve replacement or repair , or both.

3. Condition-based maintenance (CBM) where maintenance decisions are based on the current condition of the equipment, thus avoiding unnecessary maintenance and performing maintenance activities only when they are needed to avoid failure. CBM relies on condition monitoring techniques such as oil analysis, vibration analysis, and other diagnostic techniques for making maintenance decisions.

Implementing these strategies involves familiar maintenance activities like inspection , replacement, and repairs. In this section, we discuss representative models and modeling techniques dealing with maintenance strategies including, replacement, preventive maintenance, imperfect preventive maintenance , condition-based maintenance and inspection decisions. Replacement models are discussed first. They are of particular interest since they are among the earliest models developed. They were extended to include repairs and also later to incorporate condition monitoring information.

2.1 Replacement.

In this section, two classical preventive replacement policies, the constant interval (block) replacement and age-based replacement policies [42] are presented.

2.1.1 Constant Interval Replacement Policy.

In this case preventive replacements occur at fixed intervals of time, while failure replacements are carried out when necessary. The problem is to determine the optimal interval between preventive replacements so that the total expected cost per unit time of replacing the equipment is minimized. Clearly, a long interval incurs high failure replacement cost and a small interval incurs high preventive replacement cost. The optimal interval balances these two costs.

The following notation is used to develop the constant interval replacement model. Let the preventive replacement cost and failure replacement cost be denoted by C_p and C_f, respectively. Also, let $f(t)$ and $F(t)$ be the probability density function and cumulative distribution function of the equipment's failure times. As described above, preventive replacements are carried out at constant intervals of length t_p, regardless of the age of the equipment. Failure replacements occur as many times as needed in the interval $(0, t_p)$.

The total expected cost per unit time $ETC(t_p)$ is given by:

$$ETC(t_p) = \frac{\text{total expected cost in the interval } (0, t_p)}{\text{length of interval } (0, t_p)} = \frac{C_p + C_f N(t_p)}{t_p}$$

$$(1.1)$$

The total expected cost in the interval $(0, t_p)$ is the sum of the cost of a preventive replacement, namely C_p and the expected cost of failure replacements. The latter cost is given by $C_f N(t_p)$ where $N(t_p)$ is the expected number of failures in the interval $(0, t_p)$.

2.1.2   Age-Based Replacement Policy.   A drawback of the time-based replacement policy is that there is a possibility of performing a preventive replacment shortly after failure replacement which may lead to waste of resources. In an age-based replacement policy, a preventive replacement of the equipment is carried out once the age of the equipment has reached a specified age t_p. If the equipment fails prior to t_p, a failure replacement is performed and the next preventive replacement is rescheduled t_p time units later. Here also the problem is to find the optimal preventive replacement interval t_p that minimizes the total expected cost per unit time $ETC(t_p)$.

In this case there are two possible scenarios:

1. The equipment reaches its planned replacement age t_p without failing. This happens with probability $R(t_p)$.

2. The equipment fails before the planned replacement time t_p. This happens with probability $F(t_p)$.

The total expected cost in the interval $(0, t_p)$ is the sum of the cost of a preventive replacement and the expected cost of failure replacements and is given by:

$$C_p R(t_p) + C_f F(t_p)$$

The expected cycle length is equal to the length of a preventive cycle (t_p) times the probability of a preventive cycle $(R(t)$ plus the expected length of a failure cycle times the probability of a failure cycle $(F(t))$.

The expected length of a failure cycle, $M(t_p)$, can be determined by finding the mean of a distribution truncated at t_p. $M(t_p)$ is called the truncated mean. It is given by:

$$M(t_p) = \int_{-\infty}^{t_p} \frac{t f(t) dt}{F(t_p)} \tag{1.2}$$

Therefore

$$ETC(t_p) = \frac{C_p R(t_p) + C_f F(t_p)}{t_p R(t_p) + M(t_p) F(t_p)} \tag{1.3}$$

Minimizing this cost function with respect to t_p yields the optimal age-based preventive replacement interval.

The above models assume that a failure is detected instantly and replacements are done instantaneously.

These basic models have been extended in several ways. Many extensions and variations are discussed in Chapter 7 of this book.

Other extensions consider the replacement of a group of components. Cho and Parlar [22] provide a good review of the multi-component literature, including group replacement models of various types. In his published thesis [107], Wildeman provides an extensive review of the literature on multi-component maintenance models with economic dependence.

2.2   Preventive Maintenance.

For a complex system, the maintenance action is not necessarily the replacement of the whole system, but is often the repair or replacement of a part of the system. Hence, the maintenance action may not renew the system completely, i.e. make it as good as new.

In this section we discuss two preventive maintenance policies for a repairable system. For more details see Nguyen and Murthy [75]. These policies are motivated by the following practical considerations:

- Continuing to repair a system is often costly compared with replacing it after a certain number of repairs.

- The replacement (repair) cost of a failed system is usually greater than the replacement (repair) cost of a nonfailed system.

2.2.1   Preventive Maintenance Policy I:.

Replace the system after $(k-1)$ repairs. For a system subject to $(i-1)$ $(i < k)$ repairs, it is repaired at the time of failure or preventively maintained at age T_i (i.e. T_i hours from the last repair or replacement) whichever occurs first.

The following notation is used to develop the mathematical model for PM Policy I.

C_p:	preventive maintenance cost
C_b:	additional cost due to breakdown
C_r:	replacement cost
$f_i(t)$:	probability density function of the failure time of an equipment that has been subjected to $(i-1)$ repairs and PMs
$F_i(t), R_i(t)$:	Cumulative distribution and reliability functions
$h_i(t)$:	hazard function

The expected cost of repair and PM for a system that has been subjected to $(i-1)$ repairs and PMs is

$$
\begin{aligned}
CPM(T_i) &= (C_p + C_b)F_i(T_i) + C_pR_i(T) \\
&= C_p + C_bF_i(T_i)
\end{aligned}
$$

Hence the expected total cost of a replacement cycle is

$$
EC(k, T_1, ..., T_k) = (k-1)C_p + C_r + C_b \sum_{i=1}^{k} F_i(T_i).
$$

The expected length of a replacement cycle is

$$
L(k, T_1, ..., T_k) = \sum_{i=1}^{k} \int_0^{T_i} R_i(T_i)dt.
$$

The total expected cost per unit time is

$$
ETC(k, T_1, ..., T_k) = \frac{(k-1)C_p + C_r + C_b \sum_{i=1}^{k} F_i(T_i)}{\sum_{i=1}^{k} \int_0^{T_i} R_i(T_i)dt} \tag{1.4}
$$

The problem is to find the optimal number of repairs k and ages $T_1, T_2, ...,$ T_k that minimize (1.4).

2.2.2   Preventive Maintenance Policy II:.   Replace the system after $(k-1)$ repairs. For a system subject to $(i-1)$ $(i < k)$ repairs, it is always repaired at age T_i (i.e. T_i hours from the last PM). In case of failure, a minimal repair is performed.

It is assumed that preventive maintenance or breakdown repair would bring the system to the as good as new condition. A *minimal repair* is a maintenance action that does not affect the failure rate of the system. Let C_m denote the cost of a minimal repair and $h(t)$ denote the hazard function.

Similar to the development of the model of Policy I, the expected total cost per unit time for Policy II is given by:

$$
ETC(k, T_1, ..., T_k) = \frac{(k-1)C_p + C_r + C_m \sum_{i=1}^{k} \int_0^{T_i} r_i(t)dt}{\sum_{i=1}^{k} T_i} \tag{1.5}
$$

Again, the problem is to find the optimal number of repairs k and ages $T_1, T_2, ..., T_k$ that minimize (1.5).

2.3   Imperfect Preventive Maintenance.

In many preventive maintenance models, the system is assumed to be as good as new after each preventive maintenance action. However, a more realistic situation is one in which the failure pattern of a preventively maintained system changes. One way to model this is to assume that, after PM, the failure rate of the system is somewhere between as good as new and as bad as old. This concept is called imperfect maintenance.

Several assumptions have been made to analyze imperfect PM [45]:

1. The state of the system after PM is "bad as old" with probability p and " good as new" with probability $1 - p$ [19,69].

2. PM reduces the failure rate, but does not restore the age to zero [73,74].

3. The age of the system is reduced at each PM intervention introducing the concept of an improvement factor [44, 56, 58, 74]

4. The system has different time to failure distribution between PMs [73]

5. The age of the system is reduced and the failure rate of the system is adjusted incorporating an improvement factor [44, 56, 58, 74]

6. The system degrades with time. PM restores the failure function to the same shape while the level remains unchanged and the failure rate function is monotone [21].

To illustrate this type of models, consider the following PM policy [74]

PM is done at fixed intervals h_k ($k = 1, 2..., N - 1$) and the unit is replaced at PM N. If the systen fails between PMs, it undegoes only minimal repair . The PM is imperfect. The age after PM k reduces to $b_k t$ when it was t before PM (b_k is constant and $0 \le b_k \le 1$).

The problem is to find the optimal interval lengths h_k ($k = 1, 2..., N - 1$) and the optimal number of PM N when a replacement is done so that total expected cost per unit time is minimized. Let y_i denote the age of the system right before the ith PM, C_{pm} be the cost of a scheduled PM, C_r be the cost of replacement, and C_m be the cost of minimal repair at failure. Also, let $r(t)$ denote the hazard rate function.

The imperfect preventive maintenance model is developed under the following assumptions:

1. The system undergoes PM activities at times $h_1, h_1 + h_2, ...$ where h_i, $i = 1, 2, ..., N-1$ is the length of the ith interval and is replaced at the end of Nth interval.

2. The PM is done at fixed intervals h_k $(k = 1, 2..., N-1)$ and the unit is replaced at PM N. If the systen fails between PMs, it undergoes only minimal repair . The PM is imperfect. The age after PM k reduces to $b_k t$ $0 = b_o < b_1 < b_2 < ... < b_N < 1$ when it was t before PM. The system becomes $t(1 - b_k)$ units of time younger after PM k.

3. Time returns to zero after replacement, i.e., the system becomes as good as new.

4. The hazard rate function $r(t)$ is continuous and strictly increasing.

5. The times for PM, minimal repair , and replacement are negligible.

6. A cycle starts with a new system and ends after N intervals with a replacement.

Let y_k be the effective age of the system right before PM k. Note that $y_k = h_k + b_{k-1}y_{k-1}, k = 1, 2, ..., N$. Hence, during the kth PM interval the age of the system changes from $b_{k-1}y_{k-1}$ to y_k. Note that $h_k = y_k - b_{k-1}y_{k-1}$.

It is more convenient to express the cost function in terms of y_k than in terms of h_k. The expected total cost per cycle is again the sum of the expected cost of minimal repairs the expected cost of PMs, and the cost of replacement. Thus

$$C(y_1, y_2, ..., y_N) = C_m \sum_{i=1}^{N} \int_{b_{k-1}y_{k-1}}^{y_k} r(t)dt + (N - 1)C_{pm} + C_r. \quad (1.6)$$

The expected cycle length, $E(T)$, is given by:

$$E(T) = \sum_{i=1}^{N-1}(1 - b_i)y_i + y_N. \quad (1.7)$$

The problem is to find the optimal number of PMs, N, and the optimal interval lengths $h_i, i = 1, ..., N$ that minimize the total cost per unit time

$$
\begin{aligned}
ETC(y_1, y_2, ..., y_N) &= \frac{C(y_1, y_2, ..., y_N)}{E(T)}, \\
&= \frac{C_m \sum_{i=1}^{N} \int_{b_{k-1}y_{k-1}}^{y_k} r(t)dt + (N - 1)C_{pm} + C_r}{\sum_{i=1}^{N-1}(1 - b_i)y_i + y_N}
\end{aligned}
$$

Recently, Pham and Wang [80] provided a review of imperfect preventive maintenance models. The reader is referred to this extensive review for more information on imperfect PM models.

2.4   Condition-Based Maintenance. Condition-based maintenance consists of deciding whether or not to maintain a system according to its state using condition monitoring techniques. There are three types of decisions that need to be made:

1. Selecting the parameters to be monitored, which depends on several factors such as the type of equipment and technology available.

2. Determining the inspection frequency.

3. Establishing the warning limits that trigger appropriate maintenance action, which could be static or dynamic.

Representative models for condition-based maintenance, based on Cox's proportional hazards model [27], are discussed in this section. Other modeling schemes such as the state space model and the Kalman filter for prediction [25] can also be used as modeling tools.

Cox's proportional hazards model differs from other models in the way the reliability of equipment is estimated. In this case, equipment reliability is not estimated based only on its age as is usually the case. In addition to age, reliability is estimated also based on concomitant information such as the information obtained from condition monitoring techniques (vibration analysis, oil analysis, etc.) Essentially the procedure is to modify the failure rate function $h(t)$ by multiplying it by a function $g(z(t))$ where $z(t)$ is the concomitant variable (sometimes called covariate or explanatory variable). Thus the modified failure rate function becomes:

$$h(t, z(t)) = r(t)g(z(t)),$$

which is the proportional hazards model. A commonly used form for $g(z(t))$ is

$$g(z(t)) = e^{\gamma z(t)},$$

where γ is the regression coefficient. In fact, Cox's proportional hazards model is a multiple regression model. In the case of several covariates (several monitored variables), γ and $z(t)$ can be made vectors.

$$h(t, z(t)) = h_o(t)e^{\gamma_1 z_1(t) + \gamma_2 z_2(t) + \cdots},$$

where $h_o(t)$ is called the baseline hazard rate which is the hazard rate that the system would experience if the effects of all covariates are equal to zero. The advantage of this method is that the hazard rate function is determined based on both the age of equipment and it condition. This makes it very useful in a condition-based maintenance context.

There are situations when the hazard rate increases with time. Then it is reasonable to schedule a preventive replacement when the failure rate becomes too high. To determine the optimal failure rate level (threshold level), it is necessary to predict the future behavior of covariates. Makis and Jardine [61, 62] have assumed a widely accepted Markov process model to describe the stochastic behavior of covariates and have presented optimal replacement decision models. One of these models was included in a condition-based-maintenance software package [43]. The reader is also referred to the chapter by Kumar in this book and to the survey by Kumar and Klefsjo [53].

2.5   Inspection. The inspection problem in maintenance tends to be centered around determining the optimal interval between checks on equipment to assess its status.

The problem of determining an optimal inspection policy has been discussed widely in the literature [82]. Some of these models combine inspection and replacement decisions. It is assumed that minimal repairs at a modest cost will bring the production process to the in-control state. However, the residual life after these repairs will depend on the age of the system. After a certain numbers of repairs a replacement is performed. The problem is to determine an optimal inspection policy and the optimal replacement time that will minimize the expected cost per unit time. Models in these area differ in the following features:

1. Failure process mechanism (e.g. Markovian versus non-Markovian)

2. Types of repairs (e.g. minimal repairs)

3. Assumption on inspection interval (e.g. constant interval length, constant interval hazard)

Examples of this model include the work of Barlow Proschan [10], Munford and Sahani [68], Munford [67], and more recently Banerjee and Chuiv [8], among others.

Examples of models involving inspection decisions only include the work of Baker [6], Ben-Daya and Hariga [13], and Hariga [37]. More information on these models can be found in the chapter by Hariga and Al-Fawzan in this book. In many cases, the inspection problem is jointly considered with repair and/or replacement decisions.

Other situations in which the inspection problem arises is when the purpose of the inspection is to find out whether the system has shifted to an out of control state, in which case a restoration action of the system to the in-control state is needed. Lee and Rosenblatt [55] consider an economic lot-sizing problem involving inspections s aimed at assessing the status of the production process. In a similar context, Rahim [83] and Ben-Daya [11] considered quality control inspections to assess the status of the production system.

The inspection problem arises also in the context of CBM. As discussed earlier, CBM consists of deciding whether or not to maintain a system according to its state. The state of the system can be represented by an indicator variable that measures its deterioration. Once indicators, such as bearing wear and gauge readings which are used to describe the state have been specified, inspections are made to determine the values of these indicators. Appropriate maintenance action may then be taken, depending on the state. The state of the system is represented by the value taken by a random variable X. This random variable is observed at times $t_i(i = 1, 2, ...)$, and x_i is the value of X at time t_i. This variable is equal to zero if the system is new. The greater x_i, the closer the system to a breakdown. The system breaks down if the value of X exceeds a given limit L. Taylor [103] proposed one of the earliest research work in this area. The model assumes that the system is subject to shocks that follow a Poisson distribution. The magnitude of system deterioration resulting from a shock follows a negative exponential distribution and the magnitudes of detrioration are independent of each other. The maintenance and repair costs are assumed to be constant. Under these assumptions, Taylor showed that the optimal policy is a control limit policy (CLP). That is, there exists a limit x^* such that if the state of deterioration of the system is greater that x^*, but less than a breakdown limit, then a maintenance is required. However, no action is needed if the state of deterioration is below x^*.

Bergman [17] introduced a more general model in which the state of deterioration of the system is a nondecreasing random variable. He also proves that the CLP is an optimal policy. Zuckerman [109] provided the properties of the system (conditions) under which CLP is optimal.

Several inspection models [7] use the delay time concept. Delay time is defined as the time lapse from when a fault could first be noticed until the time when its repair can be delayed no longer because of unacceptable failure consequences [24]. A repair can be undertaken during the delay time period to avoid breakdown.

A basic delay time inspection model is presented in the remainder of this section. This model is based on the following assumption:

1. The probability density function of the delay time τ, $f(\tau)$ is known.

2. Inspections take place at regular intervals of time of equal length T and each inspection costs C_i. Its duration is T_p.

3. Inspections are perfect.

4. The time of origin of the fault is uniformly distributed over time since the last inspection and is independent of τ.

5. Faults arise at the rate of λ per unit time.

6. A breakdown repair costs C_f and has an average duration T_f and an inspection repair costs C_p and has an average duration T_p.

The problem is to determine the length of the inspection interval T that minimizes the expected downtime per unit time or expected cost per unit time.

Suppose that a fault arising in the interval $(0, T)$ has a delay time in the interval $(\tau, \tau + d\tau)$ with probability $f(\tau)d\tau$. A breakdown repair will be performed if the fault arises in the interval $(0, T - \tau)$, otherwise an inspection repair will be carried out. The probability of the fault arises before $(T - \tau)$, given that a fault will occur is $(T - \tau/T)$ because of Assumption 4. The probability of a breakdown repair is given by

$$p(T) = \int_0^T \frac{T - \tau}{\tau} f(\tau)d\tau \qquad (1.8)$$

The expected downtime per unit time $D(T)$ is given by

$$D(T) = \frac{\lambda T T_f p(T) + T_p}{T + T_p} \qquad (1.9)$$

The expected cost per unit time $C(T)$ is given by:

$$C(T) = \frac{\lambda T [C_f p(T) + C_p(1 - p(T))] + C_i}{T + T_p} \qquad (1.10)$$

These models have been used in preventive maintenance, condition monitoring and other settings. For a review of delay time models , the reader is referred to the paper by Baker and Christer [7].

3.   MAINTENANCE PLANNING AND SCHEDULING

Maintenance Planning and Scheduling form one of the most important elements of the maintenance function. They are prerequisite for effective maintenance control. The maintenance planning and scheduling models can be classified based on function or horizon. The functional classification results in planning models and scheduling models. The horizon classification results in long, medium and short term models.

The long and medium term models address maintenance capacity planning and spare parts provisioning for maintenance. The short term planning models focus on resource allocation monitoring and control.

The long term scheduling models focus on the allocation of resources and scheduling production units for planned maintenance such models are used in scheduling maintenance for electric generators in large electric utilities or production units in a process plant. Medium range scheduling models are used to schedule large maintenance jobs or shut down maintenance. Heuristic rules are used for short term maintenance scheduling as reported in Worral and Mert [108]. In this section synopsis of the optimization and simulation models developed for these problems will be presented.

3.1   Maintenance Capacity Planning.

The general maintenance capacity planning is defined as the determination of the optimal level of resources (workers, skills, spare parts, equipment and tools) required to meet the forecasted maintenance load . Two major subproblems of this capacity planning have been studied in the literature, which are:

1. To determine the optimal mix of the skills of trades that are needed to meet the maintenance load from available sources.

2. To determine the optimal spare parts provisioning policy to meet maintenance requirements. The models for the second subproblem will be presented in the next section .

Several optimization models appeared in the literature for the above two problems. Samples of these models are given in Roberts and Escudero [85] and Duffuaa and Al-Sultan [31]. It is very complex to treat the general maintenance capability planning problem in a single optimization models including all aspects of the problem. It is to be noted many variables in this problem are stochastic in nature. However, several simulation models have been developed for the general problem or simplified version of it by Duffuaa and Raouf [32] and Al-Zubadi and Christer [4].

The subproblem that deals with the determination of the optimal mix of trades to meet the maintenance has been formulated as a linear program (LP) after relaxing the integrality of some variables. The LP formulation and simulation models that address capacity planning problems are given in Chapter 2.

3.2   Allocation of Resources.

An important aspect of planned maintenance is the ability to allocate resources to different plant areas in order to complete planned work orders and manage the maintenance backlog in an effective manner. Taylor [102] developed a LP model for allocating labor crews to different prioritized work orders in a cost effective manner. The model gives consideration to different craft areas such as electrical, masonry and special allocable projects crews. The model ensures that high priority jobs are completed first. The details of this model are given in Chapter 2.

3.3   Scheduling Preventive Maintenance. This type of problem deals with scheduling major planned preventive maintenance for a number of production units. The maintenance program and tasks are already planned for each unit and must be performed in a specified time window. This problem is typical in electric utilities and process plants. Several optimization models appeared in the literature for this problem and is an active area of research. The objective function for most models in the literature is to minimize cost depending on the type of industry. The constraints that are considered in this problem include:

1. Maintenance completion constraints. This constraint ensures that whenever maintenance started on a unit it must be completed without interruption.

2. Preventive maintenance (PM) window constraints. This constraint ensures that the PM for each unit is done in the specified window

3. Crews availability constraints.

4. Precedence constraints. This type of constraint requires that maintenance is performed on critical units prior to others.

5. Production schedule constraints. This type of constraint ensures delivery of products.

6. Reliability constraint. The type of constraint ensures that the amount to be produced is attained with a certain reliability level.

A sample of the models in this area appeared in Dopazo and Merril [30] and Escudero [34]. Also an invited review of the work in this area was given by Kralj and Pedrovic [51]. Chapter 3 of this book provides a typical model in the area of electric power generation with an account of the approaches used to solve such models.

3.4   Scheduling Maintenance Jobs. This type of scheduling problem deals with the scheduling of available maintenance jobs at the beginning of each horizon. It falls under the category of medium or short term maintenance scheduling problem. One of the first models for this problem was formulated by Roberts and Escudero [85]. The model by Roberts and Escudero is a deterministic mixed integer programming model. However the general maintenance job scheduling problem is far more complex than the formulation given by Roberts and Escudero [85]. This has been realized by Duffuaa and Al-Sultan [31] and a more realistic statement of the problem was given, together with a stochastic programming model that formulates it. The details of this model with an example are given in Chapter 2.

4. SPARE PARTS PROVISIONING

The availability of spare parts and material is critical for maintenance systems. Insufficient stocks can lead to extended equipment down time. However, excessive stocks lead to large inventory costs and increase system operating costs. An optimal level of spare parts material is necessary to keep equipment operating profitably and at a higher level of overall equipment effectiveness.

In order to appreciate the potential of the problem of spare parts and material provisioning in maintenance and realize the benefits of optimization models in this area, the reader is referred to Sherbrooke [91]. Sherbroohe estimated that in 1968 the cost of recoverable spare parts in the United State Air Force (USAF) amounted to ten billion dollars. This cost is about 52% of the total cost of inventory for that year. It is worth noting that all these parts are used for maintaining the USAF equipment.

The literature on general inventory models is vast and can not be covered with justice in this section. However in this section the important models dealing with spare parts provisioning for maintenance are reviewed. Then two important models, one dealing with irreparable items and the other with repairable items are briefly stated. The emphasis is on models that link stock levels of spares and material to maintain system effectiveness.

Generally the literature in this area can be divided into two major categories:

1. Irreparable item inventory models.

2. Repairable (recoverable) items inventory models.

Two reviews have appeared in the literature in this area. The first review by Nahmis [70] covering most of the work that appeared prior to 1981. He organized his review in the following manner:

- The $(S - 1, s)$ ordering policies and their importance in managing repairable items.

- The METRIC (multi-echelon technique for recoverable item control) model and its extensions.

- Continuous review models for deterministic and random demand.

- Periodic review models under random demand.

The second review by Cho and Parlor [22] focused on maintenance models for multi-unit systems. They included a section on inventory and maintenance models. The rest of the section is organized into two subsections dealing with models for irreparable and repairable items.

4.1   Irreparable Item Inventory Models.

The models in this subsection focus on irrepairable items. One of the earlier work in this area is by Henin [39]. He formulated a model for determing optimal inventory policies for a single part. Later Henin and Perrakis [40] generalized the model for many components and spare parts. Optimal ordering and replacement policies for Markovian systems are modeled by Kawai [49, 50].

The work of Osaki and his co-researchers is notable in this area. They have constructed a variety of mathematical models under different sets of conditions. The reader is refered to Kaio and Osaki [47, 48], Osaki and Nakagawa [76], Thomas and Osaki [103, 104] who have introduced an ordering policy in which ordering and replacement times are jointly optimized. Park et al. [78] have generalized some of these models to the case of random lead time. A comprehensive review in this area is given by Dohi, Kaio and Osaki in Chapter 7.

Sheikh et al. [90] presented a model for determining the number of spare parts needed during an operation period t. The spare parts stocking policy is based on a reliability approach. This model is presented in Chapter 4 of this book. The above model is practical and easy to implement provided the assumptions about the distributions are validated and good estimates for K and μ are available. Recently Kabir and Al-Olayan [46] developed a similar model for jointly optimizing age replacement and spare parts provisioning. In the past age replacement were made assuming spare parts were available. However, this assumption is questionable in many practical cases.

4.2   Repairable-Item Inventory Model.

The field of repairable-item inventory systems with returns has been gradually recognized by a number of researchers and practitioners. The reason is that, expensive repairable spares such as aircraft engines and locomotives motors constitute a large part of the total investment in many organizations. These spares are critical to the operations and their unavailability can lead to excessive downtime costs. In many situations, repairing failed items certainly has cost advantages over purchasing new items.

In repairable-item inventory systems with returns, two types of inventories (i.e., the serviceable and repairable inventories) are considered. The serviceable inventory is depleted by outside demand for serviceable units in a repair facility. The repairable inventory, on the other hand, is replenished by returns of repairable units from customers and is depleted by repairing or junking repairable units in the repair facility. Phelps [81] considered the situation in which the demand process for serviceable units and the return process of failed units are perfectly correlated. This means that a return of a unit simultaneously generates a demand for a unit. They both assumed the inventory levels are reviewed periodically.A similar repairable-item inventory problem was considered by Heyman [41]. The objective here is to determine the optimum level of the servicable inventory that minimizes the total of repair , disposal, purchasing and holding costs.

Several models for optimizing the inventory and servicing of repairable items have appeared in the literature. These models include:

- The M machine K repairmen queuing model. This model has been applied by Albright [2] for a fixed state-state dependent ordering policy for a system with M identical machines. The objective was to minimize the long run expected discounted cost.

- Fault tolerance (FT) repairable items inventory models. This model has been introduced by Lawrence and Schafer [54] and Smith and Schafer [97] to represent the situation of N identical machine where it is assumed the system is operative as long as m-out-of-N (where $0 < m \leq M$) components are working. The FT model is used to determined the optimal inventory policy which minimizes the expected cost subject to the constraint of total inventory investment.

- Multi-dimensional Markovian models. These models have been used to formulate complex repairable items inventory systems with returns Albright and Soni [3] using a continuous ordering policy of the (s, S) type. Such systems consist of a storage facility for serviceable units and a repair facility with a finite number of identical repairmen having exponential repair times. Demand for serviceable units occurs when any item being leased fails or new customers are generated. An item is returned when a customer brings a failed item for repair or a lease expires. Any of the above events is assumed to occur independently according to Poisson processes with different mean rates. The system is assumed to have as many as

M items while the total customer in the system is limited to N. A certain population of failed items is assumed to be irreparable, and thus new items must be brought from an outside source.

■ Mult-echelon inventory repariable items inventory models. These type of models are used to represent multi-echelon repairable-item inventory systems where several bases (each of which may consists of several sub-areas) are supported by a central depot. Units (which may or may not be identical) in each base are subject to random failure. A failed unit may be repaired at the same base; otherwise, it must be sent to the central depot for repair. Each base receives scheduled trans-shipment from the central depot. The main objective of such a system is to develop an optimal replenishment policy for each base and the central depot. Based on the capacity availability of the repair facilities in a system, multi-echelon repairable-item inventory models can be classified into two general groups: METRIC-based models and non-METRIC models. The METRIC-based models assume a Poisson (compound Poisson) process with a constant failure rate and infinite repair capacity (i.e., ample service). These two assumptions make control of repair levels irrelevant, and thus reduce the degree of difficulty of obtaining solutions to the problems. In the non-METRIC models, the ample service assumption is relaxed. Here, the optimization decision is not only on the levels of spares but on the levels of repair . Since the advent of the METRIC model by Sherbrooke [91] many METRIC-based multi echelon inventory models have been developed (see Sherbooke [92], Simon [95], Muckstadt [63, 64, 65], Muckstadt and Thomas [66]).

■ Multi-indenture and multi-echelon inventory repairable items inventory models. These types of models represent a hierarchical maintenance supply support system. In this type of system each location may have repair and stockage capabilities. In this type of model an $(S-1, S)$ rule is used for all items in all locations. In this rule whenever a unit is lost from stock, a replacement unit is ordered from the next higher location in the support system. As part of the ordering rule, it is assumed that each demand for one to be replaced is accompanied by a failed unit. In each location a failed item has a repair cycle which depends on the availability of spare parts. If an item can not be repaired at a location it is sent to the next higher level repair facility. If an item is available at the next higher repair facility a certain time is involved which the resupply time. Clark [23] presented a multi-indenture , multi-echelon

known as the A_o optimal inventory model.The model generalized
the work of Roseman and Hoekstra [86], Sherbrooke [91], Muckstadt [63, 64, 65], Slay and O'Malley [96] and Fitzgerald [35]. The
objective of the model is to maximize operational availability of
each equipment subject to budget constraints on spare stockage.
Since this model generalizes several models and is important in
this area, the model is stated below.

The model uses the following notation:

A_o	operational availability
$MTBE$	mean time between failures
$MTTR$	mean time to repair
$MSRT$	mean supply response time
A_{eu}	fraction of time equipment e is available for use at location u.
M_{iu}	mean time to return a failed unit of item i at location u to a serviceable condition,
D_{iu}	mean supply response time ($MSRT$ for item i at location u),
τ_{iu}	mean time to repair ($MTTR$) for item i at location u when all needed repair parts are available, 0 if location u does not operate the equipment,
S_{iu}	stock level of item i at location u,
λ_{iu}	expected number of demands per time unit upon inventory for item i at location u,
$p(x; \lambda_{iu}T_{iu})$	probability of x units of stock reduction for item i at location u,
T_{iu}	mean stock replenishment time (time to replace an inventory loss through repair or procurement for item i at location u).
γ_{iu}	probability that a demand for item i at location u results in a loss (discard or sent elsewhere for repair) which must be replaced through resupply.
L_{iu}	average resupply time assuming stock is available at the resupply source,
L'_{iu}	average additional resupply time due to expected shortages at the resupply source,
R_{iu}	average repair cycle assuming availability of spares for items within i at the next lower indenture of the parts breakdown, and
R'_{iu}	average additional repair cycle due to expected shortages of spares for items within i at the next

lower indenture.

The expected operational availability of an equipment can be calculated as follows:

$$A_o = \frac{\text{Up-time}}{\text{Up-time} + \text{Down-time}} = \frac{MTBF}{MTBF + MTTR + MSRT} \quad (1.11)$$

The next equation relates the mean time to return a failed equipment to use to mean repair time and mean supply response time. It is stated

$$M_{iu} = D_{iu} + \tau_{iu} \quad (1.12)$$

The next equation relates the mean supply response time to demand and mean stock replenishment time and is stated as:

$$D_{iu} = \frac{1}{\lambda_{iu}} \sum_{x > S_{iu}} (x - S_{iu}) P(x; \lambda_{iu}, T_{iu}) \quad (1.13)$$

The above equation captures the concept that demand is uncertain and shortages will occur with some probability according to stock level S_{iu}. It provides the expected number of units short at any time for a given stock level S_{iu}. The next equation relates mean replenishment time to mean supply time and average repair cycle

$$T_{iu} = \gamma_{iu}(L_{iu} + L'_{iu}) + (1 - \gamma_{iu})(R_{iu} + R'_{iu}) \quad (1.14)$$

Equation (11) specifies the degree to which resupply time versus repair time contributes to mean replenishment time. For totally consumable items $\gamma_{iu} = 1$, and the only consideration is the time it takes to obtain resupply from a higher source. For totally repairable item $\gamma_{iu} = 0$. This indicates that the model can handle both consumable and repairable items. The next equation relates supply time between a level and the next higher level in the supply chain.

$$L'_{iu} = D_{iv} \quad (1.15)$$

where v is the higher level support facility providing supply of item i to location u. The next equation states that the extra repair time for item i at location u due to storage of lower level parts is given by the weighted average of the mean time required to return a part to a serviceable condition through replacement or further repair

$$R'_{iu} = \frac{\sum_{j \in i} \lambda_{ju} \mu_{ju}}{\sum_{j \in i} \lambda_{iu}} \quad (1.16)$$

where j is an item within i at the next lower indenture level of the parts broken.

The following equation provides a formula for operational availability for the end equipment e at location u. This is equivalent to A_o.

$$A_{eu} = \frac{1}{1 + \lambda_{eu}\mu_{eu}} \tag{1.17}$$

Equation (14) can be easily obtained by setting λ_{eu} to be $\frac{1}{MTBF}$ and use equation (9) for μ_{eu}. The model described above establishes structural relationships among stock levels for items, locations, repair time, supply response time and relates them to end equipment operational availability. The model can be solved to determine stock levels to achieve operational availability goals. Clark [23] provided a scheme for determining these stock levels.

Due to the complexity of the models in this area and random nature of the variables in these systems, several simulation studies have been reported in the literature. For simulaion studies in this area the reader is referred to Scudder [88], Buyukhurt and Parlar [20], Dhakar et al. [29], Kumar et al. [52] and Madu and Kuei [57].

5. INTEGRATED MODELS

Production, quality and maintenance are three important aspects in any industrial process and are interrelated problems. Understanding and solving these problems require adequate modeling of the dependencies between production, quality and maintenance. In the past, production, quality and maintenance have been treated as three separate problems. Because of the interdependence between them, many attempts have been made to develop models that take into consideration more than one of the three aspects. This line of research has witnessed a considerable growth in the last decade.

Maintenance as a separate problem received a lot of attention from researchers. Pierskalla and Voelker [82] presented a survey of the literature on maintenance models up to 1976. Other surveys include Sherif and Smith [93], Valdez and Feldman [106], Cho and Parlar [22]. The reader is referred also to the classical book by Jardine [42]. Many reliability books contain chapters dealing with maintenance issues, see for example [33]. Maintenance activities interfere with the production schedule and with the deterioration pattern of the production process. The absence of a unified production-maintenance models results in limited understanding of the existing relationships between production and maintenance parameters and the trade-offs between alternative maintenance and production policies. There is a growing interest in developing production models considering maintenance requirement, as we will see later.

The maintenance quality relationship is also well documented [12]. In most case, statistical process control cannot be separated from preventive maintenance. As stated by Tagaras [99], information obtained in the course of statistical process control determines possible restoration actions to be taken and thus affects preventive maintenance schedule. Similarly, maintenance activities interfere with the deterioration pattern of the production process, thus changing the process control requirements. Many models addressing statistical process control and maintenance simultaneously have appeared in the literature.

Interest in integrated models addressing all three aspects at the same time is also present in recent literature. This presents a natural development built on the above directions of research.

In this section, we provide a brief discussion of representative integrated models of production quality and maintenance. For an elaborated survey of these models, the reader is referred to Ben-Daya and Rahim [16].

5.1   Production and Maintenance.
The problem of integrating production and maintenance has been generally approached in the literature in two different ways. Some authors approached this problem by determining the optimal preventive maintenance schedule in the production system and others by taking maintenance as a constraint to the production system. Under the first approach, Rosenblatt and Lee [87] have developed a model to determine the EPQ of a single product and the schedule of inspection for process maintenance. Groenevelt et al. [36] have also developed a model which studies the problem of selecting the economic lot size for an unreliable manufacturing facility with a constant failure rate and general randomly distributed repair times.

Tse and Makis [105] studied a general EPQ model with random machine failures and planned preventive replacement. They considered two types of failure and a general cost structure. When a major failure occurs (Type I failure), the production unit is replaced, and the interrupted lot is aborted. When a minor failure occurs (Type II failure), the machine is restored to the operating condition via a minimal repair and the production can be resumed immediately at a cost lower that the set up cost. The objective was to find the optimal lot size and optimal preventive maintenance replacement time.

Ashayeri [5] proposed a production and maintenance model for the process industry using a different modeling approach. A mixed integer linear programming model is developed to simultaneously plan preventive maintenance and production in a process industry environment. The model schedules production jobs and preventive maintenance jobs, while minimizing costs associated with production, backorders, corrective and preventive maintenance.

Under the second approach, Rishel and Christy [84] have studied the impact of incorporating alternative scheduled maintenance policies into the material requirements planing (MRP) system. Brandolese *et al.* [18] have considered the problem of planning and management of a multi-product and one-stage production system made up of flexible machines operating in parallel. They developed a model to find the optimal schedule of both production and maintenance intervention which were considered as capacity-consuming jobs to be scheduled on the production resources.

Several authors looked at the effect of maintenance of Just-In-Time (JIT) and flexible manufacturing systems (FMS) [1, 98].

5.2   Maintenance and Quality.   Ben-Daya and Duffuaa [12] discussed the relationship between maintenance and quality and proposed a broad framework for modeling the maintenance-quality relationship.

Statistical process control is one of the most common methods for controlling product quality. As mentioned earlier, for many processes, it is not possible to separate process control from preventive maintenance because of their interrelationships. Tagaras [99] provided an integrated cost model for the joint analysis and optimization of process control and maintenance operations. The proposed model can be viewed as an improvement and extension of the models in Pate et al. [79] which addressed the inspection-maintenance problem with more emphasis on maintenance. Tagaras's model allows for an arbitrary number of out of control states and different sampling and maintenance intervals. preventive maintenance is performed periodically on a subset of the quality control inspection epochs.

The effect of maintenance on the economic design of $\bar{x}$-control charts for general time to shift distribution with increasing hazard rate was also studied by Ben Daya and Rahim [15].

Tapiero [100] formulated a problem of continuous quality production and maintenance of a machine. He assumed that quality is a known function of the machine's degradation states. He used different applications of a specific quality function to obtain analytical solutions to open-loop and feedback stochastic control maintenance problems.

Makis [59] studied the EPQ model for a production process subject to random deterioration. The process is monitored through imperfect inspections. The objective is to find the optimal lot size and optimal inspection schedule in a production run.

5.3   Joint Models for Maintenance, Production and Quality.

As a natural extension to the above models, papers appeared in the literature addressing production, quality and maintenance in a single model. Some of the models integrating production and quality and reviewed above considered corrective maintenance. In this section, we consider two examples of research dealing with preventive maintenance and the modeling of its effect on production quantity and quality.

Ben Daya [11] has extended Rahim's [83] model by introducing imperfect preventive maintenance concept into the determination of EPQ and the economic design of a $\bar{x}$-control chart for a class of deteriorating processes where the in control period follows a general probability distribution with an increasing failure rate. Ben-Daya and Makhdoum [14] investigated the effect of PM policies on EPQ and design of the $\bar{x}$-control chart.

Makis and Fung [60] have studied the problem of the joint determination of the lot size, inspection interval and preventive replacement time.

6.   FUTURE RESEARCH DIRECTIONS

The research into production technologies should take into consideration the maintenance dimension. The significance and complexity in maintenance in modern production and manufacturing systems indicate that maintenance should be viewed as a more important area of study and research. Several challenging problems in various maintenance areas require further attention and research including:

1. Condition monitoring and condition-based maintenance.

2. Inventory and maintenance models.

3. Integrated models.

4. Maintenance planning and scheduling.

5. Bridging the gap between theory and practice through intelligent maintenance optimization systems.

Technology is providing the maintenance function with an ever increasing range of diagnostic methods and equipment. This will enhance the ability of implementing condition-base maintenance strategies. Models dealing with condition monitoring and condition-based maintenance have witnessed a tremendous growth in recent years. Models that integrate economic considerations and risk estimation using condition monitoring methods to identify optimal maintenance strategies are in great demand. Although some work has been done in this direction, more work is needed. The opportunities for new developments in this area are tremendous.

Spare parts management can be improved a great deal by taking into account the failure pattern of the equipment requiring the parts. Simply relying on historical usage data may not be as effective. Integrated inventory and maintenance models have been developed in the past. However, more work is needed. In particular, proposed models should be realistic so that they can be used to improve spare parts management and reducing costs. Also, this will lead to improvements in equipment availability.

In the past, production, quality and maintenance have been treated as three separate problems. Because of the interdependence between them, many attempts have been made to develop models that take into consideration more than one of the three aspects. This line of research has witnessed a considerable growth in the last decade. This trend will continue. New and alternative ideas for modeling the interdependence between production, quality, and maintenance are needed to generate models that can be used for their joint optimization.

Maintenance planning and scheduling grow in importance as maintenance becomes a more significant portion of total production costs in more capital intensive industries using complex technologies. As discussed in Part II of this book, some work has been done in maintenance planning and scheduling, but much more is needed. Models taking into account the stochastic nature of maintenance jobs, effect of information delays, among other factors need to be developed to provide tools for effective planning and scheduling.

Maintenance information systems must be integrated with optimization models for effective utilization of these models. This requires cooperation between people in the area of maintenance and optimization and information systems designers.

Future developments in maintenance research must also strive to bridge the gap between theory and practice. The success of this type of research should be measured by its relevance to practical situations and by its impact on the solution of real maintenance problems. The developed theory must be made accessible to practitioners through information technology tools.

References

[1] Albino, V, Carella, G., and Okogbaa, O.G. Maintenance Policies in Just-In-Time Manufacturing Lines. *International Journal of Production Research.* **30** (1992) 369-382.

[2] Albright, A. Evaluation of Costs of Ordering Policies in Large Machine Repair Problems. *Naval Research Logistics Quarterly,* **31** (1984) 387-398.

[3] Albright, A. and Soni, A. Markovian Multi-Echelon Repairable Inventory System. *Naval Research Logistics,* **35** (1988) 49-61.

[4] Al-Zubaidi, H. and Christer, A.H. Maintenance manpower modelling for a hospital Buliding complex. *European Journal of Operational Research,* **99** (1995) 603-618.

[5] Ashayeri, J., Teelen A., and Selen, W. A Production and Maintenance Planning Model for the Process Industry. *International Journal of Production Research,* **34** (1996) 3311-3326.

[6] Baker M.J.C., How often should a machine be inspected?, *International Journal of Quality and Reliability Management,* **7**, 1990, 14–18.

[7] Baker, R.D. and Christer, A.H. Review of Delay-Time OR Modeling of Engineering Aspects of Maintenance. *European Journal of Operational Research,* **73** (1994) 407-422.

[8] Banerjee, P.K. and Chuiv, N.N. Inspection Policies for Repairable Systems. *IIE Transactions,* **28** (1996) 1003-1010.

[9] Barlow, R.E. and Hunter, L.C. Optimum Preventive Maintenance policies. *Operations Research,* **8** (1960) 90-100.

[10] Barlow, R.E. and Proschan, F. *Mathematical Theory of Reliability.* Wiley, New York, 1965.

[11] Ben-Daya, M. Integrated Production, Maintenance, and Quality Model for Imperfect Processes. *IIE Transactions,* **31** (1999) 491-501.

[12] Ben-Daya, M. and Duffuaa, S.O. Maintenance and Quality: the Missing Link. *Journal of Quality in Maintenance Engineering,* **1** (1995) 20-26.

[13] Ben-Daya M. and Hariga M., A maintenance inspection model : optimal and heuristic solutions, *International Journal of Quality and Reliability Management,* **15** (1998) 481-488.

[14] Ben-Daya, M. and Mahkdoum, M. Integrated Production and Quality Model under Various Preventive Maintenance Policies. *Journal of the Operational Research Society,* **49** (1998) 840-853.

[15] Ben-Daya, M. and Rahim, M.A. Effect of Maintenance on the Economic Design of $\bar{x}$−Chart. *The European Journal of Operational Research*, **120** (2000) 131-143.

[16] Ben-Daya, M. and Rahim. M.A. Integrated Production, Quality and Maintenance Models: a Review. In *Integrated Modeling in IQM: Production Planning, Inventory, Quality and Maintenance*, Kluwer Academic Publishers, Forthcoming.

[17] Bergman, B. Optimal Replacement Under a General Failure Model. *Advances in Applied Probability*, **10** (1978) 431-451.

[18] Brandolese, M., Francei, M., and Pozetti, A. Production and Maintenance Integrated Planning. *International Journal of Production Research*, **34** (1996) 2059-75.

[19] Brown, M. and Proschan, F. Imperfect Repair. *Journal of Applied Probability.* **20** (1983) 851-859.

[20] Buyukkurt, M.D., and Parlar, M. (1989). A Comparison of Allocation Policies in a Two-Echelon Repairable-Item Inventory Model. Submitted.

[21] Canfield, R.V. Cost Optimization of Periodic Preventive Maintenance. *IEEE Transactions on Reliabiity.* **R-35** (1986) 78-81.

[22] Cho, D. and M. Parlar. A Survey of Maintenance Models for Multi-Unit systems. *European Journal of Operational research* **51** (1991) 1-23.

[23] Clark, A.J. Experience with Multi-Indentured, Multi-Echelon Inventory Model. *TIMS Studies in Management Science, edited by L.B. Schwarz*, **16** (1981) 299-330.

[24] Christer, A.H. and Waller, W.M. Delay-time Models of Industrial Maintenance Problems. *Journal of Operational Research Society*, **35** (1984) 401-406.

[25] Christer A.H. and Wang, W. A State Space Condition Monitoring Model for Furnace Erosion Prediction and Replacement. *1995 IFRIM Workshop Proceeding*, San Farnacisco, July 1995.

[26] Chu, C., Proth, J.M., and Wolf, F. Predictive Maintenance: The One-Unit Replacement Model. *International Journal of Production Economics*, **54** (1998) 285-295.

[27] Cox, D.R and Oakes, D. *Analysis of survival data.* Chapman & Hall, London, 1984.

[28] Dekker, R. On the Use of Operations Research Models for Maintenance Decision Making. *Microelectronics and Reliability* **35** (1995) 1321-1331.

[29] Dhakar, T.S., Schmidt, C.P., and Miller, D.M. Base Stock Level Determination for High Cost Low Demand Critical Repairable Spares. *Computers Ops Res.*, **21** (1994) 411-420.

[30] Dopazo, J. F., and Merrill, H.M. Optimal Generator Maintenance Scheduling using Integer Programming. *IEEE Transactions on Power Apparatus and Systems*, **94** (1975) 1537-1545.

[31] Duffuaa, S.O. and Al-Sultan, K.S. A Stochastic Programming Model for Scheduling Maintenance Personnel. *Applied Mathematical Modelling*, **25** (1999) 385-397.

[32] Duffuaa, S.O. and Raouf, A. A Simulation Model for Determining Maintenance Staffing in an Industrial Environment. *Simulation*, **8** (1993) 93-99.

[33] Elsayed, E. A. *Reliability Engineering.* New York, Addison Wesley Longman, 1996.

[34] Escudero, L. F. On Maintenance Scheduling of Production Units. *European Journal of Operational Research* **9** (1982) 264-274.

[35] Fitzgerald, J.W. Three Echelon LRU Search Algorithm. Working Paper, Air Force Logistics Command, Wright-Patterson Air Force Base, Ohio, 1975.

[36] Groenevelt, H.A., Seidmann, A. and Pintelon, L. Production Lot Sizing with Machine Breakdown. *Management Science*, **38** (1992) 104-123.

[37] Hariga, M. A Maintenance Inspection Model for a Single Machine with General Failure Distribution. *Microelectronics and Reliability*, **36** (1996) 353-358.

[38] Hax, A.C. and D. Candea, *Production and Inventory Management*, Prentice Hall, Englewood Cliffs, N.J., 1984.

[39] Henin, C.G. Optimal Replacement Policies for a Single Loaded Sliding Standby. *Management Science*, **18** (1972) 706-715.

[40] Henin, C.G., and Perrakis, S. Optimal Replacement Policies with Two or More Loaded Sliding Standbys. *Naval Research Logistics Quarterly*, **30**(1983) 583-599.

[41] Heyman, D.P. Optimal Disposal Policies for a Single-Item Inventory System with Returns. *Naval Research Logistics Quarterly*, **25** (1977) 581-596.

[42] Jardine, A.K.S. *Maintenance, Replacement and Reliability.* Pitman, London, 1973.

[43] Jardine A.K.S., Makis, V., Banjevic, D., Braticevic, D., and Ennis, M. A Decision Optimization for Condition-Based Mainteanance. *Journal of Quality in Mainteanance Engineering*, **4** (1998) 115-121.

[44] Jayabalan, V. and Chaudri, D. Optimal Mainteanance and Replacement Policy for a Deteriorating System with Increased Mean Downtime. *Naval Resarch Logistics.* **39** (1991) 67-78.

[45] Jayabalan, V. and Chaudri, D. An Heuristic Approach for Finite Time Maintenance Policy. *International Journal of Production Economics.* **27** (1992) 251-256.

[46] Kabir, A.B.M.Z., and Al-Olayan, A.S. A Stocking Policy for Spare Part Provisioning under Age Based Preventive Replacement. *European Journal of Operational Research,* **90** (1996) 171-181.

[47] Kaio, N., and Osaki, S. Optimum Ordering Policies with Lead Time for an Operating Unit in Preventive Maintenance. *IEEE Transactions on Reliability,* **27** (1978) 270-71.

[48] Kaio, N., and Osaki, S. Optimum Planned Maintenance Policies with Lead Time", *IEEE Transactions on Reliability,* **30** (1981) 79.

[49] Kawai, H. An Optimal Ordering and Replacement Policy of a Markovian Degradation System under Complete Observation. *Journal of Operational Research Society of Japan,* **26** (1983) 279-290.

[50] Kawai, H. An Optimal Ordering and Replacement Policy of a Markovian Degradation System under Incomplete Observation. *Journal of Operations Research Society of Japan,* **26** (1983) 293-306.

[51] Kralj, B. L. and Pedrovic K. M. Optimal Preventive Maintenance Scheduling of Thermal Generating Units in Power Systems : A Survey of Problem Formulations and Solution Methods. *European Journal of Operational Research,***35** (988) 1-15.

[52] Kumar, N., Vrat, P., and Sushil. A Simulation Study of Unit Exchange Spares Management of Diesel Locomotives in the Indian Railways", *International Journal of Production Economics,* **33** (1994) 225-236.

[53] Kumar, D. and Klefsjo, B. Proportional Hazards Model: a Review. *Reliability Engineering and System Safety,* **44** (1994) 177-188.

[54] Lawrence S.H., and Schafer M.K. Optimal Maintenance Center Inventories for Fault-Tolerant Repairable Systems. *Journal of Operations Management,* **4** (1984) 175-181.

[55] Lee, H.L., and Rosenblatt J.J. Simultaneous Determination of production Cycles and Inspection Schedules in a Production System. *Management Science,* **33** (1987) 1125-1136.

[56] Lie, C.H. and Chun, Y.H. An Algorithm for Preventive Maintenance policy. *IEEE Transactions on Reliability.* **R-35** (1986) 71-75.

[57] Madu, C.N., and Kuei, C.H. Simulation Metamodels of System Availability and Optimum Spare and Repair Units. *IIE Transactions*, **24** (1995)

[58] Malik, M.A.K. Reliable Preventive Maintenance Scheduling. *AIIE Transactions*. **11** (1979) 221-228.

[59] Makis, V. Optimal Lot Sizing and Inspectionn Policy for an EMQ Model with Imperfect Inspections. *Naval Research Logistics* **45** (1998) 165-186.

[60] Makis, V. and Fung, J. Optimal Preventive Replacement, Lot Sizing and Inspection Policy for a deteriorating Production System. *Journal of Quality in Maintenance Engineering* **1** (1995) 41-55.

[61] Makis, V. and Jardine, A.K.S. Optimal replacement in the proportional hazards model. *INFOR* **30** (1991) 172-183.

[62] Makis, V. and Jardine, A.K.S. Computation of Optimal policies in replacement models. *IMA Journal of Mathematics Applied in Business & Industry* **3** (1992) 169-175.

[63] Muckstadt, J.A. A Model for a Multi-Item, Multi-Echelon, Multi-Indenture Inventory System. *Management Science*, **20** (1973) 472-481.

[64] Muckstadt, J.A. Some Approximations in Multi-Item Multi-Echelon Inventory Systems for Recoverable Items. *Naval Research Logistics Quarterly*, **25** (1978) 377-394.

[65] Muckstadt, J.A. A Three-Echelon, Multi-Item Model for Recoverable Items. *Naval Research Logistics Quarterly*, **26** (1979) 199-221.

[66] Muckstadt, J.A., and Thomas, L.J. Are Multi-echelon inventory methods worth implementing in systems with low-demand-rate items. *Management Science*, **26** (1980) 493-494.

[67] Munford, A.G. Comparison Among Certain Inspection Policies. *Management science*, **27** (1981) 260-267.

[68] Munford, A.G. and Sahani, A.K. A Nearly Optimal Inspection Policy. *Operational research Quarterly*, **23** (1973) 373-379.

[69] Murthy, D.N.P. and Nguyen, D.G. Optimal Age-Policy With Imperfect Preventive Maintenance. *IEEE Transactions on Reliabiity*. **R-30** (1981) 80-81.

[70] Nahmias, S. Managing Repairable Item Inventory Systems: a Review. *TIMS Studies in the Management Sciences*, **16** (1981) 253-277.

[71] Nahmias, S. *Production and Operations Research*, McGraw Hill, Singapore, 1997.

[72] Nakagawa, T. Mean Time to Failure with Preventive Maintenance. *IEEE Transactions on Reliability.* **29** (1980) 341.

[73] Nakagawa, T. Periodic and Sequential Preventive Maintenance Policies. *Journal of Applied Probability.* **23** (1986) 16-21.

[74] Nakagawa, T. Sequential Imperfect Preventive Maintenance Policies. *IEEE Transactions on Reliability.* **37** (1988) 295-298.

[75] Nguyen, D.G. and Murthy, D.N.P. Optimal Preventive Maintenance Policies for Repairable Systems. *Operations Research.* **29** (1981) 1181-1194.

[76] Osaki, S., and Nakagawa, T. Two Models for Ordering Policies. *Notes in Decision Theory*, No.34, University of Manchester, Manchester, 1977.

[77] Osaki, S., Kaio, N., and Yamada, S. A summary of optimal ordering policies. *IEEE Transactions on Reliability*, **30** (1981) 272-277.

[78] Park, Y.T., and Park, K.S. Generalized Spare Ordering Policies with Random Lead Time", *European Journal of Operational Research*, **23** (1986) 320-330.

[79] Pate-Cornell, M.E., Lee, H.E., and Tagaras, G. Warnings of Malfunction: the Decision to Inspect and Maintain Production Processes on Schedule or on Demand. *Management Science*, **33** (1987) 1277-1290.

[80] Pham, H. and Wang, H. Imperfect Maintenance. *European Journal of Operational Research*, **94** (1996) 425-438.

[81] Phelps, E. Optimal Decision Rules for Procurement Repair or Disposal of Spare Parts. RM 2920-PR-, RAND Corp., Santa Monica, CA, 1962.

[82] Pierskalla, W.P. and Voelker, J.A. A Survey of Maintenance Models: the Control and Surveillance of Deteriorating Systems. *Naval Research Logistics Quarterly* **23** (1976) 353-388.

[83] Rahim, M.A. Joint Determination of Production Quantity Inspection Schedule, and Control Chart Design. *IIE Transactions*, **26** (1994) 2-11.

[84] Rishel, T.D. and Christy, D.P. Incorporating Maintenance Activities into Production Planning: Integration at the Master Schedule versus Material Requirements Level. *International Journal of Production Research*, **34** (1996) 421-446.

[85] Roberts, S. M., and Escudero, L.F. Minimum Problem Size Formulation for the Scheduling of Plant Maintenance Personnel. *Journal of Optimization Theory and Applications*, **39** (1983) 345-362.

[86] Rosenman, B. and Hoekstra D. A Management System for High-Value Army Components. U.S. Army Advanced Logistics Research Office, Frankfort Arsenal, Report No. TR64-1, Philadelphia, Pennsylvania, 1964.

[87] Rosenblatt, M.J. , and Lee, H.L. Economic Production Cycles with Imperfect Production Processes. *IIE Transactions*, **18** (1986) 48-55.

[88] Scudder, G.D. Priority Scheduling and spares stocking policies for a repair shop: the multiple failure case. *Management Science*, **30** (1984) 739.

[89] Scudder, G.D., and Hausman, W.H. Spares Stocking Policies for Repairable Items with Dependent Repair Items. it Navel Research Logistic, **29** (1982) 303-322.

[90] Sheikh, A.K., Callom, F.L., and Mustafa S.G. Strategies in spare parts management using a reliability engineering approach. *Engineering Costs and Productions Economics*, **21** (1991) 51-57.

[91] Sherbrooke, C.C. METRIC: a Multi-Echelon Technique for Recoverable Item Control. *Operations Research*, **16** (1968) 122-141.

[92] Sherbrooke, C.C. A Evaluator for the Number of Operationally Ready Aircraft in a Multi-Echelon Availability models. *Operations Research*, **19** (1971) 618-635.

[93] Sherif, Y.S. and Smith, M.L. Optimal Maintenance Models for Systems Subject to Failure: a Review. *Naval Research Logistics Quarterly*, **28** (1981) 47-74.

[94] Silver, E.A., D.F. Pyke and R. Peterson, *Inventory Management and Production Planning and Scheduling*, John Wiley and Sons, Inc., New York, 1998.

[95] Simon, R.M. Satisfactory Properties of a Two-Echelon Inventory Modes for Low Demand Items. *Operations Research*, **19** (1971) 761-773.

[96] Slay, F.M., and O'Malley, T.J. An Efficient Optimization Procedure for a Levels-of-Indenture Inventory Model. AF-605 (draft), Logistics Management Institute, Washington, D.C., 1978.

[97] Smith C.H., and Schafer, M.K. Optimal Inventories for Repairable Redundent Systems with Aging Components. *Journal of Operations Management*, **5** (1985) 339-350.

[98] Sun, Y. Simulation for Maintenance of an FMS: an Integrated system of Maintenance and Decision Making. *International Journal of Advanced Manufacturing Technology*. **9** (1994) 35-39.

[99] Tagaras, G. An Integrated Cost Model for the Joint Optimization of Process Control and Maintenance. *Operations Research* **39** (1988) 757-66.

[100] Tapiero, C.S. Continuous Quality Production and Machine Maintenance. *Naval Resarch Logistics Quaterly*, **33** (1986) 489-499.

[101] Taylor, H.M. Optimal Replacement under Additive Damage and other Failure Models. *Naval Resarch Logistics Quaterly*, **22** (1975) 1-18.

[102] Taylor, R. J., A linear programming model to manage the maintenance backlog *Omega, International journal of Management Science,***24-2**(1995) 217-227.

[103] Thomas, L.C., and Osaki, S. A note on ordering policy. *IEEE Transactions on Reliability*, **27** (1978) 380-381.

[104] Thomas, L.C., and Osaki, S. An optimal ordering policy for a spare unit with lead time. *European Journal of Operational Research*, **2** (1978) 409-419.

[105] Tse, E. and Makis, V. Optimization of Lot Size and the Time of Replacement in a Production System Subject to Random Failure. In *Proceedings of Automation'94*, Taipei, Taiwan, **4** (1994) 163-169.

[106] Valdez-Flores, C. and Feldman, R.M. A Survey of Preventive Maintenance Models for Stochastically Deteriorating Single Unit Systems. *Naval Research Logistics Quarterly.* **36** (1989) 419-446.

[107] Wildeman, R. *The Art of Grouping Maintenance*, Thesis Publishers, Amsterdam, 1996.

[108] Worral, B.M., and B. Mert. Applications of Dynamic Scheduling Rules in Maintenence Planning and Scheduling. *Proceedings of the 5th International Conference on Production Research (ICPR)*, August 12-16, 1979, pp. 260-264.

[109] Zuckerman, D. Replacement Models under Additive Damage. *Naval research Logistics Quarterly* , **24** (1977) 549-558.

II

MAINTENANACE PLANNING & SCHEDULING

Chapter 2

MATHEMATICAL MODELS IN MAINTENANCE PLANNING AND SCHEDULING

Salih O. Duffuaa
Systems Engnieering Department
King Fahd University of Petroleum and Minerals
Dhahran-31261, Saudi Arabia.

Abstract   This chapter presents various advanced mathematical models in the area of maintenance planning and scheduling (MPS) that have high potential of being applied to improve maintenance operations. First, MPS is classified based on function and horizon. The classes resulted include maintenance capacity planning , resource allocation,and maintenance job scheduling . Each class requires a model that captures its characteristics. The models presented in this chapter are based on linear, nonlinear and stochastic programming . Models limitations as well as direction for further research are outlined in the chapter.

Keywords: Linear programming , stochastic programming , planning and scheduling, capacity planning resource allocation.

1.   INTRODUCTION

The purpose of this chapter is to present recent advanced models and approaches for maintenance planning and scheduling . Maintenance Planning and Scheduling are the most important elements of the maintenance function and are prerequisite for effective maintenance control. Their importance has been recognized and emphasized in Duffuaa and Al-Sultan [7] and Duffuaa et al [5]. Some of the reasons for their importance are:

- In an era of automation and new production systems, such as Just-In-Time (JIT), operating with minimum stocks of finished products and work in process have made interruptions to production costly. Such interruptions can be minimized with effective planning and scheduling models.

- Effective planning and scheduling improve the utilization of production capacity and maintenance resources.

- Effective planning and scheduling models aid in provisioning the needed spare parts and material in advance and thus to minimize the cost of maintenance.

- They contribute to the high quality of maintenance work by adopting the best methods and procedures and assigning the most qualified trades for the jobs.

Maintenance planning and scheduling models can be classified based on function or horizon. The functional classification results in planning models and scheduling models. The horizon classification results in long, medium and short term models.

The long and medium term models address maintenance capacity planning and spare parts provisioning for maintenance. The short term planning models focus on resource allocation monitoring and control.

The long term scheduling models focus on the allocation of resources and scheduling production units for planned maintenance. Such models are normally developed for scheduling maintenance for electric generators in large electric utilities or production units in a process plant. Medium range scheduling models are used to schedule large maintenance jobs or shut down maintenance. Heuristic rules are used for short term maintenance scheduling . Figure 1.1 depicts the classes of maintenance planning and scheduling problems.

The sections in this chapter are organized around this classification. Section 2 outlines models for maintenance capacity planning while Section 3 presents models that are formulated for allocation of resources to accomplish planned work orders. Section 4 highlights approaches and models for planning and scheduling preventive maintenance for a number of production units. Section 5 presents models for scheduling planned and emergency maintenance jobs and section 6 concludes the chapter by summarizing model limitations and suggesting areas for further research.

2.   MAINTENANCE CAPACITY PLANNING

The general maintenance capacity planning determines the optimal level of resources (workers, skills, spare parts, equipment and tools) re-

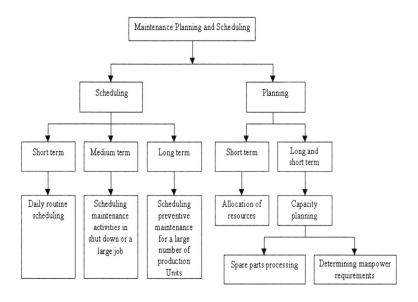

Figure 2.1   Classification of maintenance planning and scheduling.

quired to meet the forecasted maintenance load (Duffuaa et al)[5]. Two major subproblems of this type of capacity planning have been studied in the literature, which are:

1. To determine the optimal mix of the skills of trades that are needed to meet the maintenance load from available sources.

2. To determine the optimal spare parts provisioning policy to meet maintenance requirements.

Several models have been proposed in the literature to deal with the above two problems. A sample of these models appeared in Robert and Escudero [12], Duffuaa and Al-Sultan [7], Duffuaa and Raouf [8] and Al-Zubaidi and Christer[1]. It is very complex to treat the general maintenance capacity planning problem in a single optimization model including all aspects of the problem. It is to be noted also, that many variables in this problem are stochastic in nature. This led to the development of several simulation models for the general problem or simplified versions of it. Such models appeared in Duffuaa and Raouf [8] and Al-Zubadi and Christer[1]. The subproblem that deals with the determination of the optimal mix of trades to meet the maintenance requirements was formulated as a Linear Programming (LP)after relaxing the integrality of some variables. The formulation given below deals with mechanical work but can be easily modified for other types of work:

x_{ijkt} =   Number of man-hours from the mechanical trade of skill level i; $i = 1, 2, ..., I$, from source $j, j = 1, 2, ..., J$, made available to perform mechanical work of grade k, $k = 1, 2, ..., K$, in period $t, t = 1, 2, ..., T$;

c_{ij} =   Hourly cost of a mechanical trade of skill level i from source j ;

r_{ktl} =   Cost of backlogging one man-hour of grade k work from period l to period t ;

b_{ktl} =   Number of manhours of grade k work backlogged from period l to period t ;

p_{ijk} =   Productivity of a mechanical trade of skill level i from source j when performing mechanical work of grade k ;

F_{kt} =   Forecasted mechanical load of grade k in period t ;

B_{kt} =   Backlog of grade k work at period t in terms of manhours;

UB_k =   Upper limit for a healthy backlog for grade k work ;

LB_k =   Lower limit for a healthy backlog for grade k work ;

U_{ijt} =   Upper limit on the availability of skill i mechanical trade from source j in period t .

The LP model for determining the required number of mechanical manhours of different skills made available from different sources to perform all grades of mechanical work, consists of an objective function and a set of constraints. The objective function is to minimize manpower cost and the cost due to backlogging maintenance work. The constraints include the work balance constraints, a reasonable ratio between in-house manhours and overtime, and limits on labor availability. The model is stated as :

$$\text{Min} \sum_i \sum_j c_{ij} \left(\sum_k \sum_t x_{ijkt} \right) + \sum_k \sum_t \sum_l r_{ktl} b_{ktl} \qquad (2.1)$$

subject to:

- Backlog balance constraints

$$\sum_{l=1}^{t-1} b_{ktl} = B_{k,t} \qquad k = 1, 2, ..., K, \ t = 1, 2, ..., T \qquad (2.2)$$

- Work balance constraints

$$\sum_i \sum_j p_{ijk}x_{ijkt} + B_{kt} = F_{kt} + B_{k,t-1} \quad k = 1, 2, ..., K, \ t = 1, 2, ..., T$$

(2.3)

- Limit on overtime man-hours in terms of in-house regular man-hours (taken as 25%); 1 represents regular in-house and 2 represents overtime man-hours.

$$x_{i2kt} - 0.25x_{i1kt} \le 0 \quad k = 1, 2, ..., K, \ t = 1, 2, ..., T, \ i = 1, 2, ..., I$$

(2.4)

- Limits on man-hour availability

$$\sum_k x_{ijkt} \le U_{ijt} \quad k = 1, 2, ..., K, \ t = 1, 2, ..., T, \ i = 1, 2, ..., I$$

(2.5)

- Lower and upper limits on backlog on different work grade

$$LB_k \le B_{kt} \le UB_k \quad k = 1, 2, ..., K, \ t = 1, 2, ..., T \qquad (2.6)$$

The preceding model can be solved by any linear programming algorithm such as the simplex-based code in the Linear Interactive and Discrete Optimization (LINDO) package (Schrage [13]) or the one in the International Mathematical Software Library (IMSL) or the Optimization Software Library (OSL).

Stochastic simulation offers a viable approach for modeling maintenance system. Through such models the general problem for maintenance capacity planning can be addressed. Duffuaa et al [4] developed a conceptual model for maintenance that has been proposed and implemented using AweSim, the commercial version of SLAM II simulation language. The model has six modules. The modules are:

1. Load Module

2. Planning and Scheduling Module

3. Spare Parts Module

4. Equipment Module

5. Quality Control Module

6. Measure of Permanence Module

The above modules have been integrated in a simulation model and used to study various issues regarding effective maintenance management. The issues studied include manpower utilization, effect of different priority systems and effect of different spare parts service levels.

3.   ALLOCATION OF RESOURCES

An important aspect of planned maintenance is the ability to allocate resources to different plant areas in order to complete planned work orders and manage the maintenance backlog in an effective manner. A cost based linear programming model has been developed by Taylor [31] to optimize the allocation of labor crews to different prioritized work orders. The model gives consideration to different craft areas such as electrical, masonry and special projects crews and ensures that priority one jobs are completed first. The model is stated as follows:

B = base maintenance cost per hour;

R_{ij} = hours used by the Regular Crew in craft area i to satisfy priority j work orders, i =1, 2, ... n, j = 1,2,... m;

P_{ij} = hours used by the Projects Crew in craft area j to satisfy priority j work orders;

U_{ij} = unscheduled hours in craft area i for priority j work orders ;

V_{ij} = artificial cost to stimulate completing high priority work orders first;

D_i = artificial cost to assign Regular before Project Crews;

S_i = supply of Regular Crew labor-hours available in craft area i;

T = supply of Project Crew labor-hours available ;

Q_{ij} = labor-hours required in craft area i for completion of priority j work orders;

r = fraction of work that must be done by Regular Crew before using Projects Crew ;

k = fraction of total labor-hours that can be added as overtime;

n = number of areas under consideration;

k = number of priorities;

$$\text{Minimize} \sum_{i=1}^{n} \sum_{j=1}^{m} (B + V_{ij})R_{ij} + (B + V_{ij} + D_i)P_{ij} \qquad (2.7)$$

subject to

$$\sum_{j=1}^{m} R_{ij} \leq S_i \qquad \text{for all } i$$

$$\sum_{i=1}^{m} \sum_{j=1} P_{ij} \leq T$$

$$\sum_{i=1}^{n}(R_{im} + P_{im}) \leq k\left(T + \sum_{i=1}^{n}S\right)$$

$$R_{i1} \geq r(R_{i1} + P_{i1})\text{for all}\, i \text{ and } 0 \leq r \leq 1$$

$$R_{i1} + P_{i1} + R_{in} + P_{in} = Q_{i1} \text{ for all } i$$

$$R_{ij} + P_{ij} + U_{ij} = Q_{ij} \text{ for all } i\, ; j = 2, 3, ..., m-1$$

$$R_{ij}, P_{ij}, U_{ij} \geq 0$$

The above model has been tested and used to allocate crews on a weekly basis in an aluminum smelting plant.

4. SCHEDULING PREVENTIVE MAINTENANCE

This type of problem deals with scheduling major planned preventive maintenance for a number of production units. This problem is an example of a long term scheduling problem. The maintenance program and tasks are already planned for each unit and must be performed in a specified time window. This problem is typical in electric utilities and processing plants. Several optimization models appeared in the literature for this problem and it is an active area of research. The objective function for most models in the literature is to minimize cost depending on the type of industry. The constraints that are considered include:

1. Maintenance completion constraints. This constraint ensures that whenever maintenance is started on a unit must be completed without interruption.

2. preventive maintenance (PM) window constraints. This constraint ensures that the PM for each unit is done in the specified window.

3. Crews availability constraints

4. Precedence constraints . These types of constraints require maintenance to be performed on critical units prior to less critical ones.

5. Production schedules constraints. This type of constraint ensures delivery of products.

6. Reliability constraint. The type of constraint ensure that a certain production level is attained with certain reliability .

A sample of the models in this area appeared in Dopazo and Merril[9] and Escudero [10]. In addition, on invitation, a review of the work in this area was given by Kralj and Pedrovic [11]. Chapter 3 of this book provides a typical model in the area of electric power generation with an account of the approaches used to solve such models.

5.   SCHEDULING MAINTENANCE JOBS

This type of scheduling problem deals with the scheduling of available maintenance jobs at the beginning of each horizon. It falls under the category of medium or short term maintenance scheduling problem. One of the first models for this problems was formulated by Roberts and Escudero [12]. The model in Roberts and Escudero [12] is a deterministic mixed integer programming model.

In general the problem of scheduling maintenance jobs is far more complex than the problem formulation given by Roberts and Escudero [12]. This has been realized by Duffuaa and Al-Sultan [7] and a realistic statement of the problem is given together with a stochastic programming model that formulates it. Duffuaa and Al-sultan [7] described the maintenance scheduling problem as follows. At the beginning of each horizon there is a set of known jobs to be scheduled in the next horizon and while these jobs are being performed there is a probability that a set of anticipated jobs (emergency jobs) might occur. The known jobs come from planned maintenance that include: (1) preventive and routine maintenance, (2) scheduled overhauls and some emergency maintenance jobs on hand. Therefore a realistic schedule must take into consideration both known (on hand) and anticipated jobs in order to minimize interruptions and delays. It can be seen that the above statement casts the problem in a stochastic framework.

The scheduling of maintenance jobs involves both deterministic constraints resulting from the maintenance jobs on hand and stochastic constraints resulting from uncertain future jobs (emergency jobs). Stochastic programming with recourse (SR) is a formulation that handles current constraints and uncertain future events (Dantzig [3] and Vajda [15]). Therefore it seems to be a natural model for scheduling maintenance jobs. The stochastic programming model with simple recourse (SR) is stated as follows:

$$\text{Max } cx + E\left[\inf q^+y^+ + q^-y^- \mid y^+ - y^- = R - Tx, v^+, v^- \geq 0\right] \quad (2.8)$$

subject to

$$Ax \leq b$$
$$x \geq 0$$

where R is a vector of random variables with known distribution function and E the expected value operator; $A(m_1 \times n_1), b(m_1 \times 1), c(1 \times$

$n_1), q^+(1 \times m_2), q^-(1 \times m_2)$ and $T(m_2 \times n_1)$ given matrices. Another way of stating the SR model is as follows

$$\text{Max} \quad cx + E\left[\inf \; q^+ y^+ + q^- y^-\right] \tag{2.9}$$

subject to

$$
\begin{aligned}
Ax &\leq b \\
Tx + y^+ - y^- &= R \\
x \geq 0, y^+, y^- &\geq 0
\end{aligned}
$$

The above is the formulation for the simple recourse problem. For more on stochastic programming with resource, see Dantzig [3] and Vajda [15]. Next we present a recourse formulation for the maintenance scheduling problem.

5.1 Problem Formulation.

In this section a model is formulated for scheduling maintenance jobs. The model extends the minimum formulation given by Roberts and Escudero [12] by incorporating the stochastic component of the maintenance scheduling problem. Prior to stating the model, the following notations are defined :

i	=	subscript, for the ith skill;
j	=	subscript, for the jth job;
k	=	subscript, for the kth hour clock time;
l	=	subscript, for the lth starting time for a job;
t	=	subcript, for the tth hour from the start of a job, incremental time;
c_{jkl}	=	values associated with Y_{jkl};
h_{ijt}	=	hours required for the ith skill on the tth job, at the rth hour from the start of the job;
H	=	horizon of scheduling , in hours;
n	=	number of known jobs to be scheduled (deterministic part);
N	=	total number of jobs (both known and expected);
S	=	total number of skill types available;
S_{ik}	=	skill hours available for ith skill, kth hour clock time;
Y_{jkl}	=	binary variable for jth job,kth hour clock time, lth starting time for the job; starting time for the job; if $Y_{jkl} = 1$, the job is scheduled; if $Y_{jkl} = 0$, the job is not scheduled;
Z_{jl}	=	binary variable for jth job, lth starting time for the job; $Z_{jl} = 1$, the job scheduled; $Z_{jl} = 0$, the job is not scheduled;
θ_j		job duration, in terms of number of clock hours for job j;
$\{I_j\}$	=	index set for skill types required for job j;

$\{I_k\}$ $=$ index set of skill types that are used in hour k;

$\{J_{ik}\}$ $=$ index set for jobs that use skill i in hour k, for all t and l subscripts;

$\{K_i\}$ $=$ index set of hours where skill i is used;

n_j $=$ duration of job j in hours;

R_{ijk} $=$ reserved manpower of skill i for anticipated job j in hour k(hour k is in the future);

P_j^k $=$ probability job of type j occurs in hour k

d_{ijk}^+ $=$ positive deviation from the manpower of skill i reserved for anticipated job j in hour k;

d_{ijk}^- $=$ negative deviation from the manpower of skill i reserved for anticipated job j in hour k;

μ_i^+ $=$ penalty of having extra manpower reserved;

μ_i^- $=$ penalty of having less manpower reserved.

The model formulated for the maintenance job scheduling problem incorporating deterministic and stochastic components is given below:

$$[ll]\text{Max} \quad \sum_j \sum_k \sum_l c_{jkl} Z_{jl} - \sum_{i=1}^S \mu_i^+ \left[\sum_{j=n+1}^N \sum_{k=1}^H P_j^k d_{ijk}^+ \right]$$

$$- \sum_{i=1}^S \mu_i^- \left[\sum_{j=n+1}^N \sum_{k=1}^H P_j^k d_{ijk}^- \right] \qquad (2.10)$$

subject to

$$\sum_{i\in\{I_j\}} \sum_{t=1}^{\theta_j} h_{ijt} Y_{ill} \leq \sum_{i\in\{I_j\}} \sum_{t=1}^{\theta_j} h_{ijt}, \quad j = 1, ..., n \text{ and } \ell = 1, ..., H - \theta_j + 1$$

Skill hour balances

$$\sum_{j\in\{J_{ik}\}} h_{ijt} Y_{jll} + \sum_{j\in\{J_{ik}\}} R_{ijk} \leq S_{ik}, \quad i = 1, 2, ..., S, \ k = 1, 2, ..., H$$

Selection equations

$$\sum_{l=1}^{H-\theta_j+1} Z_{jl} = 1 \ j = 1, 2, ..., n$$

$$Y_{ill} - Z_{jl} = 0 \ j = 1, 2, .., n\ell = 1, ..., H - \theta_j + 1$$

Goal equations

$$R_{ijk} + d_{ijk}^+ - d_{ijk}^- = h_{ijk} \ j = 1, 2, ..., n, l = 1, ..., H - \theta_j + 1, i = 1, 2, ..., S$$

y_{ill}, Z_{jl} are zero-one variables $d^+_{ijt}, d^-_{ijt}, R^-_{ijk} \geq 0$.

The recourse formulation is a trade off between complete anticipation models and complete adaptation models . In the first, decisions are made right at the beginning of the horizon and the plan is not altered by the outcomes of the subsequent period, while in the latter, there will be adjustments from one period to another depending on the outcome of that particular period. It is clear that the recourse formulation tries to strike a balance between the two models.

To demonstrate the utility of the above model the following example taken from Duffuaa and Al-Sultan [7] is presented. The horizon is taken to be an 8 hour shift and the block for each job shows the skill requirements for that job. For example, job 1 requires 2 mechanics in the first hour, 1 mechanic in the second hour, 3 electricians in the third hour and 1 plumber in the fourth hour. M represents mechanic, E electrician and P plumber. Jobs one through four are the jobs on hand at the beginning of the horizon.

Job 1	2M	M	2E	P

Job 2	M	P	P	E	M

Job 3	P	E	M

Job 4	E	P	2E	M

Stochastic component (expected jobs in the horizon) are jobs five to nine.

	1	2	3	4	5	6	7	8	Probability
Job 5		2M	E						0.5
Job 6			2M	E					0.25
Job 7					2M	E			0.25
Job 8	P	2E							0.3
Job 9				P	E				0 7

The deterministic part in the example consists of jobs at hand at the beginning of the horizon. The stochastic part shows the jobs expected in the coming hours. For example, job 4 is expected to arrive at hour 2 with probability 0.5 and requires 2 mechanics in the first hour and an electrician in the second hour. Similarly the hour corresponding to the first block is the hour when the job is expected to arrive. The last column provides the probability with which it will arrive at that hour. The maintenance department has two mechanics and two electricians and one plumber. c_{jkl} is 100 for jobs 1 and 2 and 20 for jobs 3 and 4. μ_j^+ is taken to be 100 and μ_j^- is taken to be 20 for all j.

The above example is formulated using the proposed approach in this section and the resulting program is as follows:

$$
\begin{aligned}
\text{Max} \quad & 100Z_{11}+100Z_{12}+100Z_{13}+100Z_{14}+100Z_{15}+100Z_{21} \\
& +100Z_{22}+100Z_{23}+100Z_{24}+20Z_{31}+20Z_{32}+20Z_{34}+20Z_{35} \\
& 20Z_{36}-20Z_{41}-20Z_{42}-20Z_{43}-20Z_{44}-20Z_{45}-10D_{352}^+ \\
& -50D_{352}^--10D_{253}^+-50D_{253}^--5D_{363}^+-25D_{363}^- -5D_{264}^+ \\
& -25D_{264}^--5D_{375}^+-25D_{375}^--5D_{276}^+-25D_{276}^- -6D_{182}^+ \\
& -30D_{182}^--6D_{283}^+-30D_{283}^--1.4D_{194}^+-70D_{194}^- -14D_{295}^+ \\
& -70D_{295}^-
\end{aligned}
$$

$$
\begin{array}{rll}
2) & D_{352}^+-D_{352}^-+R_{352} & = 2 \\
3) & D_{252}^+-D_{253}^-+R_{253} & = 1 \\
4) & D_{363}^+-D_{363}^-+R_{363} & = 2 \\
5) & D_{264}^+-D_{264}^-+R_{264} & = 1 \\
6) & D_{375}^+-D_{375}^-+R_{375} & = 2 \\
7) & D_{276}^+-D_{276}^-+R_{276} & = 1 \\
8) & D_{182}^+-D_{182}^-+R_{182} & = 1 \\
9) & D_{283}^+-D_{283}^-+R_{283} & = 2 \\
10) & D_{194}^+-D_{194}^-+R_{194} & = 1 \\
11) & D_{295}^+-D_{295}^-+R_{295} & = 2 \\
12) & Z_{11}+Z_{12}+Z_{13}+Z_{14}+Z_{15} & = 1 \\
13) & Z_{21}+Z_{22}+Z_{23}+Z_{24} & = 1 \\
14) & Z_{31}+Z_{32}+Z_{34}+Z_{35}+Z_{36}+Z_{33} & = 1 \\
15) & Z_{41}+Z_{42}+Z_{43}+Z_{44}+Z_{45} & = 1 \\
16) & 2Z_{11}+Z_{21} & \leq 2 \\
17) & Z_{11}+2Z_{12}+Z_{22}+Z_{352} & \leq 2 \\
18) & Z_{12}+2Z_{13}+Z_{23}+Z_{363} & \leq 2 \\
19) & Z_{13}+2Z_{14}+Z_{24}+Z_{32}+Z_{41} & \leq 2 \\
20) & Z_{14}+2Z_{15}+Z_{21}+Z_{42}+Z_{33}+Z_{25}+R_{375} & \leq 2 \\
21) & Z_{15}+Z_{22}+Z_{34}+Z_{43} & \leq 2
\end{array}
$$

22) $Z_{23}+Z_{35}+Z_{44}$ ≤ 2
23) $Z_{24}+Z_{45}$ ≤ 2
24) Z_{41} ≤ 2
25) $Z_{31}+Z_{42}$ ≤ 2
26) $2Z_{11}+Z_{32}+2Z_{41}+Z_{43}+R_{253}+R_{283}$ ≤ 2
27) $2Z_{12}+Z_{21}+2Z_{42}+Z_{44}+R_{33}+R_{264}$ ≤ 2
28) $2Z_{13}+Z_{22}+Z_{34}+2Z_{43}+Z_{45}+R_{295}$ ≤ 2
29) $2Z_{14}+Z_{23}+Z_{35}+2Z_{44}+R_{276}$ ≤ 2
30) $2Z_{15}+Z_{24}+Z_{36}+2Z_{45}$ ≤ 2
31) Z_{31} ≤ 1
32) $Z_{21}+Z_{32}+Z_{41}+R_{182}$ ≤ 1
33) $Z_{21}+Z_{22}+Z_{42}+Z_{33}$ ≤ 1
34) $Z_{11}+Z_{22}+Z_{23}+Z_{34}+Z_{43}+R_{194}$ ≤ 1
35) $Z_{12}+Z_{23}+Z_{24}+Z_{35}+Z_{44}$ ≤ 1
36) $Z_{13}+Z_{24}+Z_{36}+Z_{45}$ ≤ 1
37) Z_{14} ≤ 1
38) Z_{15} ≤ 1

The above program was fed to the package LINDO. The results are summarized below.

$$Z_{14} = 1, \ Z_{14} = 1, \ Z_{14} = 1, \text{ and } Z_{14} = 1.$$

This means job number one starts at the beginning of hour 4 and finishes by the end of hour number 7. Job number two starts at the beginning of hour number 1 and is completed by the end of hour number five. Job number 3 starts at the beginning of hour number one and is completed by hour number three and job number 4 starts at the beginning of hour number five and is completed by the end of hour number eight, at the end of the horizon. A brief schedule is given for the deterministic jobs

Horizon

Job	1	2	3	4	5	6	7	8
1				2M	M	2E	P	
2	M	P	P	E	M			
3	P	E	M					
4				E	P	2E	M	

In addition, the solution provides the manpower reserved for anticipated jobs. All required manpower has been reserved for jobs 5, 6 and 9 at the times they are expected. No manpower is reserved for job 7 and only part of the needed manpower is reserved for job 8. Decisions can be made on how to utilize the reserved manpower in case the jobs do not arrive.

6. CONCLUSION AND FURTHER RESEARCH

In this chapter models for maintenance planning and scheduling have been presented. The models fall in two classes. The first class consists of models developed for planning and the second class for scheduling . The models presented have not been widely used in industry for several reasons. The reasons include:

1. Unavailability of data.

2. Lack of awareness about these models.

3. Some of these models have restrictive assumptions.

Further research is needed to relax some of the assumptions in these models to enhance their applicability and use. In this regard simulation offers a good modelling tool for using these models in 'what if' type analysis. Efforts also need to be made in the data collection area to provide needed data for such models. Data required include time standards, failure data, job types and their occurrence and maintenance load.

References

[1] Al-Zubaidi, H., and Christer, A.H. Maintenance Manpower Modelling for a Hospital Buliding Complex. *European Journal of Operational Research*, **99**(3) (1995) 603-618.

[2] Ashayer,J., Teelen,A., and Selen,W. A Production Planning Model for Process industry. *International Journal of Production Research*, **34** (12)(1996) 3311-3326.

[3] Dantzig, G. *Linear Programming and Extensions* . Princeton University Press, 1962.

[4] Duffuaa,S. O., Bendaya, M., Al-Sultan, K. S. and Andijani,A. A. A simulation Model for Maintenance Systems in Saudi Arabia. *2nd Progress Report on the KACST Project # AR-16-85,*1999.

[5] Duffuaa, S. O., Raouf, A. and Campbell, J. D. *Planning and Control of Maintenance Systems Modeling and Analysis.* John Wiley, 1999.

[6] Duffuaa, S.O., and Al-Sultan,K.S. A Mathematical Model for Effective Maintenance Planning and Scheduling. *Proceedings of the*

Second Scientific Symposium on Maintenance, Planning and Operation, Riyadh , 184-193, 1993.

[7] Duffuaa, S.O. and Al-Sultan, K.S. A Stochastic Programming Model for Scheduling Maintenance Personnel. *Applied Mathematical Modelling*, **25** (1999) 385-397.

[8] Duffuaa, S.O. and Raouf,A. A simulation Model for Determining Maintenance Staffing in an Industrial Environment. *Simulation*, **8** (1993) 93-99.

[9] Dopazo, J. F., and Merrill,H. M. Optimal Generator Maintenance Scheduling Using Integer Programming *IEEE Transactions on Power Apparatus and Systems*, **94** (1975) 1537-1545.

[10] Escudero, L. F. On Maintenance Scheduling of Production Units. *European Journal of Operational Research*, **9** (1982) 264-274.

[11] Kralj, B. L. and Pedrovic K. M. Optimal preventive maintenance scheduling of thermal generating units in power systems : A survey of problem formulations and solution methods. *European Journal of Operational Research*, **35** (1988) 1-15.

[12] Roberts, S. M., and Escudero,L. F. Minimum Problem Size Formulation for the Scheduling of Plant Maintenance Personnel. *Journal of Optimization Theory and Applications*,**39-3** (1983) 345-362.

[13] Schrage, L. *Linear and Quadratic Programming with Lindo*. The Scentific Press, Palo Alto, California, 1991.

[14] Taylor, R. J. A Linear Programming Model to Manage the Maintenance Backlog *Omega, International journal of Management Science*, **24-2** (1995) 217-227.

[15] Vajda, S. *Probablistic programming*, Academic Press Inc., 1972.

[16] Worral, B. M., and Mert,B. Applications of Dynamic Scheduling Rules in Maintenance Planning and Scheduling,*Proceedings of the 5th International Conference of Production Research (ICPR)*, 260-264, 1979.

Chapter 3

ELECTRIC GENERATOR MAINTENANCE SCHEDULING

Ibrahim El-Amin
Department of Electrical Engineering
King Fahd University of Petroleum and Minerals
Dhahran 31261
Saudi Arabia

Abstract   The complexity of electric power systems has resulted in the development of software-based techniques for solving operational problems. Generator maintenance scheduling (GMS) is one of those problems. The objectives of this paper are to present a review of the problem formulation, model development and the solution techniques. GMS is a large-scale, nonlinear and stochastic optimization problem with many constraints and conflicting objective functions. It is formulated in such a way as to define the time sequence of maintenance for a set of generating units so that all operational constraints are satisfied and the objective function obtains a minimum value. GMS constraints are related to the maintenance technology of the generating units, power system requirements and manpower and material resources. The technology defines the most desirable period, duration, sequence and uninterrupted duration of the maintenance work. It must be understood that GMS is a multiobjective optimization problem. This paper shows that the most widely used approach involves the minimization of the fuel and maintenance costs . The reserve and reliability criteria are also used by assigning some weighting coefficients. There are many algorithms and approaches that are suitable for the solution of the GMS problem. These are grouped in a number of categories such as heuristic search algorithms, mathematical programming methods, expert systems , fuzzy logic approach, simulated annealing , maintenance scheduling and tabu search methods. Each method has been successfully applied to a specific problem or network. However, it is important to note that there is no general consensus or agreement about the most suitable method. Future research may address the issues and impact of the independent power producer (IPP) on GMS problems. Constraints on transmission systems also have to be included and considered in GMS formulation.

Keywords: Generator Maintenance Scheduling , Power system reserve margin, Reliability , Transmission constraints, Tabu search , Genetic algorithm, Fuzzy systems, Optimization

1. INTRODUCTION

The complexity of electric power systems has warranted the development of techniques for solving operational and planning problems. Generator maintenance scheduling (GMS) is one of those problems. GMS is a large-scale, nonlinear and stochastic problem. It is important in system design, planning and operation. Today power systems have a large number of units with little reserve margins. The de-regulation of power utilities and the introduction of the independent power producers (IPP) have added more dimensions and complexity. The reliability of system operation and production costs are affected by the maintenance of generating facilities. The GMS task is to arrange the generating units for maintenance in such a way that the production costs are minimized and that certain levels of system security and adequacy are met. The term maintenance is taken in this paper to mean preventive maintenance. An optimal GMS increases system-operating reliability, reduces generation cost, and extends generator lifetime. Additionally, it could potentially defer some capital expenditure for new plants and allow critical maintenance work to be performed. A sub-optimal GMS contributes to higher production costs and lower system reliability. The maintenance schedule affects many short and long-term planning functions. For example, unit commitment, fuel scheduling, reliability calculations and production costing all have the maintenance schedule as an input. A sub-optimal schedule could adversely affect each of these functions. GMS is a large-scale, nonlinear and stochastic optimization problem with many constraints. The number of independent variables in the mathematical model is determined by the number of units and by the number of time-stages considered in the planning horizon. Non-linearity is manifested by the relationship between fuel consumption and output power. This is often a quadratic relationship. The relationship between the cost of unsupplied energy and demand is also nonlinear.

The stochastic nature of the GMS problem is derived from demand uncertainty, the generator's forced outage rate and others.

The astronomical number of possible maintenance schedules and the complexity of constraints and objectives make the maintenance problem a difficult one. However, the GMS problem can be efficiently formulated and solved by powerful operation search tools and algorithms provided that the characteristics and peculiarities of the electric power system are recognized and acknowledged. The generated electric power cannot be stored, and at every instant generation must be equal to demand. The transmission network has a limited capability . There is also a need to ensure an adequate amount of reserve capacity. The operation of interconnected power systems and the IPP upon the reserve of each network influences the GMS and the operational policy. The new competitive environment of the electricity market makes the loss of sales due to maintenance a very serious task for system operators.

The problem is to define the time sequence of maintenance for a set of generating units so that all operational constraints are satisfied and the objective function obtains a minimum value.

This paper presents an overview of the GMS problem and the state of the art methods of solving it. The GMS problem has been the subject of numerous publications and is well covered in the technical literature [1, 2, 5, 7, 17-19, 23, 26, 27]. The increased use of artificial intelligence tools in power systems has resulted in several publications that address the GMS problem.

The GMS problem has received considerable attention in the past. Dopazo and Merril formulated the problem as a mixed integer programming model (MIPM) [5]. Escudero et al. also formulated the GMS problem as MIPM but they considered a different objective function [7]. They utilized a combination of the implicit enumeration and branch-and-bound techniques. In 1982 Yamayee reviewed the literature and concluded that the existing methods made a number of assumptions and, in general, they solved only relatively small problems [26]. Yellen and Al-Khamis demonstrated the use of the decomposition approach for solving the GMS problem [2, 27]. Kralj and Rajakovic formulated a multi-objective mixed integer-programming model and used the branch-and-bound approach for solving it [18].

Recently the Bender decomposition method was used for solving the GMS problem when it was integrated with transmission system maintenance and other constraints. In this method, the problem is broken into two sub problems: a deterministic, mixed integer decision problem and a stochastic linear operation sub problem. This transforms the very complex problem into two very simple lower dimension problems.

Expert systems (ES) and rule-based approaches have also been used for the solution of GMS. This avoids the complexities and uncertainties of the maintenance schedule. ES should have powerful knowledge base rules before they can provide acceptable optimal GMS solutions [19].

GMS is a multi-objective optimization problem, and the use of fuzzy set theory is promising. It is combined with dynamic programming to simplify many of the complicated factors and uncertainties. The simulated annealing (SA) , maintenance scheduling (GA) and tabu search (TS) methods have all been used for the solution of GMS [23]. Here GMS is formulated as a mixed integer programming problem. The model is mostly deterministic. The purpose here is to reduce the computational time.

An extensive literature survey and review of the GMS problem is presented in [1,17].

2.   MAINTENANCE GOALS AND MODEL FORMULATION

2.1   Introduction.   An optimal GMS solution is expected to achieve many goals. Chief among them are the following:

- increased generator availability, thus reducing the probability of forced outage of the unit;

- improved efficiency, thus extending the lifespan of the equipment;

- minimized fuel consumption , maintenance costs, and the cost of the generated electrical energy;

- maximized revenues and reduced system losses;

- achieving the required system security and reliability levels;

- postponing the installation of new generating units.

The complexity of the GMS model is influenced by many elements. The choice of independent variables plays an important role. These variables are normally integer variables associated with the starting time of a maintenance interval.

GMS constraints are related to the maintenance technology of the generating units, the power system requirements, and the manpower and material resources . The technology defines the desirable maintenance period, duration, sequence of unit maintenance, and the uninterrupted duration of the maintenance work. The constraints also include preferred maintenance periods and required reliability levels. The use of the loss-of-load-probability (LOLP) and the net reserve are common.

The constraints on the resources are attributed to the crews and their specialities, and the simultaneous maintenance of units of a particular group . The constraints on materials such as spare parts and special tools are also considered. Most of the constraints, however, may be expressed in the form of linear equalities and inequalities in terms of the independent variables.

2.2 The GMS Constraints.

2.2.1 Maintenance Completion Constraints.

The maintenance completion constraint ensures that once a unit is removed from the system for maintenance, it completes the maintenance without interruption and it does so in a time period that is exactly equal to its maintenance duration of M_i weeks. The maintenance completion constraint also ensures that a unit is maintained just once during the planning horizon. Let:

N	: total number of units in the system;
T	: length of the maintenance planning horizon;
E_i	: earliest time (week) that maintenance on unit i
s	can start;
L_i	: latest time (week) that maintenance on unit i
	can start;
$[E_i, ..., L_i]$	: maintenance window for unit i; and,
M_i	: maintenance duration of unit i.

Let:

$$Z_{ik} = \left(\begin{array}{ll} 1 & \text{if unit } i \text{ begins maintenance in the } k\text{th week,} \\ 0 & \text{elsewhere,} \end{array} \right. \tag{3.1}$$

where $k \in [E_i, ..., L_i]$ and $i = 1, ..., N$.

Since unit i has to begin maintenance just once during the planning horizon, we have:

$$\sum_{k=E_i}^{L_i} Z_{ik} = 1, i = 1, ..., N \tag{3.2}$$

Let:

$$x_{it} = \left(\begin{array}{ll} 1 & \forall y = k, ..., k + M_i - 1 \\ 0 & \text{elsewhere} E_i \leq t < k; (k + M_i - 1) < t \leq L_i \end{array} \right. \tag{3.3}$$

To ensure that a unit i is in the maintenance state for just M_i consecutive periods, we have the following equation:

$$\sum_{t=k}^{k+M_i-1} x_{it} = M_i, i = 1, ..., N \quad E_i \leq k \leq L_i \tag{3.4}$$

The maintenance completion constraint can be expressed as:

$$(1 - Z_{ik}) Q \geq \left[\sum_{t=k}^{k+M_i-1} x_{it} - M_i \right] \geq (Z_{ik} - 1) Q \qquad (3.5)$$

$$\sum_{k=E_i}^{L_i} Z_{ik} = 1 \quad i = 1, ..., N$$

$$E_i \leq k \leq L_i \qquad (3.6)$$

where Q is a large number.

2.2.2   Crew Constraints. The crew constraint depends on the available manpower and it states that the same crew cannot maintain two units simultaneously .

Let units i_1 and i_2 have the same maintenance crew assigned to them. Both units cannot be maintained at the same time.

If the maintenance duration of unit i_1 (M_{i_1}) is longer than that of unit i_2 (M_{i_2}), the following equation ensures that unit i_2 will not be in maintenance for a period equal to that of unit i_1:

$$\sum_{t=k}^{k+M_{i_1}-1} x_{i_2,t} = 0$$

$$E_{i_1} \leq k \leq L_{i_1} \qquad (3.7)$$

The crew constraint is expressed in terms of the x_{it} variables of the second unit i_2 as follows:

$$(1 - Z_{i_1,k}) Q \geq \left[\sum_{t=k}^{k+M_{i_1}-1} x_{i_2,t} \right] \geq (Z_{i,k} - 1) Q \qquad (3.8)$$

2.2.3   Precedence Constraints. There are situations that require the maintenance of the most critical generating unit first before committing the other units in the system for maintenance. The GMS model handles this requirement by using the precedence constraint . The precedence constraint specifies the sequence in which maintenance on the units has to be performed.

Assume that unit i_2 can only be maintained after the completion of the maintenance of unit i_1 whose duration is M_{i_1}. The following equation, in terms of x_{i_2} explains the constraint:

$$\sum_{t=E_{i2}}^{k-1} x_{i_2,t} + \sum_{t=k}^{k+M_{i1}-1} x_{i_2,t} = 0 \tag{3.9}$$

The first term ensures that unit i_2 shall not be in a maintenance state before unit i_1 begins its maintenance. The second term guarantees that i_2 can begin maintenance only after i_1 has completed its maintenance.

$$(1 - z_{i_1,k}) Q \geq \left[\left(\sum_{t=E_{i2}}^{k-1} x_{i_2,t} + \sum_{t=k}^{k+M_{i_1}-1} x_{i_2,t} \right) \right] \geq (z_{i_1,k} - 1) Q \tag{3.10}$$

2.2.4   Capacity Constraints. This constraint ensures that the capacity on maintenance does not exceed the gross reserve during any stage of maintenance (t),

$$\sum_{i=1}^{N} x_{it} R_i \leq G_t \tag{3.11}$$

where

R_i : rating of unit i;
D_t : peak demand in period t;
IC: installed capacity $= \sum_{i=1}^{N} R_i$;
G_t : gross reserve in period $t = IC - D_t$.

2.2.5   Reserve Constraints. For the secure operation of a power system, it is necessary that the system always has a specified generation reserve margin . This is a deterministic constraint that defines the level of security in a power system. The reserve constraint is expressed as:

$$\sum_{i=1}^{N} R_i (1 - x_{it}) \geq D_t + R_{min} \quad t = 1, ..., T \tag{3.12}$$

where R_{min} is the minimum reserve margin kept for reliability consideration .
The value of the margin may be chosen on the basis of the size of the largest unit in the system or as a constant percentage of peak demand.

2.2.6   Reliability Constraints. The security of the power system is measured in a stochastic manner to reflect the outage of the generators . It is measured by defining a reference value of LOLP or an index of energy not served. It may be expressed as:

$$E\left\{\sum_i r_{i,t}\right\} \leq \epsilon, \tag{3.13}$$

where $E\{\}$ = the expected value;
$\quad \epsilon$ = the acceptable level of energy not served; and,
$\quad r_{i,t}$ = a vector which corresponds to the energy that has not served a time t in period T.

The inclusion of the reliability constraint would require a complex simulation, as it cannot be expressed in an analytical form. The effect of this constraint is normally reflected in the objective function or indirectly through equation (3.12).

2.2.7   Manpower Constraints.   The manpower hours required to carry out scheduled maintenance must be less than the maximum manpower hours available from the utility . This is expressed as:

$$\sum_i^{NT} MH_{pi} \leq MH_p^{max}(T), \tag{3.14}$$

where,

MH_{pi} $\quad = \quad$ manpower hours needed to maintain generator i;
$MH_p^{max}(T)$ $\quad = \quad$ maximum available manpower hours; and,
NT $\quad = \quad$ number of generator units maintained at the same time interval T.

There are several constraints that are related to the power networks and a number of others that reflect interchange and pool transactions.

2.2.8   Geographical Constraints.   This constraint is used to limit the number of maintained generators in a region.   The purpose is to avoid the loss of large amounts of transmission and the reduction of the reserve capacity in a region. It is expressed as:

$$N_m(T) \leq N_m^{max}(T), \tag{3.15}$$

where,

$N_m(T)$ $\quad = \quad$ number of maintained generators in region m; and,
$N_m^{max}(T)$ $\quad = \quad$ maximum number of generators to be maintained in region m.

2.2.9   Interchange Capability Constraint. This specifies that the maximum amount of flow shall not exceed the transmission capability of the interconnection tie lines in a multi-area system. It is stated as follows:

$$|fl_{a,t}| \leq \overline{fl_a} \tag{3.16}$$

where $fl_{a,t} = $ tie line outflow from area a; and,
$\overline{fl_a} = $ allowed maximum flow for $fl_{a,t}$.

2.3   Objective Criteria for an Optimization Model of the GMS Problem.
It is essential to point out that GMS is a multi-objective optimization problem. It is a compromise between adequate reliability, minimum fuel costs, and maximum usage of the available resources but without constraint violations. It is left to the planner to decide which optimality criterion is suitable for the system requirements. The most widely used criterion is the minimization of fuel and maintenance costs. The reserve and reliability criteria are also used by assigning some weighting coefficients to each .

2.3.1   The Economic Criterion. The cost components include those of fuel, equipment, and the personnel required to carry out the maintenance . It may also include the cost of unserved energy. The fuel cost is a variable quantity that depends on the output power of the machine. The relationship is mostly non-linear. It is usually represented by a second order polynomial in terms of the output power. The maintenance cost, on the other hand, is fixed for each particular generator. The fuel cost is much larger than the maintenance cost. The cost of unserved energy is stochastic and depends on the utility policy for load shedding and the acceptable levels of unserved energy. It is usually very small when compared to the other cost components. It is mostly neglected in the GMS problem. The sum of the fuel and maintenance costs is referred to as the operating cost of the utility.

The GMS model seeks to minimize the total generator operating cost over the operational planning period.

The objective function is thus stated as follows:

$$Min \sum_{i=1}^{N} \sum_{t=1}^{T} H \left(a_i + b_i p_{it} + c_i p_{it}^2 \right) (1 - x_{it}) + \sum_{i=1}^{N} \sum_{t=1}^{T} H p_{it} v_i \left(1 - x_{it} \right) \tag{3.17}$$

where,

p_{it} :   generator output (MW) of unit i in period t;
a_i, b_i, c_i :   fuel cost coefficients;

v_i :   variable operation and maintenance cost for unit i in $ /MWh;

H :   number of hours in a period.

The first term represents the cost of producing the energy over the planning horizon. The second term is maintenance cost of each unit

2.3.2   The Reserve Criterion. The reserve criterion has many forms . Chief among them are:

- The minimum net reserve

- The sum of weighted net reserves in all the maintenance intervals

- The sum of squares of net generation reserve

The reserve criterion tends to level the reserve and is given by

$$Min \sum_{t=1}^{T} S_{avg} - S_t \qquad (3.18)$$

where:

S_t :   net reserve in period t; and,

S_{avg} :   average net reserve.

The net reserve is

$$\sum_{t=1}^{T} S_t = \sum_{t=1}^{T} G_t - \sum_{i=1}^{N} R_i x_{i,t} \qquad (3.19)$$

The average reserve is

$$S_{avg} = \left[\frac{\sum_{t=1}^{T} G_t - \sum_{t=1}^{T} \sum_{i=1}^{N} R_i x_{i,t}}{|T|} \right] \qquad (3.20)$$

2.3.3   The Reliability Criterion. The objective function may also be stated as the minimization of the sum of squares of the reserve generation :

$$Z = Min \left\{ \sum_{t} \left(\sum_{i} p_{it} - \sum_{it} \sum_{it} x_{it} p_{it} - D_t \right)^2 \right\} \qquad (3.21)$$

Table 3.1   4-Unit Problem Data

i	R_i	E_i	L_i	M_i	a	b	c	v_i
1	200	1	5	4	78	7.97	0.00482	0.2
2	200	1	7	2	80	7.80	0.00462	0.2
3	300	1	7	2	7.65	7.65	0.00465	0.4
4	90	1	8	1	8.40	8.40	0.00610	0.5

Table 3.2   Weekly Peak Load and Gross Reserves

t	1	2	3	4	5	6	7	8
D_t	249	265	276	279	256	307	187	295
G_t	541	525	514	511	534	483	603	495

2.3.4   Illustration. The development of the GMS model is illustrated by considering a power system of 4 units. Each unit has to be scheduled for maintenance just once during a planning horizon of 8 weeks. The forecasted weekly peak demands and other relevant details for the 4-unit system are listed in the tables 3.1 and 3.2 respectively.

For the given problem:
N : total number of units = 4;
T : number of weeks in the planning horizon = 8; and,
IC : total installed capacity = 790 MW.

The objective is to minimize the total operating cost of the power system as given by equation (3.17), repeated here for the sake of clarity.

$$Min \sum_{i=1}^{N}\sum_{t=1}^{T} H\left(a_i + b_i p_{it} + c_i p_{it}^2\right)(1 - x_{it}) + \sum_{i=1}^{N}\sum_{t=1}^{T} H p_{it} v_i\left(1 - x_{it}\right)$$

$$(3.22)$$

Unit 1 : $E_i = 1, L_1 = 5$ and $M_1 = 4$ weeks.

Unit No. 1 has 5 possible starting periods (E_1+L_1-1). For each of the possible starting times, the following decision variables $z_{11}, z_{12}, z_{13}, z_{14},$ and z_{15} are associated. For example, if unit 1 begins maintenance in week 1, then only $z_{11} = 1$ and the rest are all equal to zero. Unit 1 will be out of the system for a period equal to its maintenance duration M_1 of 4 weeks. This is indicated by the elevation of the binary variables x_{11}, x_{12}, x_{13} and to the value of 1.

$$\left.\begin{array}{ccccc}
(1-z_{11})\,Q & \geq & [x_{11}+x_{12}+x_{13}+x_{14}-4] & \geq & (z_{11}-1)\,Q \\
(1-z_{12})\,Q & \geq & [x_{11}+x_{12}+x_{13}+x_{14}-4] & \geq & (z_{12}-1)\,Q \\
(1-z_{13})\,Q & \geq & [x_{11}+x_{12}+x_{13}+x_{14}-4] & \geq & (z_{13}-1)\,Q \\
(1-z_{14})\,Q & \geq & [x_{11}+x_{12}+x_{13}+x_{14}-4] & \geq & (z_{14}-1)\,Q \\
(1-z_{15})\,Q & \geq & [x_{11}+x_{12}+x_{13}+x_{14}-4] & \geq & (z_{15}-1)\,Q
\end{array}\right\} (3.23)$$

Out of the 5 possible combinations listed above for unit 1, only one should be active during the planning horizon. This is ensured by the following constraint:

$$z_{11} + z_{12} + z_{13} + z_{14} + z_{15} = 1 \tag{3.24}$$

The constraint, given above, ensures that unit 1 is maintained just once during the planning horizon.

Unit 2: $E_2 = 1$, and $M_2 = 2$ weeks.

Unit 2 has 7 possible starting periods $(E_2 + L_2 - 1)$. For each of the possible starting times, the following decision variables $z_{21}, z_{22}, z_{23}, z_{24}, z_{25}, z_{26}$ and z_{27} are associated. For example, if unit 2 begins maintenance in week 5, then $z_{25} = 1$ and the unit will be out of the system for a period equal to its maintenance duration of M_2 of 2 weeks. This is indicated by the elevation of the corresponding binary variables x_{25} and x_{26} to the value of 1.

We have, for unit 2, the following constraints:

$$\left.\begin{array}{ccccc}
(1-z_{21})\,Q & \geq & [x_{21}+x_{22}-2] & \geq & (z_{21}-1)\,Q \\
(1-z_{22})\,Q & \geq & [x_{22}+x_{23}-2] & \geq & (z_{22}-1)\,Q \\
(1-z_{23})\,Q & \geq & [x_{23}+x_{24}-2] & \geq & (z_{23}-1)\,Q \\
(1-z_{24})\,Q & \geq & [x_{24}+x_{25}-2] & \geq & (z_{24}-1)\,Q \\
(1-z_{25})\,Q & \geq & [x_{25}+x_{26}-2] & \geq & (z_{25}-1)\,Q \\
(1-z_{26})\,Q & \geq & [x_{26}+x_{27}-2] & \geq & (z_{26}-1)\,Q \\
(1-z_{27})\,Q & \geq & [x_{27}+x_{28}-2] & \geq & (z_{27}-1)\,Q
\end{array}\right\} (3.25)$$

The constraint:

$$z_{21} + z_{22} + z_{23} + z_{24} + z_{25} + z_{26} + z_{27} = 1 \tag{3.26}$$

ensures the selection of just one of the seven possible combinations for unit 2. The maintenance completion constraints for units 3 & 4 can be written in a similar fashion.

Unit 3: $E_3 = 1, L_3 = 7$ and $M_3 = 2$ weeks

$$\left.\begin{array}{ccccc}
(1 - z_{31})\,Q & \geq & [x_{31} + x_{32} - 2] & \geq & (z_{31} - 1)\,Q \\
(1 - z_{32})\,Q & \geq & [x_{32} + x_{33} - 2] & \geq & (z_{32} - 1)\,Q \\
(1 - z_{33})\,Q & \geq & [x_{33} + x_{34} - 2] & \geq & (z_{33} - 1)\,Q \\
(1 - z_{34})\,Q & \geq & [x_{34} + x_{35} - 2] & \geq & (z_{34} - 1)\,Q \\
(1 - z_{35})\,Q & \geq & [x_{35} + x_{36} - 2] & \geq & (z_{35} - 1)\,Q \\
(1 - z_{36})\,Q & \geq & [x_{36} + x_{37} - 2] & \geq & (z_{36} - 1)\,Q \\
(1 - z_{37})\,Q & \geq & [x_{37} + x_{38} - 2] & \geq & (z_{37} - 1)\,Q
\end{array}\right\} \qquad (3.27)$$

The constraint:

$$z_{31} + z_{32} + z_{33} + z_{34} + z_{35} + z_{36} + z_{37} = 1 \qquad (3.28)$$

ensures the selection of just one of the seven possible maintenance constraint combinations for unit 3.

Unit 4: $E_4 = 1, L_4 = 8$ and $M_4 = 1$

The maintenance completion constraints for unit 4 are:

$$\left.\begin{array}{ccccc}
(1 - z_{41})\,Q & \geq & [x_{41} - 1] & \geq & (z_{41} - 1)\,Q \\
(1 - z_{42})\,Q & \geq & [x_{42} - 1] & \geq & (z_{44} - 1)\,Q \\
(1 - z_{43})\,Q & \geq & [x_{43} - 1] & \geq & (z_{43} - 1)\,Q \\
(1 - z_{44})\,Q & \geq & [x_{44} - 1] & \geq & (z_{44} - 1)\,Q \\
(1 - z_{45})\,Q & \geq & [x_{45} - 1] & \geq & (z_{45} - 1)\,Q \\
(1 - z_{46})\,Q & \geq & [x_{46} - 1] & \geq & (z_{46} - 1)\,Q \\
(1 - z_{47})\,Q & \geq & [x_{47} - 1] & \geq & (z_{47} - 1)\,Q \\
(1 - z_{48})\,Q & \geq & [x_{48} - 1] & \geq & (z_{48} - 1)\,Q
\end{array}\right\} \qquad (3.29)$$

The constraint:

$$z_{41} + z_{42} + z_{43} + z_{44} + z_{45} + z_{46} + z_{47} + z_{48} = 1 \qquad (3.30)$$

ensures the selection of just one of the eight possible maintenance constraint combinations for unit 4.

Crew Constraints

A single maintenance crew is assigned to maintain units 1 & 2. As a result, units 1 & 2 cannot be committed for maintenance simultaneously at any stage during the planning horizon.

The maintenance duration of unit 1 is greater than that of unit 2. As per the formulation, the crew constraint is expressed in terms of variables of unit 2 as follows:

$$\left.\begin{array}{ccccc}
(1 - z_{11})\,Q & \geq & [x_{21} + x_{22} + x_{23} + x_{24}] & \geq & (z_{11} - 1)\,Q \\
(1 - z_{12})\,Q & \geq & [x_{22} + x_{23} + x_{24} + x_{25}] & \geq & (z_{12} - 1)\,Q \\
(1 - z_{13})\,Q & \geq & [x_{23} + x_{24} + x_{25} + x_{26}] & \geq & (z_{13} - 1)\,Q \\
(1 - z_{14})\,Q & \geq & [x_{24} + x_{25} + x_{26} + x_{27}] & \geq & (z_{14} - 1)\,Q \\
(1 - z_{15})\,Q & \geq & [x_{25} + x_{26} + x_{27} + x_{28}] & \geq & (z_{15} - 1)\,Q
\end{array}\right\} \quad (3.31)$$

Equation (3.31) ensures the selection of just one of the 5 possible crew constraint combinations listed above for the given planning horizon.

Precedence Constraints

The precedence constraint for the given problem is that unit 1 should be maintained before unit 2. In other words, unit 2 can be committed for maintenance only after the maintenance of unit 1 has been completed. Since unit 2 has the lesser maintenance duration compared to unit 1, the required precedence constraint is expressed in terms of the x_{it} variables of unit 2.

$$\left.\begin{array}{rcl}
(1 - z_{11})\,Q & \geq & [x_{21} + x_{22} + x_{23} + x_{24}] \\
& \geq & (z_{11} - 1)\,Q \\
(1 - z_{12})\,Q & \geq & [x_{21} + x_{22} + x_{23} + x_{24} + x_{25}] \\
& \geq & (z_{12} - 1)\,Q \\
(1 - z_{13})\,Q & \geq & [x_{21} + x_{22} + x_{23} + x_{24} + x_{25} + x_{26}] \\
& \geq & (z_{13} - 1)\,Q \\
(1 - z_{14})\,Q & \geq & [x_{21} + x_{22} + x_{23} + x_{24} + x_{25} + x_{26} + x_{27}] \\
& \geq & (z_{14} - 1)\,Q \\
(1 - z_{15})\,Q & \geq & [x_{21} + x_{22} + x_{23} + x_{24} + x_{25} + x_{26} + x_{27} + x_{28}] \\
& \geq & (z_{15} - 1)\,Q
\end{array}\right\} \quad (3.32)$$

Capacity Constraints

$$\sum_{i=1}^{4} x_{it} R_i \leq G_t \qquad (3.33)$$

$$\left.\begin{array}{r}
200x_{11} + 200x_{21} + 300x_{31} + 90x_{41} \leq 541 \\
200x_{12} + 200x_{22} + 300x_{32} + 90x_{42} \leq 525 \\
200x_{13} + 200x_{23} + 300x_{33} + 90x_{43} \leq 514 \\
200x_{14} + 200x_{24} + 300x_{34} + 90x_{44} \leq 511 \\
200x_{15} + 200x_{25} + 300x_{35} + 90x_{45} \leq 534 \\
200x_{16} + 200x_{26} + 300x_{36} + 90x_{46} \leq 483 \\
200x_{17} + 200x_{27} + 300x_{37} + 90x_{47} \leq 603 \\
200x_{18} + 200x_{28} + 300x_{38} + 90x_{48} \leq 495
\end{array}\right\} \quad (3.34)$$

Reserve Constraints

$$\sum_{i=1}^{4} R_i (1 - x_{it}) \geq D_i + R_{min} \qquad t = 1, ..., 8 \qquad (3.35)$$

For the given problem, a minimum reserve equal to 20% of the maximum weekly peak demand is kept for reliability considerations .

Maximum $(D_t) = 307$ MW. $R_{min} = 0.20(307) = 62$ MW.

The reserve constraints ensure that the total available power from the units, which are not committed to maintenance, is greater than the demand plus the reserve.

$$\left.\begin{array}{r}
200 (1 - x_{11}) + 200 (1 - x_{21}) + 300 (1 - x_{31}) + 90 (1 - x_{41}) \geq 311 \\
200 (1 - x_{12}) + 200 (1 - x_{22}) + 300 (1 - x_{32}) + 90 (1 - x_{42}) \geq 327 \\
200 (1 - x_{13}) + 200 (1 - x_{23}) + 300 (1 - x_{33}) + 90 (1 - x_{43}) \geq 338 \\
200 (1 - x_{14}) + 200 (1 - x_{24}) + 300 (1 - x_{34}) + 90 (1 - x_{44}) \geq 341 \\
200 (1 - x_{15}) + 200 (1 - x_{25}) + 300 (1 - x_{35}) + 90 (1 - x_{45}) \geq 318 \\
200 (1 - x_{16}) + 200 (1 - x_{26}) + 300 (1 - x_{36}) + 90 (1 - x_{46}) \geq 369 \\
200 (1 - x_{17}) + 200 (1 - x_{27}) + 300 (1 - x_{37}) + 90 (1 - x_{47}) \geq 249 \\
200 (1 - x_{18}) + 200 (1 - x_{28}) + 300 (1 - x_{38}) + 90 (1 - x_{48}) \geq 357
\end{array}\right\} \quad (3.36)$$

3.   SOLUTION METHODS

Many analytical and direct search optimization methods cannot be used for the GMS problem due to its features and dimensionality. There are, however, many algorithms and other approaches that are suitable for the solution of the GMS problem. They are grouped in a number of categories:

1. heuristic search algorithms;

2. mathematical programming methods;

3. expert systems ;

4. evolutionary search techniques such as: simulated annealing , genetic algorithm, and tabu search methods;

5. fuzzy logic approaches.

Heuristic techniques provide a GMS solution which is based on a trial and error concept. Each unit is considered separately. Although they provide clear and simple concepts with low computational time, they cannot guarantee the optimal solution. Most search methods are based on equalizing the net system reserve.

3.1   Mathematical Programming Methods.   Mathematical programming methods are widely used to solve the GMS problem. These include implicit enumeration , dynamic programming (DP) , branch and bound , and the decomposition methods[2, 5, 7, 18, 26, 27]. The use of mathematical programming methods for the solution of the GMS problem is not presented in this paper. These exact techniques solve small size problems successfully. However, for large problems, they tend to take a longer computational time. In many cases, they are considered impractical. Interested readers may refer to [1,17].

3.2   The Application of the Tabu Search Algorithm to GMS.   In this section, the implementation of the tabu search approach for electric generator maintenance scheduling is described. The details of the TS algorithm are well documented and will not be given here [1, 8-13, 16].

3.2.1   Initial Solution.   The initial solution is randomly generated using the following steps :

1. For generator $i, i = 1, ...N$, with maintenance window $[E_i, L_i]$, generate a random number r from the uniform distribution between 0 and 1 $[r \sim U(0, 1)]$. The maintenance starting period (MSP) for this generator is :

$$MSP = \text{Int.} \left[E_i + r \left(E_i, L_i \right) \right]$$

2. After determining the MSP for all generators, a schedule is determined and is checked to see if it satisfies all other constraints. If not, one of the MSPs is modified depending on which constraint has been violated. Again the modification is done randomly.

3.2.2   The Implementation of TS to the GMS Problem.

1. Read in the problem data. Initialize the tabu lists and the iteration counter.

2. Generate randomly the values of the maintenance starting periods (MSP) of the generators from their respective maintenance windows i.e.

$$\{z_{ik}; k \in [E_i, ..., L_i], ..., i = 1, ..., N\}$$

Based on the z_{ik} values, compute the corresponding x_{it} values.

3. Check whether the generated MSPs satisfy the maintenance completion constraints . If the maintenance completion constraints are satisfied, proceed to step 4, otherwise go to step 2.

4. Check whether the generated MSPs satisfy the crew and the precedence constraints . If the constraints are satisfied, proceed to step 7, otherwise go to step 5.

5. Check which of the crew or precedence constraints are violated .

6. If a precedence or crew   constraint, involving a pair of generators $i1$ and $i2$, are violated, randomly generate the maintenance start period, $z_{i2,k}$, of the second unit from its maintenance window, keeping the MSP $z_{i1,k}$ of the first unit fixed as its former value. Go to step 4.

7. Check whether the resource and the reserve constraints are satisfied. If any of the constraints are violated, go back to step 2.

8. If the constraints are satisfied, then a candidate solution (a set of z_{ik}'s) is obtained.

9. Check whether the z_{ik} values comprising the candidate solution are present in the respective tabu lists.

10. If the z_{ik} values are present, i.e., the solution is not tabu, store the values of z_{ik} in the respective tabu lists using the first-in-first out (FIFO) mechanism. Compute and store the objective function value for the candidate solution. The candidate solution is now the current solution. Update the iteration counter and go to step 14. Otherwise go to step 11.

11. If the z_{ik} values comprising the candidate solution are already present in the tabu lists, then the solution represented by the set of is tabu.

12. When a candidate solution is found to be tabu, check if it passes the aspiration criterion test. Compute the objective function value for the tabu solution. If the resulting objective function value is better

than the best obtained so far, then override the tabu status of the solution. Accept the candidate solution and store it in the tabu lists. The candidate solution now becomes the current solution. Update the iteration counter and go to step 14.

13. If the tabu candidate solution does not pass the aspiration criterion test, then discard it. Go to step 16.

14. Form a new candidate list using the current solution, i.e., generate neighbour solutions for the solution just accepted into the tabu lists. Go to step 15.

15. Pick the best candidate member in the candidate list for admission to the tabu list. Go to step 9.

16. Pick the next best member in the candidate list. Go to step 9.

17. Check for the stopping criteria. If they are satisfied, terminate the tabu search process. Otherwise, go back to step 15.

3.2.3   The Application of the TS to the GMS Problem.   The TS method was used to solve the GMS for a number of networks [6]. The tests included 5-unit, 10-unit and a 20-unit systems. The results were compared with the implicit enumeration methods. The results were identical and were carried out in a shorter time.

3.3   The Application of the Genetic Algorithm to the GMS Problem.   The maintenance scheduling (GA) is a search algorithm based on the concept of natural selection and genetic inheritance. It searches for an optimal solution by manipulating a population of strings that represent different potential solutions. Each solution corresponds to a sample point from the search space. Details of the GA are found in several references and will not be given here [4].

3.3.1   Initialization and Problem Encoding.   To apply the GA to the GMS problem, the schemes of genetic operation, the structure of strings, the encode/decode techniques and the fitness function must be decided. The maintenance schedule is referred to as a chromosome. It has a fixed binary length. One chromosome represents the maintenance schedules of all units for several years. The schedule of each unit is a row in the list of chromosomes. The encoding of the problem, with an appropriate representation, is a crucial aspect of the implementation of the GA to solve an optimization problem. In the binary representation, the GMS problem is encoded as a one-dimensional binary array. This binary string (chromosome) consists of sub-strings each of which contain the variables over the whole scheduling period for a particular unit. The size of the GA search space for this type of representation is

$$\sum_{2i=1}^{N} (L_i - M_i - E_i + 2) \tag{3.37}$$

For each unit $i = 1, 2, ..., N$, the maintenance window constraint forces exactly one variable in $\{x_{it}, t \in T\}$ to be one and the rest to be zero.

3.3.2   The Implementation of the GA for a GMS Solution.

The following steps describe the implementation of the GA to solve the GMS problem:

1. Read the system data, and initialize the GA parameters: population size, chromosome length, and crossover and mutation ratios.

2. Prepare randomly the maintenance starting period of the generating units using equation (3.37).

3. Calculate the fitness function of each chromosome. The fitness function represents the objective function. However, to take care of the various constraints, a penalty function approach is adopted. The penalty value for each constraint violation is proportional to the amount by which the constraint is violated. The evaluation function is the sum of penalty values for each constraint violation and the objective function itself with some weighting coefficients.

$$\text{The evaluation} = \sum_{c} \omega_c V_c + \omega_o F, \tag{3.38}$$

where ω_c and ω_o are the weighting coefficients; V_c is the amount of the violation of constraint c; and, F is the objective value.

4. Check for acceptance of the evaluation function as compared to the previous calculated values. If all the evaluation functions satisfy the stopping criteria, go to step (3.8).

5. Perform the GA crossover operator to generate a new offspring.

6. Calculate the fitness function of the new offspring.

7. Replace the old population by the new population.

8. Go to step 3.

9. End.

3.3.3   The Application of the GA to the GMS Problem.
The GA method was used to solve a 21-unit GMS system [4]. The results show that the GA is effective in solving the GMS problem. The encoding of the GA plays an important role in its effectiveness. The integer coding is robust and easy to program.

3.4   The Hybridized Use of GA, SA and TS Methods.
The convergence performance of the GA can be improved by adopting the acceptance probability of simulated annealing (SA) [15]. In order to escape from the local optimum, the TS is also added. In the TS, the best candidate solution in the neighborhood is repeatedly selected as a new current solution.

3.4.1   Implementation.
The implementation of the hybrid GA, SA, and TS method to the GMS problem is similar to the GA procedure:

1. Read the system data, and initialize the GA parameters: population size, chromosome length, & crossover and mutation ratios.

2. Generate the GA chromosome string candidates.

3. Calculate the fitness function.

4. If the newly calculated function is lower than the current solution, perform a crossover to the new GA string. Proceed to step 8.

5. If the newly calculated function is greater than the current solution and the SA temperature is greater than a certain random number, perform a crossover of the GA string and proceed to step 8. The SA temperature is related to the difference between fitness functions.

6. The old GA string remains as the current solution.

7. Repeat steps 3, 4 and 5 for the entire population

8. Search the neighborhood of the current solution by the TS process. The neighborhood is defined by subtracting from and adding to the value of the starting period of each unit's maintenance of the current solution by one.

9. Select the best fitness function as the new trial solution.

10. Check for a maximum number of moves by TS before stopping. The TS aspiration criterion has to be satisfied.

It should be pointed out the hybrid method appears to be attractive for many reasons.

1. The solution region is globally searched by the two kinds of genetic operations.

2. The survival of a newly produced solution is decided by the acceptance of the probability of SA.

3. The neighborhood of an accepted solution is searched by TS.

3.4.2   Application to a GMS problem. The hybrid method of the GA, SA and TS was used to solve the GMS problem of three test systems of 12-unit, 23-unit and 35-unit respectively [15]. The accuracy of the hybrid method is better than the GA, and GA+ SA. However, the computational time is longer.

4.   THE APPLICATION OF FUZZY SYSTEMS TO THE GMS PROBLEM

4.1   Fuzzy Systems. The most often used computation procedure in a fuzzy system can be divided into three steps: fuzzification, inference and defuzzification [21]. In this system, input variables are first fuzzified through designated membership functions. An inference engine is then started to search appropriate rules from the knowledge base. A defuzzification process to find final results follows this. For the GMS problem, the objectives and constraints are also first fuzzified with respect to membership functions.

4.2   The Implementation of a Fuzzy System to the GMS Problem. The procedure of the optimization of the GMS problem is summarized below. A genetic algorithm with a fuzzy evaluation function is proposed in order to overcome some of the limitations of conventional modeling and solution methods.

1. Read system data. Generate randomly a population of individuals. The chromosome length of an individual is equal to the number of parameters of the membership functions.

2. Use binary coding to encode the parameter strings.

3. Perform the fitness computation. The error vector between the output and target vectors is used as a fitness reference.

4. Choose the individuals with the higher fitness function.

5. Perform a crossover along the chromosome. Carry out a mutation process in the bits of the string.

6. Track all the generated individuals. If the errors have improved, decode the genes of the individual and proceed to step 7. If the errors have not improved go back to step 3.

7. Fuzzify the objective and constraint functions using the members of step 6.

8. Find the optimal decision. This corresponds to the one with the highest membership value.

4.3   Application. The proposed approach was tested on a practical Taiwan Power system (Taipower) via the utility data. The network is a 31-unit system. The results demonstrate the feasibility and effectiveness of the approach for GMS applications [14].

5.   THE INTEGRATED GENERATION & TRANSMISSION MAINTENANCE SCHEDULE

The maintenance scheduling problem is to determine the period for which the generating units of an electric utility should be taken off line for planned preventive maintenance. The objective is to minimize the total operating cost while system energy, reliability requirements and a number of other constraints are satisfied . Generating units are distributed in different regions and are interconnected by transmission lines. This may lead to different composite reliability levels for a given amount of maintenance capacity outage. Furthermore, generating unit maintenance should consider transmission forced and planned outages. When transmission maintenance and other network constraints are included, the problem becomes considerably more complex and will be referred to as an integrated maintenance scheduler (IMS), which represents a network constrained generation and transmission maintenance scheduling problem [22,24]. The model is identical to the GMS model. Additional constraints are added to account for the interchange capability and system reliability constraints . The network is probablistically modeled to include the effects of generation and transmission outages.

5.1   The Solution Methodology for an IMS Problem.
The IMS problem is decomposed into a master problem and operation sub-problems. The master problem is a relaxation of the original problem. It is an integer-programming problem. It will generate a set of trial solutions. The sub-problems are treated as independent problems for each time period. The sub-problems are solved using the

maintenance schedule from the master problem. Bender decomposition is used for the solution of the IMS problem.

5.2   The Implementation of Bender Decomposition to an IMS problem.

1. Read system data and setup an initial maintenance master problem. The initial master plan takes into account the operating constraints.

2. Solve the initial master problem. If the solution is infeasible, stop. Otherwise go to step 3.

3. Solve the operation sub-problems. The sub-problems may involve reserve and reliability constraints . If the solution is feasible, go to step 6.

4. Generate infeasible Bender cuts and add to the master maintenance problem. Go to step 2.

5. Check for convergence. If yes, stop. Otherwise, go to 6.

6. Generate infeasible cuts. Proceed to step 2.

7. Stop.

5.3   The Application of the IMS to a Network.

The method was tested on a three area, 48 generator test system based on the Southern Brazilian network [24]. The results indicate the effectiveness of the method.

6.   CONCLUSIONS

The maintenance of generating units occupies an important and crucial part in electrical power system operation and planning. The generator maintenance scheduling (GMS) is a large-scale, nonlinear and stochastic optimization problem with many constraints. It defines the schedule of maintenance for a set of generating units in such a way that all operational constraints are satisfied and the objective function obtains a minimum value. The objectives are sometimes conflicting in nature. The GMS is thus a compromise between adequate reliability, minimum fuel costs, and maximum usage of the available resources but without constraint violation. This paper has shown that the most widely used criterion is the minimization of the fuel and maintenance costs.

There are many solution algorithms and approaches that are suitable for the GMS problem. They are grouped in a number of categories such as heuristic search algorithms, mahtematical programming methods, expert systems , fuzzy logic approach, simulated annealing , genetic algorithm and tabu search methods. However, it is essential to note that no method emerges as the preferred solution technique. Each method can be successfully applied to a specific problem or network. Suffice to say that more research is needed before a final judgment can be passed.

References

[1] Ahmad, A. and Kothari D.P. A Review of Recent Advances in Generator Maintenance Scheduling. *Electric Machines and Power Systems*, **26** (1998) 373-387.

[2] Al-Khamis, T. M., Yellen, J., Vemuri S., and Lemonidis L. Unit maintenance scheduling with fuel constraints. *IEEE Transactions on Power Apparatus and Systems,* **7**(2) (1992) 933-939.

[3] Dahal K.P., Aldridge C.J., McDonald J.R. Generator Maintenance Scheduling Using a Genetic Algorithm With a Fuzzy Evaluation Function. *Fuzzy Sets and Systems.* **102** (1999) 21-29.

[4] Dahal, K.P., McDonald, J.R. Generator Maintenance Scheduling of Electric Power Systems Using Genetic Algorithms with Integer Representation. *Genetic Algorithms in Engineering Systems: Innovations and Applications*, 2-4 September 1997, Conference Publication No. 446, p. 456-461, 1997.

[5] Dopazo, J.F., and Merrill, H.M. Optimal generator maintenance scheduling using integer programming, *IEEE Transactions on Power Apparatus and Systems*, **PAS-94 5** (1975) 1537-1545.

[6] El-Amin, I.M., Duffuaa, S.O., Abbas M.T. A Tabu Search Algorithm for Maintenance Scheduling of Generating Units, *Journal of Electric Power System Research*, **54**(2) (2000) 91-99.

[7] Escudero, L.F., Horton, J.W., and Scheiderich J.F. On maintenance scheduling for energy generators, *IEEE Winter Power meeting*, (1980) 264-274.

[8] Glover, F. Tabu Search - Part I *ORSA Journal on Computing*, **1** (1989) 190-206.

[9] Glover, F. Tabu Search - Part II *ORSA Journal on Computing*, **2** (1990) 4-32.

[10] Glover, F. Tabu Search: A Tutorial Technical Report, Center for Applied Artificial Intelligence, University of Colorado, Boulder, CO, 1990.

[11] Glover, F. Artificial Intelligence, Heuristic Frameworks and Tabu Search. *Management and Decision Economics*, **2** (1990) 365-375.

[12] Glover, F., and McMillan C. The general employee scheduling problem: an integration of management science and artificial intelligence. *Computers and Operations Research*, **13**(5) (1986) 563-93.

[13] Hansen, P., and Jaumard, B. Algorithms for the maximum satisfiability problem, RUTCOR Research Report RR # 43-87, Rutgers, N.J., June, 1987.

[14] Huang, S.J. Generator Maintenance Scheduling: A Fuzzy System A roach With Genetic Enhancement. *Electric Power Systems Research*, **41** (1997) 233-239.

[15] Kim, H., Hayashi, Y. and Nara, K. The Performance of Hybridized Algorithm of GA SA and TS for Thermal Unit Maintenance Scheduling, 0-7803-2759-4/95/4.00, 1995.

[16] Knox, J. The Application of Tabu Search to the Symmetric Traveling Salesman Problem, Ph.D. thesis, Graduate School of Business, University of Colorado, July 1989.

[17] Kralj, B. L. and Pedrovic K. M. Optimal preventive maintenance scheduling of thermal generating units in power systems : A survey of problem formulations and solution methods. *European Journal of Operational Research,* **35** (1988) 1-15.

[18] Kralj, B., and Rajakovic, N. Multiobjective programming in power system optimization: new a roach to generator maintenance scheduling. *International J. of Electric Power and Energyn Systems* **16**(4) (1994) 211-220.

[19] Lin, C. E., Huang, C. J., Huang, C. L., Liang C. C., and Lee, S. Y. An Expert System for Generator Maintenance Scheduling Using Operation Index, *Transactions on Power Systems*, **7**(3) (1992) 1141-48.

[20] Marwali, M.K.C., and Shahidehpour S.M. Integrated Generation and Transmission Maintenance Scheduling with Network Constraints. *IEEE Transactions on Power Systems*, **13**(3) (1998) 1063-67.

[21] Momoh, J.A., Ma, X.W., and Tomsovic K. Overview and Literature Survey of Fuzzy Set Theory in Power Systems. *IEEE Transactions on Power Systems*, **10**(3) (1995) 1676-90.

[22] Nimura, T., Ziao, M., Yokoyama, R. Flexible Generator Maintenance Scheduling Considering Uncertainties of Objectives and Parameters. CCECE'96, 0-7803-3143-5/96/4.00, 1996.

[23] Satoh, K. and Nara, K. Maintenance scheduling by using simulated annealing. *IEEE/PES summer meeting*, (Paper 90 SM 439-0 PWRS): 850-857, 1990.

[24] Silva, E.L., Morozowsski, M., Fonseca, L.G.S., Oliveria, G.C., Melo, A.C.G., and Mello, J.C.O. Transmission Constrained Maintenance Scheduling of Generating Units: A Stochastic Programming Approach. *IEEE Transactions on Power Systems*, **10**(2) (1995) 695-701.

[25] Skorin-Kapov, K. Tabu Search applied to the quadratic assignment problem, *ORSA Journal of Computing*, **2**(1) (1990) 33-45.

[26] Yamayee, Z. A. Maintenance scheduling, Description, literature survey and interface with overall operations scheduling, *IEEE Transactions on Power Apparatus and Systems*, **PAS-101**(8) (1982) 2770-2779.

[27] Yellen, J., Al-Khamis, T.M., Vemuri, S., and Lemonidis, L. A decomposition approach to unit maintenance scheduling. *IEEE Transactions on Power Apparatus and Systems*, **7**(2): (1992) 726-733.

Chapter 4

RELIABILITY BASED SPARE PARTS FORECASTING AND PROCUREMENT STRATEGIES

A. K. Sheikh, M. Younas
Department of Mechanical Engineering
King Fahd University of Petroleum and Minerals
Dhahran 31261, Saudi Arabia

and

A. Raouf
Department of Systems Engineering
King Fahd University of Petroleum and Minerals
Dhahran 31261, Saudi Arabia

Abstract   Spare parts procurement strategies based on the maintenance engineering techniques of reliability engineering are merged with materials management discipline to provide a practical method to manage and control spare parts for industry. The well-known techniques of reliability engineering are used to determine failure rates for equipment and related parts. Then this information from the maintenance discipline is linked to the data of the materials management discipline. The results of this work will provide a scientific method of spare parts forecasting based on reliability of the parts, and more rational inventory management and procurement strategies with a minimum risk of stock out. As a consequence overstocking can be eliminated, and spare parts management can be streamlined on a rationale basis.

Keywords: Spare Parts , Nonrepairable Parts , Reliability , Failure Rates , Renewal Analysis , Loss Matrix , Inventory Management , Rational Inventory , Ordering Strategies .

1.   INTRODUCTION

With the expansion of high technology equipment in industries world-wide the need for spare parts to maximize the utilization of this equipment is paramount. Sound spare parts management improves productivity by reducing idle machine time and increasing resource utilization [22]. It is obvious that spares provisioning [3, 17, 22, 23, 24] is a complex problem and requires an accurate analysis of all conditions and factors that affect the selection of appropriate spare provisioning models. In the literature there are large numbers of papers in the general area of spare provisioning. Most of these papers deal with the repairable systems, and provide a queuing theory approach to determine spare parts stock on hand to ensure a specified availability of the system.

These queuing theory based models of repairable systems provide an important contribution to spares provisioning and help to design adequate repair facilities to meet an availability target. These models [1, 4, 8, 9] have been further extended to incorporate the inventory management aspect of maintenance [11, 13, 14, 15, 25, 27, 28]. The common features of these models presented in literature are:

1. They all deal with a repairable system, where in addition to MTTF, it is necessary to have some knowledge of MTTR . The criterion for evaluating such a system is its instantaneous availability $A(t)$ or asymptotic availability, $A(\infty)$= MTTF/(MTTF+MTTR).

2. Based on a queuing theory structure with a demand rate λ, and repair rate μ , appropriate analytical models are developed. In some cases demand rate is treated interchangeably with failure rate. There is a catch in this, since the failure rate is based on operational time to failure, where the demand rate (used in inventory models) and repair rate (used in availability models) are based on calendar time. Since most of the papers do not make this distinction clear, therefore the models or their parameters need to be modified to bring all quantities into a calendar time domain.

3. These queuing theory based models primarily deal with constant failure rates , and constant repair rates. This implies exponential time to failure and time to repair models. As a process the time to failure and time to completion of the repair are being characterized by a simple homogeneous Poisson process. This assumption is very restrictive. Of course there are a number of parts such as electronic components which have a constant failure rate, but there are hundreds of other mechanical parts which do not conform to the exponential time to failure distribution (i.e., constant

failure rate), $F(t) = 1 - e^{-\lambda t}$. These mechanical parts fail due to aging with time. The aging or wear out mechanisms such as *creep, fatigue, corrosion, oxidation, diffusion,* and *wear* are all *time dependent processes*, which result in increasing failure rates of the parts characterized by the Weibull model, $F(t) = 1 - e^{-(t/\eta)\beta}$ with $\beta > 1$. Thus the models presented in the literature, which are for $\beta = 1$, cannot be used when $\beta > 1$, because the time to failure process will have to be treated as a *non-homogeneous Poisson process.*

The effectiveness of spare parts management is based on factors which require improvements in data acquisition and methods of forecasting the spare parts requirements, analyzing the data on demand of such parts, and developing proper stocking and ordering criteria for these parts. The process to obtain meaningful data begins with part identification and usage information. Usually parts can be classified as *unique* or *common, critical* or *non-critical* to the operation or equipment. From this classification the process of data collection can begin. The information required usually includes lead-time, usage rates, and population of parent equipment from which a forecast of spare parts stocking levels is determined. However, this type of data is usually inadequate to optimize the maximum/minimum quantity of spare parts to be stocked based on some "guess". Similarly forecast of data using time series analysis or other forecasting techniques [6, 12] using part utilization rate (as determined by the history of issuing of these parts from store) in general will not reflect the real situation because of the intermittent use of several of these parts. As a consequence maintenance departments normally request more parts than optimum because they have no responsibility for inventory investments. In order to be on the safe side more parts than necessary are often requested particularly in huge organizations. Material managers have to gain the assistance of the maintenance engineer to provide the failure analysis and part reliability data in order to improve spare parts availability. Once the parts or equipment reliability data is available to the maintenance engineer, both the buyer and the engineer will achieve mutually beneficial results. Quantitative techniques based on reliability theory [10, 16, 18, 19, 21] need to be used for developing the failure rates of the required parts to be purchased and/or stocked. This failure rate could be used to determine more accurate demand rates.

Once the parts have been identified, for purchase or stock, the determination of the quantities is required. The quantity, based on failure analysis, is dependent on the population of the equipment using the part(s). A reasonable estimate of the failure rate of the part can be based on historical time to failure data from data points as small as five. In the absence of such data, the expected failure rates can be obtained from the information supplied by the manufacturers, which can be coupled with other information from failure rate data banks. This can be used as an initial starting point, which can be corrected progressively as more real life in-plant data is available. It is unlikely that sufficient past history of demand exists for spare parts having irregular or lumpy demand, or for new parts, making the need for an estimate of the demand based on failure rate and reliability analysis such as the Weibull analysis necessary.

Once the usage data and its characteristics have been obtained, the next step is to develop appropriate inventory and order policy consistent with an acceptable risk of shortage (tolerable stock-out penalty), inventory investment, criticality of the system, availability (lead time of the part) and location of the industry with respect to supply sources. After making the forecast for a specified planning horizon for example say 2 to 5 years, it is necessary to develop a purchasing/procurement policy. A good parts procurement or ordering strategy must carefully balance the conflicting cost elements involved in inventory. Ideally one should follow the economic order quantity equations using the probabilistic inventory model [31].

The choice is not always so simple; there are other important factors such as cost and criticality of the part, which influence the decision about how much to order and when to order. Therefore, spare parts need to be evaluated in terms of cost and criticality [29]. The cost relates to purchase cost and is classified as low, moderate or high. The cost of not having a part which shuts down the system or process is one extreme. The other extreme is that alternatives are readily available so that a part loss will not stop the continuation of the process, for example, redundant systems are in place.

2.   SCOPE OF THE WORK

As mentioned earlier, in the broader area of spare parts management for repairable facilities, a lot of work has been done and published in OR and Management Journals. Moreover in the specific area of spare parts management of nonrepairable systems which often fail with time dependent failure rates $(\beta > 1)$, there are some *renewal theory* based prediction models available for forecasting the needs of spares in a planning horizon [10]. However, the link between these forecasts and inventory management (ordering strategies) is missing.

Increasingly the technology is developing in such a way that for several types of spare parts , subassemblies and modules, replacing them upon failure is more economical than repairing them. There are hundreds of such subsystems and parts in any major organization, representing an inventory of millions of dollars. Bearings, seals, gears, probes, tubes, electronic modules, computer parts, gaskets, filters, airplane tires, light bulbs, turbine blades, valves are some of such parts which are mostly replaced rather than repaired. In a case where some of these parts fail due to a minor repairable malfunction, then such a part will be replaced with a new one while the malfunctioning part is send to a workshop or other repair facility. After repair this repaired part is reinstalled in the system of the next opportunity. The time to malfunction will not be treated as a time to failure of the part, and will not be used to determine failure rate (and demand rate). After the parthas been reinstalled the additional time to failure will be added in the previous time to malfunction of this part, and the resulting total time will be used to determine failure rate. The time taken in dismantling a part and installing a new part has been considered to be negligible in this paper, because for reliability characterization this time is not relevant. This is a part of MTTR and is only needed if the availability analysis is to be performed. The focus of the present work is to develop rational spare forecast and inventory management strategies for non-repairable parts or systems. There is no restriction on whether parts fail at a constant rate or an increasing rate. The failure process will be characterized as a renewal process with instantaneous replacements. After obtaining a forecast for the planning horizon, ordering strategies are proposed for different categories of parts in view of their *cost* and *criticality* to the operation. Since demand rate for the spares is derived from actual part failure rate, the decisions are truly reliability based.

A scientific approach to spare parts forecasting and management along these lines was proposed in two earlier papers [29, 30]. Based on the ideas and concepts presented in these papers a formal method of determining spare parts needed in a given planning horizon, and their stocking levels, and ordering strategies , is provided with sufficient details and illustrative examples. Less formal methods of spare part inventory policy have not always worked particularly in manufacturing situations. What is needed is a well planned, integrated methodology beginning with equipment population, and the reliability characteristics such as average life, coefficient of variation, Weibull parameter and failure rates . Then one needs to link these aspects of a maintenance system with the terminology of inventory control (i.e. Economic Order Quantity (EOQ) and demand rate). Such an approach is developed to accurately forecast the spare parts requirements and to provide rational part ordering strategies , which will result in minimum inventory with a higher level of confidence that there is a reduced risk of parts being out of stock when needed.

3.   RELIABILITY CHARACTERISTICS OF THE UNIT

Reliability defines how long equipment will run without a failure. It defines the probability of the part not failing within the specified time. Since reliability is important to the maintenance effort, it must be included in spare part forecasting techniques. Buyers should recognize that parts have varying degrees of reliability and failure rates . Based on the assessment of reliability and failure rates , buyers can develop better spare parts management strategies.

Reliability is the mathematical probability that a product will function for a stipulated period of time 't' and can be expressed as a *Reliability Function* $R(t)$ of the part as follows:

$$R(t) = P(T > t) = \int_t^\infty f(t)dt \qquad (4.1)$$

where, $f(t)$ is the probability density function of the time to failure of the part. T is the random variable life whereas 't' represents its values. Another useful function to represent the failure pattern of the parts is the "*hazard function*" or "*failure rate function*" $\lambda(t)$, which is defined as

$$\lambda(t)\Delta t = P\left[t < T < t + \Delta t \,|_{T>t}\right] = \frac{P[t < T < t + \Delta t]}{P[T > t]} = \frac{f(t)\Delta t}{R(t)} \qquad (4.2)$$

or

$$\lambda(t) = f(t)/R(t) \qquad (4.3)$$

3.1   Weibull Reliability Model.   The Weibull reliability model is a most versatile model for characterizing the life of machine parts. It is a two parameter model given by the following reliability and related functions:

$$R(t) = exp[-(t/\eta)^\beta] \qquad (4.4)$$

$$f(t) = (\beta/\eta)(t/\eta)^{\beta-1} exp[-(t/\eta)^\beta] \qquad (4.5)$$

$$\lambda(t) = (\beta/\eta)(t/\eta)^{\beta-1} \qquad (4.6)$$

where $t > 0$, $\eta > 0$ and $\beta > 0$. The parameter η is the "*characteristics life*" parameter. It has the same units as t, and the reliability at $t = \eta$ is $R(\eta) = 0.368$. The mean μ and variance σ^2 are given by the following equations:

$$\mu = \eta\Gamma\left(1 + \frac{1}{\beta}\right) \tag{4.7}$$

$$\sigma^2 = \eta^2\left(\Gamma\left(1 + \frac{2}{\beta}\right) - \Gamma^2\left(1 + \frac{1}{\beta}\right)\right) \tag{4.8}$$

The parameter β is a shape parameter and is a non-dimensional quantity only dependent upon the coefficient of variation of the time to failure of the part, i.e., $\beta = \varphi(\sigma/\mu)$ (Figure 4.1).

The versatility of the Weibull model is demonstrated in Figures 4.2 and 4.3 where failure rate functions $\lambda(t)$, and probability density functions for $\beta > 1$, $\beta = 1$, and $\beta < 1$ are plotted. In service either the components fail randomly (i.e., during their useful life period) which is reflected by $\beta = 1$, i.e., the constant failure rate, or they wear out with the passage of time ($\beta > 1$) representing an increasing failure rate. Therefore, this paper will primarily focus on these two cases of the failure rate. Determination of shape parameter and characteristics life η can either be made by analyzing the historical time to failure data, or, in the case of absence of such data, one can refer to the values tabulated in Ref. [5] as starting values. A condensed table (Table 4.1) derived from Ref. [5] is also given in this paper. The initial assessment about η can be made by using the average life $\overline{T} = \mu$ as provided by the manufacturer of the part, or determined from the past experience about the average life of similar parts. These estimates are then progressively updated when in house part failure history is accumulated.

3.2   Determining Parameters β and η From Time To Failure Data.

Equation (4.4) can be transformed as $lnln(1/R(t))$ $=\beta lnt - \beta ln\eta$,which is an equation of a straight line. If historical time to failure data $t_1 < t_2 < t_3 < \ < t_i < t_N$ is available , then the reliability function $R(t_i)$ can be estimated using the well known mean rank formula [18]:

$$R(t_i) = \frac{N + 1 - i}{N + 1} \tag{4.9}$$

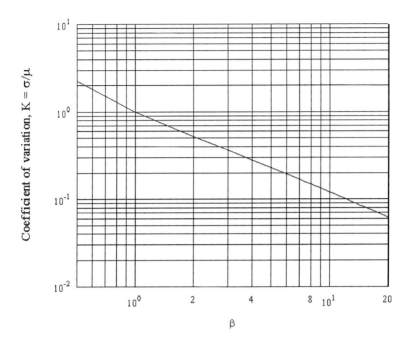

Figure 4.1 The relationship between the coefficient of variation $K = \sigma/\mu$ and shape parameter,β of the Weibull model [$\beta = 1/K$].

Table 4.1   Failure Modes and Weibull Shape Parameter β [5].

Failure Mode	Weibull Shape Parameter β
1. Deformation	1.0
2. Fracture	1.0
3. Change of Material Quality	
(a) Ageing	3.0
(b) Degradation	2.0
(c) Burning	1.0
(d) Embrittlement	1.0
4. Corrosion	2.0
5. Wear	3.0
6. Leakage	1.5

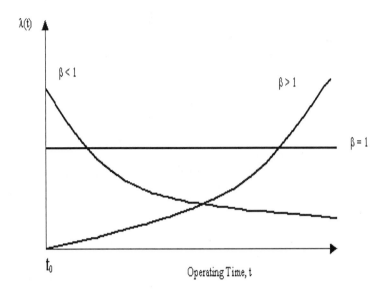

Figure 4.2 Failure rate function of Weibull model for $\beta = 1, \beta < 1$ and $\beta > 1$.

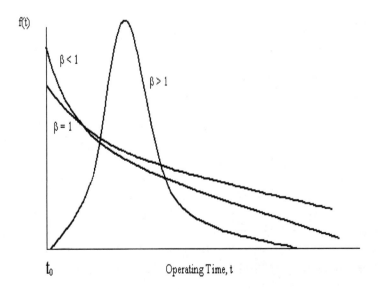

Figure 4.3 Probability density function of Weibull model for $\beta = 1$ (constant failure rate), $\beta < 1$ (decreasing failure rate) and $\beta > 1$ (increasing failure rate).

and $lnln(1/R(t_i))$ can be plotted against lnt_i , and a straight line can be fitted to this data by the linear regression approach. The parameters of the fitted line (slope and intercept) can be used to determine the Weibull parameters β and η as explained in Ref.[28].

Example 1

For illustrative purposes time to failure data of 54 nominally identical extrusion dies used in an Aluminum extrusion plant is analyzed and results are plotted in Figure 4.4. The time to failure of a die is proportional to the weight of metal being extruded until it fails. Thus the life in this particular situation is represented by kg of metal being extruded. High coefficient of determination $R^2 = .995$ indicates an excellent fit to the data.

The shape and scale parameters of these dies are $\beta = 1.83$ and $\eta = 25730$ kg.

Since $\beta > 1$, this represents *a wear out mode of failure*. Other important reliability indicators of these dies are:

$$\mu = \eta\Gamma\left(1+1/\beta\right) = 25730\Gamma\left(\langle 1+1/1.83\rangle\right) = 22850 \text{ kg}$$
$$\sigma = \eta\left(\Gamma\left(1+2/\beta\right) - \Gamma^2\left(1+1/\beta\right)\right)^{0.5}$$
$$= 25730\left(\Gamma\left(1+2/1.83\right) - \Gamma^2\left(1+1/1.83\right)\right)^{0.5} = 11073 \text{ kg}$$

Coefficient of Variation of die life can be found from Figure 4.1 as 0.57, or can be determined approximately as $K = 1/\beta = 1/1.83 = 0.546$.

4.   SPARE PARTS CALCULATIONS FOR FAILURE REPLACEMENT

4.1   Failure Replacements.   Replacements of individual units are made just after their failure. This type of replacement is primarily made for parts which will not cause any further damage to the system (machine) due to their failure in operation. If the parts are such that their actual failure in service may result in damage to the other parts of the system, then *condition monitoring* [20] is necessary and the replacement can be made just prior to the failure or somewhat earlier than that. In other words the time to replace a part under the condition monitoring scheme is when a certain manifestation of progressive damage reaches a critical level.

4.2   Failure Replacements Process Characterization by Renewal Function.   One realization of the renewal or counting process of failure replacements is given in Figure 4.5.

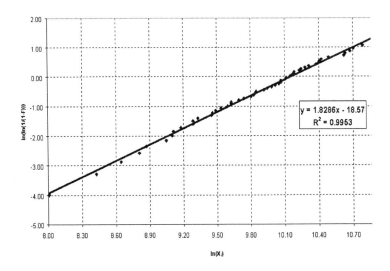

Figure 4.4 Weibull reliability analysis of 54 extrusion eies life :die life Xi is expressed as Kg of metal extruded until failure.

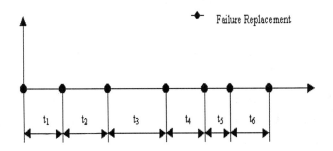

Figure 4.5 One realization of renewal or counting process of part replacement

The renewal function $H(t) = E[N(t)] = \overline{N}(t)$ defines the expected number of failures in time t and is give by the following integral equation [22]

$$H(t) = E[N(t)] = F(t) + \int_o^t H(t - x)dF(x). \qquad (4.10)$$

The renewal rate function $h(t) = dH(t)/dt$ gives the expected number of renewals per unit time, and satisfies the following integral equation [10]:

$$h(t) = dE[N(t)]/dt = f(t) + \int_o^t h(t - x)f(x)dx. \qquad (4.11)$$

The variance of number of renewals $V(N(t))$ is given by [22, 24]

$$V[N(t)] = \sigma^2[N(t)] = 2\int_o^t H(t - x)dH(x) + H(t) - H^2(t) \qquad (4.12)$$

Equations (4.10-4.12) are general equation valid for any probability model. However, by substituting the Weibull cumulative distribution function for time to failure, $F(t) = 1 - R(t)$, as

$$F(t) = 1 - exp[-(t/\eta)^\beta]. \qquad (4.13)$$

these functions can be evaluated for the Weibull Model. The renewal functions for the Weibull model for different values of β (or coefficient of variation $K = 1/\beta$) are given in Figures 4.6-4.8 [32].

4.3   Asymptotic Behavior of Renewal Functions.
Consider replacements of a part having an average time between failure as $\overline{T}$ and standard deviation of time between failures as $\sigma(T)$ (coefficient of variation of time between failures, $K = \sigma(T)/\overline{T}$) . If the operation time t of the system or machine on which this part is installed is quite long and several replacements need to be made during this period, then the average number of Failures $E[N(t)] = H(t)$ will stabilize to the following asymptotic value of the *renewal function* [10],

$$\overline{N}(t) = H(t) = E[N(t)] = \frac{t}{\overline{T}} + \frac{1}{2}(K^2 - 1) = \frac{t}{\overline{T}} + \frac{1}{2}(\frac{1}{\beta^2} - 1) \qquad (4.14)$$

and the *failure intensity* or *renewal rate function* is given by

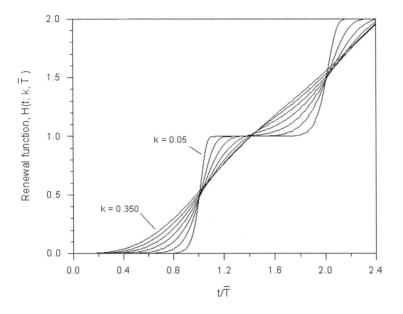

Figure 4.6 Renewal function of Weibull model.

$$h(t) = \frac{dH(t)}{dt} = \frac{dE[N(t)]}{dt} = \frac{1}{\overline{\overline{T}}} \qquad (4.15)$$

The *variance of number of failures* in time t, is

$$V[N(t)] \cong K^2 \left(\frac{t}{\overline{\overline{T}}} \right) \cong \frac{1}{\beta^2} \left(\frac{t}{\overline{\overline{T}}} \right) \qquad (4.16)$$

The *standard deviation of number of failures* in time t, is

$$\sigma[N(t)] = \frac{\sigma(T)}{\overline{\overline{T}}} \sqrt{\frac{t}{\overline{\overline{T}}}} = K \sqrt{\frac{t}{\overline{\overline{T}}}} = \left(\frac{1}{\beta} \right) \sqrt{\frac{t}{\overline{\overline{T}}}} \qquad (4.17)$$

Exponential model of time to failure is a special case, having $K = 1/\beta = 1$, and where $1/\eta$ is the failure rate of the part. Thus for an exponential model of time between failures, the equations (4.14) and (4.17) will reduce to equations (4.18) and (4.19) respectively.

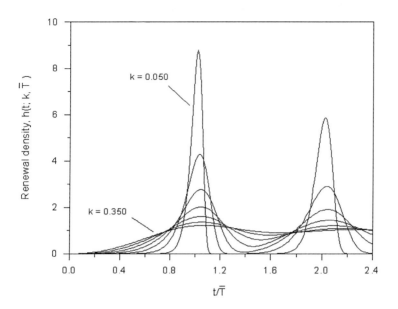

Figure 4.7   Renewal rate function of Weibull model.

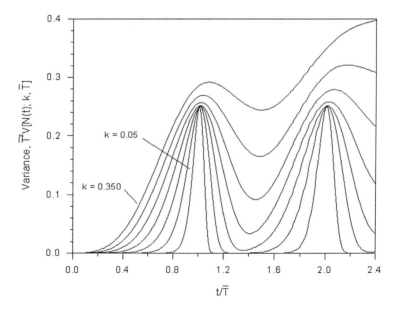

Figure 4.8   Variance of number of renewals for Weibull model.

$$H(t) = E[N(t)] = \frac{t}{\overline{T}} \tag{4.18}$$

$$\sigma[N(t)] = \sqrt{\frac{t}{\overline{T}}} \tag{4.19}$$

and equation (4.15) will remain unchanged.

4.4 Probability Distribution of $N(t)$ when t is Large.

If time t in equation (4.14-4.19) representing a planning horizon is large, then from the central limit theorem $N(t)$ is normally distributed with mean $= \overline{N(t)}$ and standard deviation $\sigma(N(t))$ [22] and is given by:

$$P(N(t) \leq N) = \Phi\left(\frac{N - \overline{N}(t)}{\sigma[N(t)]}\right) = \Phi(Z_p) = p \tag{4.20}$$

Then the number of spares N needed during this period with a probability of shortage $= 1 - p$, is given by

$$\begin{aligned}
N &= \frac{t}{\overline{T}} + \frac{1}{2}(K^2 - 1) + K\sqrt{\frac{t}{\overline{T}}}\Phi^{-1}(p) \\
&= \frac{t}{\overline{T}} + \frac{1}{2}(\frac{1}{\beta^2} - 1) + \frac{1}{\beta}\sqrt{\frac{t}{\overline{T}}}\Phi^{-1}(p)
\end{aligned} \tag{4.21}$$

where $\Phi^{-1}(p)$ is the inverse normal function and is available in probability textbooks [18],as well as in most of the spreadsheet software such as Excel. For an exponential model ($K = 1$) equation (4.21) will reduce to

$$N = \frac{t}{\overline{T}} + \sqrt{\frac{t}{\overline{T}}}\Phi^{-1}(p) \tag{4.22}$$

Example 2:

A work piece needs to be manufactured on a machining center. The milling cutter required for machining has an average life $\overline{T}$ of 160 minutes of cutting time at the adopted machining conditions. The quantity of jobs demands a total of 26 hours of machining time. With a confidence of 95%, calculate the number of milling cutters needed, $\beta = 3$ for the cutter.

Solution: $\Phi^{-1}(p) = \Phi^{-1}(0.95) = 1.65$
$\overline{T} = 160$ minutes
$\beta = 3$
$t = 26$ x $60 = 1560$ minutes
Using equation (4.21)

$$N = \frac{1560}{160} + \frac{1}{2}\left(\frac{1}{3^2} - 1\right) + \left(\frac{1}{3}\sqrt{\frac{1560}{160}}\right) \quad (1.65)$$

N=11.02 cutters, say 11 cutters.

5.   SPARE PART CALCULATIONS FOR SCHEDULED REPLACEMENTS

5.1   Scheduled Replacement Strategy.
Units or parts are replaced at scheduled times (for example, during scheduled shut-down maintenance of the machine) or periodic intervals, i.e., $t_s, 2t_s, 3t_s, ..., jt_s$ etc. However, if any failure of the units occurs during these scheduled intervals, the additional failure replacement/replacements will also be made. These additional failure replacements will be rather small in number provided there is an optimal or near optimal selection of the scheduled replacement intervals t_s.

In a planning horizon of duration $0 - t$ such that $t = N_s t_s$, where N_s represent the scheduled replacement cycles, we have N_s scheduled replacements of the part and $N_s H(t_s)$ unscheduled or failure replacements of the parts. Thus total numbers of parts needed in N_s replacement cycles are

$$N = N_s[1 + H(t_s)] \quad (4.23)$$

$H(t_s)$ can be read from Figure 4.6, or from the Tables of the Weibull renewal model such as those given by Lipson & Sheth [19]. Equation (4.23) implies that the parts are being replaced corresponding with reliability level $R(t_s) = exp[-(t_s/\eta)^\beta]$. Note N in equation (4.23) will be greater than given by equation (4.21). Therefore, the scheduled replacements can only be justified if various cost elements and value of β(or $K = \sigma/\overline{T}$) warrant it.

5.2   Comparison Of Long Run Expected Cost Of Replacements Under Scheduled And Failure Replacements.
Long run expected cost of replacement per unit part for this type of replacement policy is given by:

$$E\{C(t)\} = [C_f + C_s H(t_s)]/t_s \quad (4.24)$$

where C_s is the cost of a scheduled replacement, and C_f is the cost of a failure replacement.. Therefore, the policy of equations (4.23, 4.24) will only be superior under certain conditions of β(or $K = \sigma/\overline{T}$) and C_s/C_f. This policy can be compared with the Failure Replacement Policy, which has an expected cost

$$C_f \left[\frac{H(t)}{t} \right] = C_f \left[\frac{t}{\overline{T}} + \frac{1}{2} \left(\frac{1}{\beta^2} - 1 \right) \right] \left(\frac{1}{t} \right) = \frac{C_f}{\overline{T}} + \Delta(t) \cong \frac{C_f}{\overline{T}}. \quad (4.25)$$

Thus for superiority of the Scheduled Replacement Policy

$$C_f \left[\frac{(C_s / C_f) + H(t_s)}{t_s} \right] < \frac{C_f}{\overline{T}} \quad (4.26)$$

$$\left[\frac{C_s}{C_f} \right] < \left[\frac{t_s}{\overline{T}} - H(t_s) \right] \quad (4.27)$$

where $H(t_s)$ will depend upon parameter β. The assumption is that C_f is the same under both policies.

Example 3: Same as Problem 2, when Scheduled Replacement Policy is adopted and $C_s/C_f = 0.6$, and $\tau_s = t_s/\overline{T} = 0.92$,

Solution: $C_s/C_f = 0.6, K = 0.333, \tau_s = t_s/\overline{T} = 0.92$,
$t_s = 0.92(160) = 147$ minutes
$N_s = \frac{1560}{147} = 10.61 \cong (11 cutters = N_s)$

Thus 11 schedules replacements are made in 1560 minutes. Additional cutters due to unscheduled failures during 11 periods of replacements $= N_s[H(t_s)] = N_s H(t_s/T) = N_s[H(0.92)] = 10.61(0.6) = 6.3 \rightarrow 6$ cutters (where $H(0.92) = 0.6$ is read from Figure 4.5). Thus, the total number of cutters needed for the planning horizon of 1560 minutes of cutting time $= 11 + 6 = 17$ cutters.

Although the number of cutters needed is more than calculated in Example 2, the cost of a premature failure C_f in this case is much greater than the cost of a scheduled failure C_s, which makes consumptions of 17 cutters more economical than a failure replacement policy with its associated cost of damage or penalty due to cutter failures.

The *general rules for selecting the appropriate replacement policies in view of the above parameters* are discussed in Ref [30], and the results are summarized below:

1. If $\beta = 1$, then adopt the Failure Replacement Policy.

2. If $\beta > 1$, but $C_f > C_s$, then further investigate the possibility of adopting Scheduled Replacement Policy. Under certain conditions of β and C_s/C_f the Scheduled Replacement Policy will be better. If desired the optimal interval for scheduled replacements t_s^* can be obtained from the method shown in Ref [30].

3. If $\beta > 1$, but $C_f < C_s$, then adopt the Failure Replacement Policy.

6.   PURCHASING AND MAINTENANCE LINK

Purchasing has to work with the maintenance, reliability , or plant engineer to correlate the actual time in service of a part when the part usage or precise demand is not known. If usage history is known, based on sufficient past demand (typically 5 or more data points), then this can be used directly in the inventory models. It is unlikely that sufficient, past history of demand exists for spare parts having irregular or lumpy demand, or for new parts, making the need for an estimate of the demand based on failure rate and Weibull analysis necessary. Once the failure intensity is established there is a need to link the usage rate $\bar{U}(t)$ based on calendar time (t_c) of the part usage to the failure intensity of the part, which is based on actual operating time (t). In order to establish such a link we define the part utilization ρ as follows:

$$\rho = \frac{t}{t_c} \tag{4.28}$$

t_c is a convenient time period (often consider as one year). Since $t_c > t$, the range of ρ will be from 0 to 1 $(0 < \rho < 1)$. The *usage rate per unit of calendar time* $\overline{U(t)} = \bar{U}$ and the average operational time between failures $\bar{T}$ is related as follows:

$$\bar{U} = \frac{1}{\bar{T}}\rho \tag{4.29}$$

where $\rho \leq 1$ is the *utilization ratio (capacity ratio, or duty cycle)* of the part to take care of the idle time of the machine. For nonstop operation of the part $\rho = 1$ resulting in $\bar{U} = \frac{1}{\bar{T}}$, typical usage rate for various spare parts can be developed in house for a given plant.

The intensity of failure (i.e., $1/\bar{T}$), usage rate , and the demand rate $\bar{D}$ of the part are related to each other as follows [29]:

$$\bar{D} = N_s(t_o)\bar{U} = N_o\bar{U} = N_o\rho/\bar{T} \tag{4.30}$$

where $N_s(t_o) = N_o$ = total number of parts which are in operation at some reference time $t_o \geq 0$. Thus using equations (4.21), (4.28), and (4.30), the number of parts (N) needed in a calendar time t_c is given as follows:

$$\begin{aligned} N &= t_c\bar{D} + 0.5(K^2 - 1) + K\sqrt{t_c\bar{D}}\Phi^{-1}(p) \\ &= t_c\bar{D} + \tfrac{1}{2}(\tfrac{1}{\beta^2} - 1) + \tfrac{1}{\beta}\sqrt{t_c\bar{D}}\Phi^{-1}(p) \end{aligned} \tag{4.31}$$

6.1 Determining Demand Rate From the Knowledge of Usage Rate.

The failure intensity can also be calculated by a less sophisticated method of using historical usage rates. **Usage rate** of a spare part X_j associated with a specific machine 'j' can be calculated, by using historical information of spare part utilization in the past, with the help of the following equation;

$$\overline{U}(t) = \frac{\sum X_j}{t \sum Q_j N_j} \tag{4.32}$$

where
$\overline{U}(t)=$ Average rate of utilization per part per month (year).
$\sum X_j =$ Total net issues for spare parts X_j to be used on machines j.
$t =$ Number of month (year) for which transaction history is available.
$Q_j =$ Quantity of machines of type j using the parts X_j.
$N_j =$ Number of parts X_j installed on jth machine.

Once the usage data and factors have been obtained, the next step is to develop the appropriate inventory and order policy consistent with risk of shortage (stock out penalty), inventory investment, criticality of the system availability (lead time of the part) and location of the process with respect to supply sources.

7. SPARE PART CLASSIFICATION

There are other important factors such as cost and criticality of the part, which influence the decision about how much to order and when to order. Therefore, spare parts need to be evaluated in terms of **cost** and **criticality.**

The cost relates to purchase cost and is classified as **low, moderate** or **high.** The cost of not having a part which shuts down the system or process is one extreme. The other extreme is that alternatives are readily available so that a part loss will not stop the continuation of the process. For example, redundant systems are in place.

Criticality is based on the cost of not completing the process or assigned equipment function i.e. "the mission". Criticality can also be classified as **low, moderate** or **high.** Highly critical parts are those which are absolutely essential for mission success. Moderately critical parts are such that if they are out of stock at the time of demand, it will have only a slight to moderate effect on mission success, whereas parts of low criticality are not absolutely essential for mission success. If such parts of low criticality are not available on demand, alternate parts can be substituted, or in-plant manufacturing of such parts is

Table 4.2   Loss matrix to determine ordering policies.

Cost /Criticality ↓ →	Low L	Moderate M	High H
Low, L	l_{LL}^{***}	l_{LM}^{***}	l_{LH}^{***}
Moderate, M	l_{ML}^{**}	l_{MM}^{**}	l_{MH}^{*}
High, H	l_{HL}^{**}	l_{HM}^{*}	l_{HH}^{*}

possible, or they are instantly available in the market. To assess the criticality of parts a number of factors need to be considered, including the Pareto analysis [21, 26] of the part/equipment failure to establish ABC classification, and development of a criticality loss matrix [29].

8.   PRACTICAL SPARE PARTS MANAGEMENT

Considering **cost** and **criticality** of parts jointly, one can represent the loss to the organization due to their absence on demand, in the form of a **loss** or **regret** matrix as given in Table 4.2, where l_{iK} = loss due to the event that a part of cost, $i, (i = L, M, H)$ and criticality, k, $(k = L, M, H)$ is out of stock when demanded. The ordering strategies can be decided on the basis of this loss matrix , by grouping various closely related outcomes together (indicated by symbols, $*, **,$ and $***$ in the loss matrix).

Ordering Strategy 1:
For parts grouped as class 1, symbol $*$, *set* $[l_{HM}, l_{HH}, l_{MH}]$

Maintain a given quantity of these items in the inventory for compensation of repair and recycle times and procurement lead times, etc, and order new items on a one for one basis when failure occurs, and a spare is withdrawn from the inventory. This strategy will minimize the risks involved due to tying up too much capital and the resultant high inventory maintenance cost. Initial stock level for such parts can be calculated using equations (4.21) or (4.22) depending upon the coefficient of variation, K, and incorporating a high value of p. For example, a value of $p = 0.999$, will imply that there is a very low probability of stock out in time interval $0 - t_c^*$, where t_c^* is much less than the mission time, but greater than $[\overline{T}_c + 3\sigma_c +$ (Manufacturer's lead time + Shipment time)] i.e. $t_c^* \geq [\bar{T}_c + 3\sigma_c + (\text{Manufacturer's lead time} + \text{Shipment time})]$, where all quantities are expressed in calendar time. As an additional factor of safety the number of parts calculated using equation (4.21) or (4.22) corresponding to $t = t_c^*$, can be increased by one in the initial acquisition. Later on the orders are placed on a one by one basis.

Ordering Strategy 2:
*For parts grouped as Class 2, Symbol **, set* $[l_{ML}, l_{MM}, l_{HL}]$

In addition to the purchase cost side of the decision process, the other critical cost factors in terms of service level (risk) and safety stock need to be considered. The relevant model for ordering strategy 2 is a standard inventory model used to calculate safety stock associated with a low probability of stock out. The formula to calculate the optimum reorder quantity Q^* for this model is [21]

$$Q^* = \sqrt{\frac{2\bar{D}C_0}{C_h}} \qquad (4.33)$$

where
$\bar{D}=$ average demand per year, units/year
$C_0 =$ cost per order, \$/order
$C_h =$ Annual holding and storage cost per unit of average inventory, (\$/units/year)

This formula is consistent with the probabilistic inventory model illustrated in Figures 4.9 and 4.10. The model's variables are mentioned in Figure 4.9 and Figure 4.10, which illustrate the concept of safety stock, and associated risk of stock out (dark shaded area of the normal probability curve in Figure 4.10). Average demand during lead time (LT) is calculated from equations (4.30) and (4.14), and standard deviation of demand during lead-time is calculated from equations (4.30) and (4.16) by replacing $t = LT$, in these equations.

The effect of high stock out costs will increase the order quantity, which decreases the frequency of exposure to risk of stock out. If the cost of stock out C_e can be assigned to the number of order cycles (exposures to a stock out) then the formula for the optimum order quantity can be modified as

$$Q^* = \sqrt{\frac{2\bar{D}(C_0 + C_e)}{C_h}} \qquad (4.34)$$

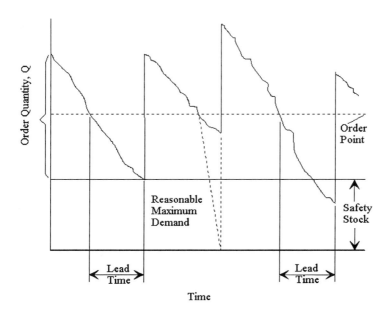

Figure 4.9   Illustration of inventory model and related variables.

Ordering Strategy 3:
*For parts grouped as Class 3. Symbol ***, set $[l_{LL}, l_{LM}, l_{LH}]$*

These parts are of low cost and varying degree of criticality. There could be two possible strategies for this type of parts:

Strategy 3(a)
If the following conditions are simultaneously satisfied:

i) spare parts can be stored for an entire mission duration without any damage during the storage.

ii) Storage cost for these parts is very small.

iii) There is a danger of non-availability of these parts during the mission time. This danger could be because of unusual circumstances. This is not inherent in assessment of part criticality.

Parts are not instantly accessible in the local market. They have to be ordered sufficiently far ahead of time.

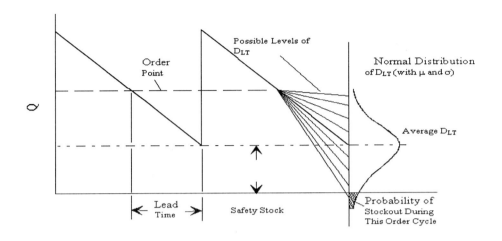

Figure 4.10   Effect of normal demand during lead time.

Then the strategy for ordering is to order them in one lot of N items to cover the entire mission period. For a specific risk of stock out say $1 - p = 0.05$, or $p = 0.95$, and t representing the mission time, the order quantity $Q = N$ can be calculated using equations (4.21) or (4.22) depending upon the value of K.

Strategy 3(b)

If one or more of the above conditions are not satisfied then make several smaller orders during the mission period, which will be identical to strategy 2, discussed in the previous section.

The ordering strategies discussed above cover both the safety stock and initial stocking level. However, the costs and risks need to be balanced in order to arrive at an inventory stocking level for specific spare parts . The following section summarizes the overall decision process in setting industrial policies.

- Spreadsheet based Weibull analysis of time to failure of parts frequently replaced in plant under consideration provides the Weibull parameters of these parts. This analysis also leads to characterization of the modes of failure.

■ Reliability models expressing operational time t, are expressed in term of calendar time t_c, using part utilization ratio, (duty cycle, or capacity ratio); $\rho = t_o/t_c$.

■ Spreadsheet based forecasting of spare parts using Weibull reliability and renewal models is made for a specified planning horizon depending upon the part replacement strategy i.e. [10, 16, 19, 21]

  − Failure replacement

  − Scheduled replacement (*constant interval replacement*)

■ The results are expressed both in terms of operational time and calendar time. Relevant charts (non dimensional graphs) could be developed to facilitate the decision making.

■ From part failure rates , the demand rate of the parts can be established and the statistical characteristics of demand (or demand rate) can be determined which provide the necessary input along with various inventory cost parameters in appropriate stochastic inventory control model(s). Optimal inventory management strategies (EOQ, Lead time, etc) are established.

■ Factors determining criticality of parts are established. Pareto analysis is used along with other relevant criteria such as a criticality loss matrix to determine the criticality of parts.

■ Part ordering strategies are streamlined depending upon the outcome of cost and criticality matrix.

Example 4.

An aluminum extrusion plant has signed a contract to supply for 3 years a certain type of extruded cross section of aluminum, at a rate of 200000 kg per year (uniformly spread over the entire year).The historical time to failure data of the die used for extruding this cross section is retrieved from the company's data files and a Weibull analysis is performed on this data. The results are plotted in Figure 4.4. The total annual production of all type of aluminum extrusion in the plant is 1000000 kg.

The calendar year, t_c is proportional to 1000000 kg. The operational time during the calendar year, t is proportional to200000 kg. The utilization ratio $\rho = t/t_c$ =200000/1000000=0.2.

From Figure 4.4, the Weibull parameters are $\beta = 1.83$ and $\eta = 25730$

kg. The cost of the die (US $5000 /die) is high, and its criticality (being vital to the operation) is also high. Thus the criticality - cost loss matrix evaluation is l_{HH}. Thus ordering strategy 1 is most appropriate in this case. Die Manufacturer's lead time is 7 weeks, shipment time is 1 week. Thus total lead time and shipment time is 2 months $=1/6$ years.

$\overline{T} = 22850$ kg.

$\overline{T}_c = \overline{T}/\rho = 22850/.2 = 114250$ kg $\rightarrow 114250$ kg $/1000000$ kg/ year $= 0.114258$ year

$3\sigma_c = 3\sigma/\rho = 38970$ kg $\rightarrow 38970$ kg $/1000000$ kg/ year $= 0.0389$ year

$t_c^* = 0.114258 + 0.0389 + 1/6 = 0.31989$ year

Operational time $t^* = \rho t_c^* = 0.063978$ year

$\rightarrow 0.06397$ year x1000000 kg/ year $= 63979$ Kg

From Figure 4.6, $H(t^*)$ can be read for $K < 0.35$. For larger values of K, equation (4.14) is valid. Using Equation (4.14) $H(t^*) = H(63979) = 2.4624$ dies $\Rightarrow 3$ dies. Therefore the ordering policy is to have an initial inventory of $3+1=4$ dies and then place an order for one die when ever a die fails. During the three years operation with $1 - p = .9999$, the total number of dies needed are forecasted using equation (4.21) , as 33 dies.

Example 5.

In an existing inventory system the following information is available for a part of **moderate** cost and **low** criticality:

Existing Inventory Policy

Item. Filter assembly.
On hand inventory: 28,
Unit cost: Co = $128 (moderate)
Holding cost: 15% of $128 = $20
Average Demand Rate: $\bar{D}$= 13 parts/year
Planning Horizon: = 1 year
Coefficient of Variation of Life: K = 1

Code: XY - WX - YZA.
Safety stock = 0,

Proposed Inventory Policy

Number of spare parts   needed in one calendar year with a risk of being out of stock $=0.05$ (i.e., $p = 0.95$), can be calculated using equation (4.29):

$N = 1$ x $13 + 19(1x13)^{0.5}(1.65) = 13 + 6 = 19$ parts.

9.   EXTENDING THE IDEAS TO REPAIRABLE SYSTEMS

The present paper deals with nonrepairable systems, but could be further extended to repairable systems. Early work in this regard reported by Cooper [7], utilizes the Erlang's loss formula to estimate the spares during the constant failure rate region ($\beta = 1$). The formula proposed by Cooper [7] is given by:

$$\overline{A}(S, M) = (M^S/M!)/\sum_{k=0}^{S}(M^k/k!) \tag{4.35}$$

$\overline{A}(S, M)$ = steady state **unavailability** of the system when the number of spares on hand is S and the number of units under repair are M. $A(S, M) = 1 - \overline{A}(S, M) =$ **availability** of the system.

The number of units under repair depends on the total number of units in service (N), the repair rate (μ), and the average lead time (l) for obtaining spares. The average number of units under repair is the product $N\mu l$. Thus, the number of spares S required to obtain a specified availability level, $A^*(S, M) = 1 - \overline{A}^*(S, M)$, can be obtained by solving equation (4.35) for S. The repair rate $\mu = C\lambda$ is a multiple of failure rate , λ , where C is a constant. To facilitate the solution of equation (4.35), graphs have been developed for different ranges of $N\mu l$ and availability $A(S, M)$ by $AT\&T$ [2]. The ideas presented in our paper can be integrated into the approach of the papers dealing with repairable systems, by relaxing the constant failure rate condition in model formation .Models can be developed for the availability of a repairable system , although the resulting equations would be quite complex. This paper can become a starting point for such a work. This work using modest mathematics represents a very important segment of the spare parts management picture. The approaches to extend this work are as follows:

- There is not much literature available in spare parts   forecasting and their interface with inventory models for aging parts. The time to failure process will become nonstationary for $\beta > 1$ this needs to be embedded in the queuing model formulations. This can be done, by using the time to failure process as a renewal process, in developing the appropriate availability equations. Mathematically this will be complex but the asymptotic results are expected to be relatively simple.

- Alternatively MTTF based on Weibull ($\beta > 1$) distribution can be approximated as $1/\lambda$ for a queuing theory based model. This approximation of replacing Weibull ($\sigma/\mu < 1$) by exponential ($\sigma/\mu = 1$) will lead to over estimation of demand rate and over-stocking will result. If $1 < \beta < 1.5$, this approach may be justified on the ground that overstocking will be less; if $\beta > 1.5$, the discrepancy will be higher.

10. CONCLUDING REMARKS

This paper provides a computer-oriented methodology to calculate reliability centered spare parts requirements for nonrepairable systems, or nonrepairable parts of a repairable system, and their rational inventory management strategies. These strategies are based upon a stochastic characterization of the time between failures of various frequently failing parts and equipment. This characterization can be used to determine the appropriate part failure rate, reliability functions and the replacement models. Then these models are linked with probabilistic inventory control models through part utilization ratio (duty cycle), $\rho = t/t_c$. In addition to that for various part failure modes the corresponding β values are documented which could be the starting point to use the methodology if in-plant historical time to failure data is not yet available.

Having assessed the cost of ordering and the cost of stock out based on the safety stock analysis, a decision process can be developed to balance the results and determine a spare parts policy. This process can take the form of the following decision rules:

1. If purchase cost is low and critical cost is low to high and the part is readily available then maintain stock at local supplier or use just in time techniques.

2. If purchase cost is low and critical cost is low to high and the part is not readily available then order large quantities of stock for life of mission.

3. If purchase cost is medium and criticality is low to medium or purchase cost is high but criticality is low, then order in batches at optimal order quantity and maintain a reasonable safety stock.

4. If purchase cost is high and criticality is medium to high, maintain an absolute minimum+1 additional spare, and order frequently on a one by one basis.

Material availability and location of the process from supply sources will play a significant role in shaping the inventory policy. Fluctuations in failure rates and population of equipment size will need to be mentioned to assure that proper inventory levels are established.

The ordering strategies discussed above cover both the safety stock and initial stocking level. However, the costs and risks need to be balanced in order to strive at an inventory stocking level for specific spare parts . The approach can easily be integrated with the existing inventory management system of any major organization, with minor modifications. *The reliability centered forecasting seamlessly integrated with inventory control strategies is a useful approach for forecasting the need of nonrepairable parts in a given planning horizon, and provide the strategies of their effective maintenance.*

ACKNOWLEDGEMENT: The authors acknowledge the support provided by King Fahd University of Petroleum & Minerals.

References

[1] Al-Bahli, A. M. Spares Provisioning Based on Maximization of Availability per Cost Ratio. *Computers in Engineering*, **24**(1), (1993) 81-90.

[2] AT&T , *Reliability Manual.* Basking Ridge, NJ;AT&T, 1983.

[3] Bartman, D., and Beckmann, M. J. *Inventory Control: Model and Methods.* Springer Verlag, 1992.

[4] Berg, M., and Posner, M.J.M. Customer Delay in $M/G/\infty$ Repair systems with Spares. *Operations Research*, **38**(2) (1990) 334-348.

[5] Bloch, H. P. and Geitner, F. K. *Practical Machinery Management for Process Plants.* **2**, Gulf Publishing Company, Houston, 1983.

[6] Bolton, W. *Production Planning and Control.* Longman Scientific & Technical, 85-96, 1994.

[7] Cooper,R.B. *Introduction to Queuing Theory.* Macmillan, New York, 1972.

[8] Dhakar, T.S., Schmidt, C.P., and Miller, D.M. Base stock level determination for high cost low demand critical repairable spares. *Computers Ops Res.*, **21** (1994) 411-420.

[9] Ebeling, C.E. Optimal Stock Levels and Service Channel Allocations in a Multi-Item-Repairable Asset Inventory system. *IIE Transactions*, **23**(2) (1991) 115-120.

[10] Gnedenko, B. V., Belyayev, Yu. K. and Solovyev, A. D. *Mathematical Methods of Reliability.* Academic Press, New York, 1969.

[11] Graves, S.C. A Multi-Echelon Inventory Model for a Repairable Item with one-for-one Replenishment. *Mgt. Sc.*, **31**(10) (1985) 1247-56.

[12] Greene, J. H. *Production and Inventory Control Handbook.* McGraw-Hill book company, 1970.

[13] Gross, D. On the Ample Service Assumptions of Palm's Theorem in Inventory Modeling. *Mgt. Sc.*, **28**(9) (1982) 1065-1079.

[14] Gross, D., Miller, D.R., and Soland, R.M. On Some Common Interests among Reliability, Inventory and Queuing. *IEEE Transactions on Reliability*, **R-34**(3) (1985) 204-208.

[15] Hall and Clark, A.J. ACIM: Availability Centered Inventory Model. *Proceedings of the Annual Reliability and Maintainability Symposium, IEEE*, (1987) 247-252.

[16] Kales, P. *Reliability for Technology.* Prentice-Hall, Inc., 1998.

[17] Langford, J. W. *Logistics: Principles and Applications.* McGraw-Hill, Inc., N.Y. 1995.

[18] Lewis, E. E. *Introduction to Reliability Engineering.* John Wiley & Sons, Inc., New York, 1996.

[19] Lipson, C. and Sheth, N. J. *Statistical Design and Analysis of Engineering Experiments.* McGraw-Hill, New York, 1973.

[20] Mann Jr, L., Saxena, A., and Knapp, G.M. Statistical-based or condition-based Preventive maintenance?. *J. of Quality in Maintenance Engineering*, **1**(1) (1995) 46-59.

[21] Monks, J. G. *Operations Management.* McGraw-Hill Book Company, 1985.

[22] Orsburn, D. K. *Spares Management Handbook.* McGraw-Hill Inc., 1991.

[23] Petrovic, R., and Pavlovic, M. A New Performance Measure for Multi-level Spare Parts Inventories. *Inventory in Theory and Practice*, Elsevier, Editor Attilia Chikan, Elsevier, 1986.

[24] Petrovic, R. Senborn, A., and Vujosevic, M. *Hierarchical Spare Parts Inventory Systems.* Elsevier, 1986.

[25] Purser,F.E. and Farmer,G.F. Optimizing Spare Parts Inventories Using RAM Techniques. *Idaho National Engineering Laboratory (EG&G Idaho),ASME* Paper 90-JPGC/Pwe-50,1990.

[26] Raza, M. K. Some quality and reliability aspects of aluminum extrusion. M.S. Thesis, King Fahd University of Petroleum & Minerals, Dhahran, 1999.

[27] Sherbrooke, C.C. METRIC: a Multi-Echelon Technique for Recoverable Item Control. *Operations Research*, **16** (1968) 122-141.

[28] Sherbrooke, C.C. *Optimal Inventory Modeling of Systems*. John Wiley, New york, 1992.

[29] Sheikh, A. K., Callom, F. L. and Mustafa, S. G. Strategies in Spare Parts Management using a Reliability Engineering Approach. *Engineering Cost & Production Economics*, **21** (1991) 51-71.

[30] Sheikh, A. K. A Statistical Method of Calculating the Spare Parts Requirement. *Symposium on Maintenance, Planning and Operation*, King Saud University, Riyadh, Saudi Arabia, March 17-19, 1990.

[31] Sipper, D. and Bulfin Jr. R. L. *Production planning, control and integration*. McGraw Hill Companies, Inc., 1997.

[32] Younas, M. Reliability Models in Fatigue Life Prediction. *M.S. Thesis*, King Fahd University of Petroleum & Minerals, Dhahran, 1984.

Chapter 5

MAINTENANCE SERVICE CONTRACTS

D. N. P. Murthy
Department of Mechanical Engineering
The University of Queensland
Brisbane, Q 4072, Australia
murthy@mech.uq.edu.au

Abstract   The maintenance of most modern products is beyond the capability of buyers for a variety of reasons. This has led to maintenance being outsourced rather than being carried out by the buyer. Many equipment manufacturers combine maintenance   service with the product and sell it as a bundle. In many other cases, the maintenance services are provided by a third party. The paper gives an overview of the literature on service maintenance contracts , the issues involved and the mathematical models to study these issues.

Keywords: : Maintenance, Warranties, Extended Warranties, Service Contract

1.   INTRODUCTION

Post-industrial societies are characterised by technology changing at an ever-increasing pace and as a result new products (consumer durable, industrial, commercial and specialised defence related products) appearing on the market with greater frequency. Also, the products are becoming more complex with, for example, the automobile of 1990 immensely more complex than the automobile of 1950. This increases the likelihood of product failure since product reliability, in general, tends to decrease as the complexity increases. Product failures occur in an uncertain manner and can be controlled through preventive maintenance actions. When a failure occurs, it needs to be rectified through corrective maintenance actions. The impact of failure in the case of consumer durable is that the buyer is deprived of its use and hence the satisfaction he drives from it. In the case of commercial and industrial products, failure impacts on the revenue generating capability and this in turn affects the final profits. For defence related products, failure can seriously impact on national security.

For simple products, often the buyer carries out the maintenance (preventive and corrective) and this is not an issue affecting the purchase decision. For complex and expensive products, maintenance becomes a major issue in the purchase decision. For consumer durables, manufacturers have used warranty to provide corrective maintenance with the manufacturer (or the dealer) repairing all failures that occur within the warranty period and often at no cost to the buyer. The warranty is tied to the sale and is factored into the sale price. Over the last decade, manufacturers have started offering extended warranty which provide the customer with coverage beyond the normal warranty period. The extended warranty is provided for an additional cost. The popularity of extended warranty has resulted in third parties (such as, financial institutions, insurance companies and independent operators) providing these services. They have been found to be highly profitable in many product markets (e.g. consumer electronics). As a result, manufacturers have scaled back considerably the basic warranty (tied to the sale) to force customers to choose extended warranties.

For most commercial and industrial businesses, it is often uneconomical to carry out the maintenance in house for a variety of reasons. These include the need for specialist work force and equipment that often require constant upgrading over the useful life of the product. In this case it is more economical to out source the maintenance in part or total. This involves a maintenance service contract between the buyer of the equipment (and recipient of maintenance service) and the service agent providing the maintenance service. As a result, the buyer of the product can be viewed as a customer of the service agent and we will use the words "buyer" and "customer" interchangeably.

This paper deals with maintenance service contracts . The aim of the paper is to give an introduction to the topic, highlight some of the issues involved and discuss some topics for future research. The outline of the paper is as follows. Section 2 gives a brief introduction to maintenance and out-sourcing of maintenance. Section 3 deals with warranties and extended warranties. Section 4 deals with a game theoretic formulation for maintenance service contract and Section 5 deals with the modeling and analysis of this formulation. Finally, the paper concludes with a brief discussion of future research in Section 6.

2.   MAINTENANCE

Maintenance can be defined as actions to (i) control the deterioration process leading to failure of a system and (ii) restore the system to its operational state through corrective actions after a failure. The former is called "preventive" maintenance and the latter "corrective" maintenance. As the name implies, corrective maintenance actions are unscheduled actions intended to restore a system from a failed state into a working state. This involves either repair or replacement of failed components. In contrast, preventive maintenance actions are scheduled actions carried out to either reduce the likelihood of a failure or to improve the reliability of a system.

2.1   Classification of Preventive Maintenance Actions.   Preventive maintenance (PM) actions are divided into the following categories:

1) Clock-based maintenance: Here PM actions are carried out at set times.

2) Age-based maintenance: Here PM actions are based on the age of the component.

3) Usage-based maintenance: Here PM actions are based on usage of the product.

4) Condition-based maintenance: Here PM actions are based on the condition of the component being maintained. This involves monitoring of one or more variables characterising the wear process (e.g., crack growth in a mechanical component).

5) Opportunity-based maintenance: This is applicable for multi-component systems, where maintenance actions (PM or CM) for a component provide an opportunity for carrying out PM actions on one or more of the remaining components of the system.

6) Design-out maintenance: This involves carrying out modifications through re-design of the component. As a result, the new component has better reliability characteristics.

2.2   Mathematical Modeling of Maintenance.   In general, most products are complex systems involving several components. Product failures occur due to failure of one or more of the components. As a result, the mathematical modeling of product failures can be done at two levels – (i) system level and (ii) component level - using stochastic point process formulations since failures occur in an uncertain manner. At the system level, one can model failures over time by an intensity function that is increasing with time to reflect the effect of degradation due to age and/or usage. At the component level,

the modeling depends on whether the component is repairable or not and the nature of repair. For non-repairable components, failures over time is best modelled by a simple renewal process when failures are detected immediately and the time to replace a failed item by a new one is negligible. More complex stochastic formulations are needed for modeling condition-based maintenance since one needs to model the degradation process over time rather than a binary characterisation (working or failed).

The literature on the mathematical modeling, analysis and optimisation of maintenance policies is very vast. Several review papers have appeared over the last 30 years. These include McCall (1965), Pierskalla and Voelker (1976), Monahan (1982), Jardine and Buzzacot (1985), Sherif and Smith (1986), Thomas (1986), Valdez-Flores and Feldman (1989). Cho and Parlar (1991) and Dekker *et al* (1997) deal with the maintenance of multi-component systems. Also, there are several books dealing with the topic, for example, Gertsbakh (1977). These deal mainly with the maintenance being carried out by the buyer and in-house.

2.3   Out Sourcing of Maintenance.

Out sourcing of maintenance implies that some or all of the maintenance actions (preventive and corrective) are carried out by an external agent under a contract. The contract specifies the terms of maintenance and the cost issues. It could be simple or complex and can involve penalty terms.

The advantages of out sourcing maintenance, in an industrial context, are as follows:

1) Better maintenance due to the expertise of the service provider.
2) Access to high level specialists on an "as needed basis".
3) Fixed cost service contract removes the risk of high costs.
4) Service agent responding to the changing needs of the customer.
5) Access to latest maintenance technology.
6) Lower capital investment on the part of the customer.
7) Managers can devote more time to other facets of the business since maintenance management involves less of their time and effort.

However, there are two disadvantages associated with out sourcing of maintenance and they are as follows:

1) Reduced in-house maintenance knowledge.
2) Getting locked in with a single service provider.

For very specialised products, the knowledge to carry out maintenance, and spares needed for replacement, can only be provided by the OEM (original equipment manufacturer). In this case, the customer is forced into a maintenance service contract with the OEM. This can result in a non-competitive market. In the USA, Section II of the Sherman Act [Khosrowpour (1995)] aims to prevent this by making it illegal for OEMs to act in this manner.

As a result, it is very important for customers to carry out a proper evaluation of the implications of out sourcing the maintenance. If done properly, out sourcing can be cheaper than in-house maintenance.

3.   WARRANTY AND EXTENDED WARRANTY

A warranty is a manufacturer's assurance to a buyer that the product or service is or shall be as represented. It may be considered to be a contractual agreement between buyer and manufacturer (or seller) which is entered into upon sale of the product or service. A warranty may be implicit or it may be explicitly stated.

In broad terms, the purpose of a warranty is to establish liability of the manufacturer in the event that an item fails or is unable to perform its intended function when properly used. The contract specifies both the performance that is to be expected and the redress available to the buyer if a failure occurs or the performance is unsatisfactory. The warranty is intended to assure the buyer that the product will perform its intended function under normal conditions of use for a specified period of time.

Many different types of warranty policies have been proposed. A taxonomy for warranty policies (for new products) has been proposed by Blischke and Murthy (1994). Different aspects of warranty have been studied extensively by researchers from diverse disciplines and the literature on the subject is considerable, see Blischke and Murthy (1996) for details. We confine our attention to a few warranty policies to highlight the maintenance aspect.

One-dimensional Free Replacement Warranty [FRW] Policy:
The manufacturer agrees to repair or provide replacements for failed items free of charge up to a time W from the time of the initial purchase. The warranty expires at time W after purchase. W is called the warranty period.

Cost Sharing Warranty [CSW] Policies:
Under the cost sharing warranty the customer and the service agent share the repair cost. The basis for the sharing can vary and as a result one can define a family of different warranty policies. We define three such policies and dealers of second hand products offer these bundled with the sale of product. The warranty period is W unless the warranty expires earlier.

Specific Parts Exclusion [SPE] Policy

Here the components of the product are divided into two sets denoted by I (denoting inclusion) and E (representing exclusion). The dealer rectifies failures of components belonging to the set I at no cost to the buyer over the warranty period. The costs of rectifying failures of components belonging to the set E are borne by the buyer. (Note: Rectification of failures belonging to the set E can be carried out either by the dealer or by a third party).

Limit on Individual Cost [LIC] Policy

The dealer rectifies all failures under warranty . If the cost of a rectification is below a limit c_I, then it is borne completely by the dealer and the buyer pays nothing. If the cost of a rectification exceeds c_I, then the buyer pays the excess (cost of rectification $- c_I$).

Limit on Individual and Total Cost [LITC] Policy

The dealer rectifies all failures under warranty . The cost to the dealer has an upper limit (c_I) for each rectification and the warranty ceases when the total cost to the dealer (subsequent to the sale) exceeds c_T or at time W whichever occurs first. The difference between the cost of rectification and that borne by the dealer is the cost to the buyer.

For more on these and other policies (for second hand products), see Chattophadyay (1999).

3.1   Warranty Analysis.

When an item is sold with warranty , the manufacturer incurs various costs (handling, material, repair, labour, etc) associated with the servicing of claims resulting due to failures under the warranty period. The cost per claim is a random variable. Since failures occur in an uncertain manner, the number of claims over the warranty period is also a random variable. As a result, the total warranty cost (i.e., the cost of servicing all warranty claims for an item over the warranty period) per unit sale is a sum of a random number of such individual costs. The cost analysis of warranties is of particular significance to manufacturers.

Expected warranty cost per unit sale

The warranty cost per unit sale is important in the context of pricing the product. The sale price must exceed the manufacturing cost plus the warranty cost (since warranty is part of the sale) or the manufacturer incurs a loss. On average, the warranty cost per item decreases as reliability increases. When a buyer has the option of choosing between different warranty policies, then this cost is of relevance.

The bulk of the literature on warranty cost analysis deals with the case where the only maintenance action used is corrective maintenance. As mentioned earlier, the cost analysis can be carried out either at the system or at the component level. In this section, we confine our attention to a simple product (e.g., coffee grinder or transistor radio) where it is either non-repairable or uneconomical to repair a failed item. As a result, a failed item needs to be replaced by a new one. This model is also applicable for complex products where the modeling is done at component level and the component under consideration is non-repairable.

Every claim under warranty results in the failed item being replaced by a new one. If one assumes that claims are made immediately after failure and that the time to replace a failed item by a new one is negligible, then the expected number of component failures during warranty , $E[N(W)]$ is given by

$$E[N(W)] = M(W) \tag{5.1}$$

where $M(.)$ is the renewal function (see, Ross (1972)) and is given by

$$M(t) = F(t) + \int_0^t M(t-x)f(x)dx \tag{5.2}$$

and $F(t)$ is the failure distribution of the item. As a result, the expected warranty cost per unit sale to the manufacturer, $E[C_m(W)]$, is given by

$$E[C_m(W)] = C_s M(W) \tag{5.3}$$

where C_s of the cost of replacing each failed item by a new one.

Blischke and Murthy (1994) deals with the cost analysis of many other types of warranty policies. An analysis of cost sharing policies can be found in Murthy and Chattophadyay (1999). Chattophadyay (1999) deals with the cost analysis of many other policies for second hand products. Sahin and Polatogu (1998) deal with the distribution for the warranty cost rather than the expected value for some simple warranty policies.

Expected demand for spares over product life cycle

We now consider the demand for spares over the product life cycle. Let L denote the product life cycle and $s(t), 0 \leq t \leq L$, denote the sales rate (i.e., sales per unit time) over the life cycle. This includes both first and repeat purchases. The total sales over the life cycle, S, is given by

$$S = \int_0^L s(t)dt \tag{5.4}$$

It is assumed that the life cycle L exceeds W, the warranty period, and that items are put into use immediately after they are purchased. Since the manufacturer must provide replacements for items that fail before reaching age W, and since the last sale occurs at or before time L, the manufacturer has an obligation to service warranty claims over the interval $[0, (L + W)]$.

The demand for spares in the interval $[t, t + \delta t)$ is due to failure of items sold in the period $[\Psi, t)$ where Ψ is given by

$$\Psi = \max\{0, t - W\} \tag{5.5}$$

It can be shown (details of the derivation can be found in Chapter 9 of Blischke and Murthy (1994)) that the expected demand rate for spares at time $t, \rho(t)$, is given by

$$\rho(t) = \int_{\psi}^{t} s(x)m(t - x)dx \tag{5.6}$$

where $m(t)$ is the renewal density function associated with the failure distribution function $F(t)$, given by

$$m(t) = f(t) + \int_{0}^{t} m(t - x)f(x)dx \tag{5.7}$$

The total expected number of spares (ETS) required to service the warranty claims over the product life cycle is given by

$$ETS = L + \int_{0}^{L+W} \rho(t)dt \tag{5.8}$$

For further details, see Blischke and Murthy (1994).

3.2   Extended Warranties.

The base warranty is an integral part of product sale and is offered by the manufacturer at no additional cost and is factored into the sale price. Extended warranty provides additional coverage over the base warranty and is obtained by the buyer by paying a premium. Extended warranties are optional warranties that are not tied to the sale process and can be offered either by the manufacturer or by a third party (e.g., credit card companies offering extended warranties for products bought using their cards).

The expected cost to the provider of extended warranties can be calculated using models similar to those for the cost analysis of base warranties. The cost of extended warranty is related to product reliability and usage intensity. The reasons for purchase of extended warranties have been analysed. Padmanabhan (1996) discusses alternate theories and the design of extended warranty policies.

3.3   Warranty and Preventive Maintenance.   As mentioned earlier, the bulk of the literature on warranty cost analysis deals with the case where the only maintenance action used is corrective maintenance. Blischke and Murthy (1994 and 1996) deal with this literature where interested readers can find further details and additional references. The use of preventive maintenance in the context of warranties has received limited attention and the literature is still disjointed. Djamaludin, Murthy and Kim (1999) deal with this topic and in this section we give a brief review of the relevant literature.

Ritchken and Fuh (1986) discuss an age replacement policy for a nonrepairable item. Any failure within the warranty period results in a replacement by a new one under a cost sharing policy. At the end of the warranty period, the item in use is preventively replaced after a period T (measured from the end of the warranty period) or on failure should it occur earlier. The optimal T is selected by minimising the buyer's asymptotic expected cost per unit time using the renewal reward theorem.

Chun and Lee (1992) consider a model where over the warranty period the item is subjected to periodic preventive maintenance actions. The preventive maintenance is imperfect in the sense that the failure rate after maintenance is lower than that before maintenance and the cost is borne by both the buyer and manufacturer. Any failures between preventive maintenance actions are repaired minimally by the manufacturer at no cost to the buyer. After the expiry of the warranty, the item is subjected to preventive and corrective maintenance actions as during the warranty periods but the costs are paid by the buyer. The period between preventive maintenance actions is optimally selected to minimise the buyer's asymptotic expected cost per unit time.

Chun (1992) deals with a model that is similar to the one in Chun and Lee (1992) but the focus is on the warranty cost to the manufacturer as opposed to the buyer.

Jack and Dagpunar (1994) deal with the model studied by Chun (1992) and they show that when the product has an increasing failure rate a strict periodic policy for preventive maintenance action is not the optimal strategy. As a result, time intervals between successive preventive maintenance actions are not identical. They derive the optimal preventive maintenance strategies to minimise the manufacturer's expected cost over the warranty period. They show that for the policy to be strictly periodic, the preventive maintenance action must result in the product being restored to as good as new.

Dagpunar and Jack (1994) deal with a model similar to that in Dagpunar and Jack (1994). The cost of each preventive maintenance action is a function of the effective age reduction resulting from the action. In this case, the optimal maintenance can result in the product not being restored to as good as new. The objective function for optimisation is manufacturer's expected warranty cost.

Sahin and Polatoglu (1996) discuss a preventive replacement policy for a repairable item following the expiration of warranty . Failures over the warranty period are minimally repaired at no cost to the buyer. The item is kept for a period T after the expiration of the warranty and replaced by a new item. Failures over this period are rectified minimally with the buyer paying the costs. They consider stationary and non-stationary strategies that minimise the long run average cost to the buyer.

Monga and Zuo (1998) deal with a model formulation where the components of a system are replaced under preventive maintenance action when their failure rate reaches some specified value. The model formulation includes warranty period and preventive maintenance action in addition to burn-in. The cost of rectifying failures under warranty is borne by the manufacturer and post warranty costs are borne by the buyer. The various decision variables (including preventive maintenance) are selected to optimise the average annual cost which is the sum of the cost to the buyer and to the manufacturer.

4.   MAINTENANCE SERVICE CONTRACTS

Since warranty is bundled with the product, the customer has no choice but to accept the terms of the warranty . In the case of extended warranty , the only choice that a customer has is whether to buy the extended warranty or not. In a maintenance service contract, the decisions of the service agent must take into account the actions of customers. The actions of customers are determined by the terms offered by the service agent. This implies a framework that takes into account these interactions.

In this section we consider maintenance service contracts where customers have to choose between different options offered by service agents and service agents take into account the optimal actions of customers in determining the optimal service contract terms. The three key elements for the study of such contracts are as follows:

1. Customers (owners of product and recipients of maintenance service)

2. Service agents (providers of maintenance service)

3. Alternate options (terms and conditions)

We discuss each of these briefly and highlight the need for a game theoretic approach to determine the optimal strategies for both customer and service agent.

4.1   Customers. We need to look at three different cases. The first case corresponds to a single customer for the product. A typical example of this is a government or a large business buying specialist products. Here the customer is more powerful than the OEM or the service provider. The customer has a strong input into product development and may dictate the subsequent post-sale support including the terms and conditions.

The second case corresponds to a relatively small number of customers for the product. These are typically, the customers for industrial and commercial products. Often, these products are either manufactured by a monopolist or by a small number of manufacturers. The terms and conditions of alternate options are determined jointly by the customer and service agent.

Finally, the third case corresponds to a large number of customers for the products (typically, consumer durables, many commercial products). The customer is not well informed about the product and lacks the knowledge to maintain it. An individual customer has very little impact on the manufacturer and very little influence on the formulation of the terms and conditions of alternate options.

4.2   Service Agent. For certain types of products, the maintenance service can only be provided by the OEM. This corresponds to a monopolistic situation. In this case, the terms of service maintenance contract might not be competitive and tend to favour the OEM. The second case corresponds to a more competitive situation where there are several service agents. In this case, the conditions and terms tend to be more competitive as long as there is no collusion between the different service providers.

4.3   Alternate Options. In this paper we confine our attention to a single service agent offering two options as part of maintenance service contracts . Both options involve only corrective maintenance actions. (Several extensions which include PM actions are discussed in Section 6.) The two options are as follows:

Option 1: For a fixed price P, the service agent rectifies all failures over the period L at no additional cost to the customer. L is the useful life of the item. If a failure is not rectified within a period τ, the agent incurs a penalty. It involves the agent compensating the customer for the equipment being non-operational for a period greater than τ. If Y denotes time for which an item is in a non-operational state before being made operational, then the penalty is given $\max \alpha\{0, (Y - \tau)\}$, where α is the penalty cost per unit time. This ensures that the agent does not deprive the customer of the use of the item for too long.

Option 2: In this case, whenever a failure occurs, the customer calls the agent to repair the failed item. The agent charges an amount C_s for each repair and there is no penalty associated with the duration for which an item stays in failed state.

4.4   Buyer's Perspective.

Under Option 1, the customer incurs a fixed cost P and is compensated if the down time (the period for which the item is in non-operational state) is greater than τ. Under Option 2, the maintenance cost incurred by a customer is a random variable since the number of failures over L is uncertain. In addition, there is no compensation for loss of revenue due to down times.

The choice between these two options is influenced by the reliability of the equipment and the parameters of the two options namely, P and α for Option 1 and C_s for Option 2. Another factor that is relevant in this context is the attitude of customers to risk. If a customer is highly risk averse, he/she is more likely to choose Option 1 in preference to Option 2. We will discuss this further in the next section where the attitude to risk is captured through a parameter of a utility function.

Finally, since customers use the product to generate income, the net profit to a customer is the difference between the revenue generated and the sum of acquisition cost and maintenance cost. If P and C_s are very large, then the net profit can be negative and this can lead to a negative value for the expected utility. In this case, the optimal action would be the following:

Option 0: Not buy the equipment. In this case there is no need for maintenance and also there is no revenue generated.

As a result, the customer decision involves choosing between the following set of actions:

A_1: Buy the product with Option 1 for maintenance,

A_2: Buy the equipment with Option 2 for maintenance.

A_0: Not buy the equipment.

4.5   Agent's Perspective.

For the agent, the revenue generated under Option 1 is a fixed amount (P) and under Option 2 is a random variable. The costs incurred are different under the two options. Under both options, there is the labour and material cost associated with each repair. Let $C_{ri}(i \geq 1)$ denote the cost for the ith repair. The total cost of repair is a random variable since the number of failures over the life L is uncertain. Under Option 1, there is the additional cost resulting from penalty payments. The penalty is dependent on the waiting and repair times after each failure. The former is influenced by the number of customers (M) serviced by the agent and the number of service channels (S) used by the agent. As a result, the total profit is a random variable under both options and affected by P, C_s, M and S.

4.6   Stackelberg Game Formulation.

The agent needs to decide on the price structure (P under Option 1 and C_s under Option 2), the total number of customers to service (M) and the number of Service channels (S) in order to maximise the expected profit. Increasing M results in greater revenue but also greater costs due to increased penalty cost due to longer waiting times. The waiting times can be reduced by increasing the number of service channels but this adds to the total cost. In addition, the profits are influenced by the actions of the customers. The agent needs to take all of these into account in determining the optimal strategy. The decision variables that the agent must select optimally are P, C_s, M and S.

A Stackelberg game theoretic formulation with the service agent as leader and the customers as followers is the most appropriate formulation to determine the service agent's optimal decisions taking into account the optimal actions of the customers. Here, the agent first evaluates the optimal actions of the customers for a given set of values for the decision variables. We confine our attention to the case where customers are identical in their attitude to risk and hence the optimal actions are the same for all. Let $A^*(P, C_s, M, S)$ denote the optimal action which maximises the expected utility. The agent then selects the decision variables optimally to maximise the expected profit taking into account the optimal actions of customers.

The optimal strategies (for both customers and the agent) depend on the information available for decision making. In the next section, we look at some of the alternate scenarios and their implications for the optimal strategies.

5.   MATHEMATICAL MODELING OF SERVICE CONTRACTS

In this section we discuss three mathematical models (Models 1 – 3) for the maintenance service contract discussed in the previous sections. Model 1 is a simple model with one agent and one customer. The agent's decision problem is to determine the optimal parameters of the two options (Options 1 and 2) discussed in the previous section. Model 2 is an extension where there are M customers and the agent has to determine the optimal value for this in addition to the parameters of the two options. Finally, Model 3 is an extension of Model with the number of service channels being the extra decision variable. We only present some salient features of the modeling and analysis. Further details of these three models can be found in the references indicated later in the section.

5.1   Model 1 [Murthy and Ashgarizadeh (1998)].

As mentioned earlier, the useful life of an item is L. Let C_b denote the customer's purchase price and R denote the revenue generated per unit time when the item is in operational state. We assume that the failure distribution of item is given by an exponential distribution with failure rate λ. The failed item is minimally repaired so that the failure rate of repaired items is also λ. This is a good model for electronic products where there is very little aging effect so that the failure rate is roughly constant. Note that, in this case there is no need for any PM actions.

Customer's Decision Problem

The consumer is risk averse and the risk aversion is modelled in terms of a utility function $U(\omega)$ given by

$$U(\omega) = \frac{1 - e^{-\beta\omega}}{\beta} \tag{5.9}$$

where β is the revenue generated by the item. Higher value of β corresponds to greater risk aversion. $\beta = 0$ implies risk neutral and in this case $U(\omega) = \omega$.

We use the following notation:

N: number of system failures over the life L

Y_i: Time for the *ith* repair $(i = 1, 2, ...N)$

$F(y)$: Distribution function for Y_i, is exponetially distributed with mean $= 1/\mu$

C_r: Cost to the agent for each repair

The revenue to the customer under the three different options are as follows:

Option A_1:

$$\omega(A_1) = RL + \sum_{i=1}^{N} [\alpha \max\{0, (Y_i - \tau)\}] - C_b - P \qquad (5.10)$$

Option A_2:

$$\omega(A_2) = RL - C_b - C_s N \qquad (5.11)$$

Option A_0:

$$\omega(A_0) = 0 \qquad (5.12)$$

Note: N and $\{Y_i\}$ are random variables. The customer chooses the optimal decision A^* from the set $\{A_0, A_1, A_2\}$ given P, and C_s to maximise the expected utility (by carrying out the expectations over the random variables). As a result, A^* is a function of P and C_s (depending on the service option).

Agent's Decision Problem

The service agent's profit depends on the customer's choice of option. It is given by

$$\pi(P, C_s; A_1) = P - C_r N - \sum_{i=1}^{N} [\alpha \max\{0, (Y_i - \tau)\}] \qquad (5.13)$$

under Option A_1; by

$$\pi(P, C_s; A_2) = \{C_s - C_r\} N \qquad (5.14)$$

under Option A_2, and by

$$\pi(P, C_s; A_0) = 0 \qquad (5.15)$$

under Option A_0. Note that the first two are random variables. The agent chooses P, and C_s optimally to maximise the expected profit by taking into account the optimal decision of the customer.

Model Analysis

The various expected values are obtained using conditional expectation. To obtain analytical results, one needs to make some simplifying assumptions. In Murthy and Asgharizadeh (1998), it is assumed that the mean time to repair is small in relation to the mean time between failures. This allows one to derive analytical results. We present the results by omitting the details and they can be found in the reference cited. The expected utilities under the three options are as follows:

$$E[U(\omega(A_1)); P, C_s)] = \frac{1 - e^{-\gamma_1}}{\beta} \qquad (5.16)$$

$$E\left[U\left(\omega\left(A_2\right);P,C_s\right)\right] = \frac{1 - e^{-\gamma_2}}{\beta} \qquad (5.17)$$

and

$$E\left[U\left(\omega\left(A_0\right);P,C_s\right)\right] = 0 \qquad (5.18)$$

where

$$\gamma_1 = \beta\left(RL - C_b - P\right) - \lambda Le^{-\mu\tau}\left[\frac{\mu}{\beta\alpha + \mu} - 1\right] \qquad (5.19)$$

and

$$\gamma_2 = \beta\left(RL - C_b\right) + \gamma L\left(1 - e^{-\beta C_s}\right) \qquad (5.20)$$

For a given P and C_s, the optimal choice between the three is easily determined by a relative comparison. The optimal decision in the $P - C_s$ plane is as shown in Figure 5.1. As can be seen, there are three different regions (denoted as Ω_i, $i = 0, 1$ and 2) with demarcation between the regions characterised by the curve Γ and two straight lines (a vertical line with $C_s = \overline{C}_s$ and a horizontal line with $P = \overline{P}$). The curve Γ and $\overline{C}_s$ and $\overline{P}$ are given by

$$P = \frac{\lambda L}{\beta}\left\{e^{\beta C_s} - e^{-\mu\tau}\left[\frac{\mu}{\beta\alpha + \mu} - 1\right] + 1\right\} \qquad (5.21)$$

$$\overline{P} = RL - C_b - \frac{\lambda L}{\beta}\left[\frac{\mu}{\beta\alpha + \mu} - 1\right]e^{-\mu\tau} \qquad (5.22)$$

and

$$\overline{C}_s = \frac{1}{\beta}\ln\left\{1 + \frac{\beta\left(RL - C_b\right)}{\lambda L}\right\} \qquad (5.23)$$

respectively.

The agent's expected profit based on the optimal action of the customer being A_1 and A_2 are as follows:

$$E\left[\pi\left(P,C_s;A^* = A_1\right)\right] = P - \lambda L\left\{C_r + \frac{\alpha e^{-\mu\tau}}{\mu}\right\} \qquad (5.24)$$

$$E\left[\pi\left(P,C_s;A^* = A_2\right)\right] = \lambda L\left\{C_s - C_r\right\} \qquad (5.25)$$

This yields

$$E\left[\pi\left(P, C_s : A^*\right)\right] = \begin{cases} \overline{P} - \lambda L\left[C_r + \left\{\alpha e^{-\mu\tau}\right\}\right]/\mu & \text{when } C_s > \overline{C}_s \\ & \text{and } P = \overline{P} \\ \lambda L\left(\overline{C}_s - C_r\right) & \text{when } C_s = \overline{C}_s \\ & \text{and } P > \overline{P} \\ 0 & \text{when } C_s > \overline{C}_s \\ & \text{and } P > \overline{P} \end{cases}$$

$$(5.26)$$

The agent's optimal decision is obtained by comparing the profits obtained from (5.24) and (5.25) provided one of them is positive. As a result, in this case the optimal pricing strategy is either $C_s > \overline{C}_s$ and $P = \overline{P}$, or $C_s = \overline{C}_s$ and $P > \overline{P}$. If the expected profit is negative for both cases, then the optimal strategy for the agent is $C_s > \overline{C}_s$ and $P > \overline{P}$ so that the customer chooses not to buy the product and hence there is no need for the service contract.

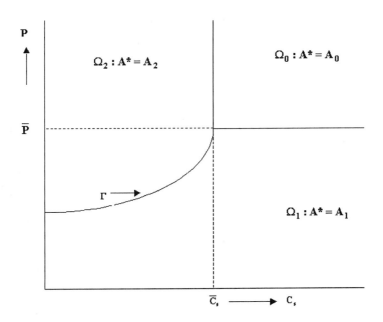

Figure 5.1 Customer's optimal strategies.

Sensitivity Analysis

The two key parameters are β (characterising the customers's attitude to risk) and λ (the failure rate of the product). It is easily shown that

$$\frac{d\overline{P}}{d\beta} < 0 \text{ and } \frac{d\overline{C}_s}{d\beta} < 0 \tag{5.27}$$

and that

$$\frac{d\overline{P}}{d\lambda} > 0 \text{ and } \frac{d\overline{C}_s}{d\lambda} < 0 \tag{5.28}$$

These characterise the sensitivity of the optimal decision strategies to changes in the parameter values.

Information Uncertainty

In the above analysis, it is assumed that both parties (agent and customer) have full information regarding the parameters of the model. Often, this is not true. Murthy and Asgharizadeh (1998) examine two cases — agent does not know the true value of β and the customer does not know the true value of λ. They model these as discrete random variables and study the impact on the optimal strategies.

5.2   Model 2 [Asgharizadeh and Murthy (1999)].

In Model 1, there was only one customer. In this model there are M customers. This is an additional decision variable that the agent needs to select optimally. Since product failures are exponentially distributed and the repair times are also exponential, the distribution for the waiting time can be obtained from the results of $M/M/1/M$ queue. If L is large so that the system has reached a steady state, then the steady state waiting time density function is given by

$$f_M(y) = \sum_{k=0}^{M-1} \hat{P}_k \frac{\mu e^{-\mu y}(\mu y)^k}{k!} \tag{5.29}$$

with

$$\hat{P}_k = \frac{(M-k)\, P_k}{\sum_{k=0}^{M-1} P_k}, \quad 0 \le k \le (M-1) \tag{5.30}$$

and

$$P_k = \frac{\left(\frac{\lambda}{\mu}\right)^k \frac{M!}{(M-k)!}}{\sum_{k=0}^{M-1} \left(\frac{\lambda}{\mu}\right)^k \left\{\frac{M!}{(M-k)!}\right\}}, \quad 0 \le k \le (M-1) \tag{5.31}$$

For a given M, following an approach similar to that used in Model 1, the customer's optimal decisions are again given by regions as shown in Figure 5.1 except that the curve Γ, $\overline{C}_s$ and $\overline{P}$ are functions of the three decision variables under the control of the agent. One needs to use a search technique to obtain the optimal M.

5.3 Model 3 [Murthy and Asgharizadeh (1999)].

In Model 2 there is only one service channel. Model 3 is an extension with S service channels. This is another decision variable that the agent must select optimally. In this case, the waiting time density function is obtained from the results of $M/M/S/M$ queue with arrival rate given by

$$\lambda_k = \begin{cases} (M-k)\lambda & \text{for } 0 \le k \le M \\ 0 & \text{for } k > M \end{cases} \tag{5.32}$$

and departure rate given by

$$\mu_k = \begin{cases} k\mu & \text{for } 0 \le k \le S \\ S\mu & \text{for } k > S \end{cases} \tag{5.33}$$

The density function for the waiting time is given by

$$
\begin{aligned}
f_{MS}(y) = {} & e^{-\mu y} \sum_{k=0}^{S-1} \hat{P}_k \\
& + \sum_{k=1}^{M-S} \hat{P}_{k+S-1} \mu (S\mu)^k \left[\frac{e^{-\mu y}}{(S\mu-\mu)^k} - \sum_{j=1}^{k} \frac{y^{(k-j)} e^{-S\mu y}}{(k-j)!(S\mu-\mu)^j} \right]
\end{aligned}
\tag{5.34}
$$

where

$$\hat{P}_{kS} = \frac{(M-k) P_{kS}}{\sum_{k=0}^{M-1} P_{kS}} \quad \text{for } 0 \le k \le (M-1) \tag{5.35}$$

$$P_{kS} = \begin{cases} \left(\frac{\lambda}{\mu}\right)^k \frac{M!}{(M-k)!k!} P_0 & \text{for } 0 \le k \le (S-1) \\ \left(\frac{\lambda}{S\mu}\right)^k \left(\frac{S^S}{S!}\right) \frac{M!}{(M-k)!} P_0 & \text{for } S \le k \le M \\ 0 & \text{for } k > M \end{cases} \tag{5.36}$$

and

$$P_0 = \left[\sum_{k=0}^{S-1} \left(\frac{\lambda}{\mu}\right)^k \left\{ \frac{M!}{(M-k)!k!} \right\} + \sum_{k=S}^{M-1} \left(\frac{\lambda}{S\mu}\right)^k \frac{S^S}{S!} \frac{M!}{(M-k)!} \right]^{-1} \tag{5.37}$$

For a given S and M, the optimal strategies are similar to those in Model 1. Again, the optimal S and M can only be obtained by a search method.

6.   CONCLUSIONS AND TOPICS OF FUTURE RESEARCH

In this paper we have given an overview of maintenance service contracts . Warranties are a special class of such contracts and the mathematical modeling and analysis of warranties has received a lot of attention. In contrast, the modeling and analysis of more general service contracts has received little attention. The only models dealing with this are ones discussed in the previous sections. These models can be extended in several ways and we discuss some of these.

1. The failure rate of the product is increasing as opposed to being constant.

2. Not all customers are identical in their attitude to risk. This can be modelled by dividing the total population into different categories with each sub-population characterised by a different risk parameter.

3. If customers are different, the agent might offer a range of sub-options for Option 1 with different values for τ and α (characterising the penalty function). This leads to different priority rules for servicing.

4. With increasing failure rate, preventive maintenance becomes important. One needs to consider the different preventive maintenance actions and these lead to a variety of new models.

5. The usage rate (or intensity) varying across the population and must be taken into account. In this case, the agent might offer service contracts that are linked to usage rate.

6. The informational aspects raise several new and challenging problems. One such is the adverse selection problem in the case of different usage rates and different maintenance service contracts .

7. We have confined our attention to a single service agent. When there is more than one service agent, one must consider different scenarios – for example, competition or different forms of collusion between the agents.

8. The logistics issues for the service agent raise several new problems that are yet to be studied.

As can be seen from the short list, maintenance service contracts offer considerable scope and challenges for new research in the future.

References

[1] Asgharizadeh, E., and Murthy, D.N.P. Service contracts : A stochastic model, *Math. and Comp. Modelling*, in print, 1999.

[2] Blischke, W.R., and Murthy, D.N.P. *Warranty Cost Analysis*, Marcel Dekker, New York, 1994.

[3] Blischke, W.R., and Murthy, D.N.P. *Product Warranty Handbook* , Marcel Dekker, New York, 1996.

[4] Chattopadhyay, G. *Modelling and Analysis of Warranty Costs for Second Hand Products*. Unpublished doctoral thesis, The University of Queensland, Brisbane, Australia, 1999.

[5] Chattopadhyay, G., and Murthy, D.N.P. Warranty cost analysis for second-hand products with cost sharing policies. *Proc. of the Fifteenth National ASOR Conference*, Gold Coast, Australia, **1** (1999) 335-347.

[6] Chun, Y.H. Optimal number of periodic preventive maintenance operations under warranty. *Reliability Engineering and System Safety* , **37** (1992) 223-225.

[7] Chun, Y.H. & Lee, C.S. Optimal replacement policy for a warranted system with imperfect preventive maintenance operations. *Microelectronics Reliability*, **32** (1992) 839-843.

[8] Cho, D. and Parlar, M. A survey of maintenance models for multi-unit systems. *European J. Operational Research*, **51** (1991) 1-23.

[9] Dagpunar, J.S. & Jack, N. Preventive maintenance strategy for equipment under warranty, *Microelectronics Reliability*, **34** (1994) 1089-1093.

[10] Dekker, R., Wildeman, R.E., and van der Duyn Schouten, F.A. A review of maintenance models with economic dependence. *Math. Methods of Oper. Res.*, **45** (1997) 411-435.

[11] Djamaludin, I., Murthy, D.N.P. and Kim, C.S. Warranty and preventive maintenance. *Proc. of the Third Australia-Japan Workshop on Stochastic Models*, Christchurch, New Zealand, (1999) 81-90.

[12] Gertsbakh, I. B. *Models of Preventive Maintenance*, North Holland, Amsterdam, 1977.

[13] Jack, N., and Dagpunar, J.S. An optimal imperfect maintenance policy over a warranty period, *Microelectronics Reliability*, **34** (1994) 529-534.

[14] Jardine, A. K. S., and Buzacott, J .A. Equipment reliability and maintenance, *European J. Operational Research* **19** (1985) 285-296.

[15] Khosrowpour, M. (ed) *Managing Information Technology Investment Outsourcing*, Idea Group Pub., U.K., 1995.

[16] Monga, A., and Zuo, M.J. Optimal system design considering maintenance and warranty , *Computers Operations Research*, **9** (1998) 691-705.

[17] McCall, J. J. Maintenance policies for stochastically failing equipment: A survey, *Management Science*, **11** (1965) 493-524.

[18] Monahan, G. E. A survey of partially observable Markov decision processes: Theory, models and algorithms, *Management Science* **28** (1982) 1-16.

[19] Murthy, D.N.P. and Asgharizadeh, E. A stochastic model for service contract, *Int. Jr. Rel. Qual. and Safety Eng.*, **5** 29-45 (1998).

[20] Murthy, D.N.P., and Asgharizadeh, E. Optimal decision making in a maintenance service operation, *Euro. Jr. Oper. Res.*, **116** 259-273 (1998).

[21] Murthy, D.N.P., and Chattopadhyay, G. Warranty cost analysis for second-hand products with cost limit policies, *Proc. of the FAIM Conference*, Tilburg, Netherlands, 1999.

[22] Murthy, D.N.P., & Padmanabhan, V. *A dynamic model of product warranty with consumer moral hazard*, Working paper, Graduate School of business, Stanford University, California, 1993.

[23] Padmanabhan, V. Usage heterogeneity and extended warranty, *Journal of Economics and Management Strategy*, **4** (1995) 33-53.

[24] Ritchken, P.H. & Fuh, D. Optimal replacement policies for irreparable warrantied items, *IEEE Transactions on Reliability*, **R-35** (1986) 621-623.

[25] Ross, S.M. *Applied Probability Models with Optimization Applications*, Holden Day, San Francisco, 1992.

[26] Sahin, I. and Polatoglu, H. Maintenance strategies following the expiration of warranty, *IEEE Transactions on Reliability*, **45** (1996) 220-228.

[27] Sahin, I. and Polatoglu, H. Manufacturing Quality, Reliability and preventive maintenance, *Production and Operations Management*, **5** (1998) 132-147.

[28] Sahin, I. and Ploatogu, H. *Quality, Warranty and Preventive Maintenance*, Kluwer, Boston, 1998.

[29] Sherif, Y. S. and Smith, M. L. Optimal maintenance models for systems subject to failure - A review, *Naval Logistics Research Quarterly* **23** (1976) 47-74.

[30] Thomas, L. C. A survey of maintenance and replacement models for maintainability and reliability of multi-item systems, *Reliability Engineering* **16** (1986) 297-309.

[31] Valdez-Flores, C. and Feldman, R. M. A survey of preventive maintenance models for stochastically deteriorating single-unit systems, *Naval Research Logistics Quarterly* **36** (1989) 419-446.

Chapter 6

SIMULATION METAMODELING OF A MAINTENANCE FLOAT SYSTEM

Christian N. Madu

Research Professor and Chair of Management Science Program, Lubin School of Business, Pace University, New York, New York

Abstract   In this chapter, we discuss the use of metamodels in analyzing maintenance float systems. Metamodels are, increasingly, being used in solving complex problems primarily because of there ease of use and tremendous appeal for practical purposes. Further, metamodels utilize the increasing power of PC-based simulations and statistical applications. Our focus here is on their application to maintenance float network problems. Maintenance float problems can be considered as part of closed queuing network problems. Such problems are very difficult to model analytically. With the use of simulation, we can better understand maintenance float problems and with metamodels, we may be able to provide some generalizations to the results obtained through simulation.

Keywords: Maintenance float, simulation, metamodeling.

1.   INTRODUCTION

Many real life problems are often difficult to model analytically. Sometimes, the application of mathematical models to such problems may lead to series of assumptions that may often narrow the scope of the model that is developed. Simulation on the other hand, is an effective tool in modeling complex systems. The use of simulation allows us to consider all the original problem's parameters and therefore, solve the problem in its original form. Although simulation could be effective in problem solving, a major drawback is the lack of generalization of its results. Unlike mathematically derived models, it is difficult to generalize simulation results beyond the scope of the original problem solved. This is a major drawback since it may necessitate conducting new simulation for similar problems when problem parameters are changed. There are however, advantages to the use of simulation. As we have already stated, a problem can be solved in its original form through simulation without concern of the influence of problem parameters on the tractability of the model. Further, simulation is a useful tool when mathematical modeling fails and when the real life model is non-existent. Thus, simulation can be used to build a prototype of the actual model. Through the prototype, we can study the behavior of the actual system, conduct sensitivity analysis, and investigate alternative scenarios about the system. However, it is not easy to conduct a sound simulation study. Law and McComas [1989] point out that in order to conduct a sound simulation study, attention should be focused on the following activities: problem formulation, data and information collection, probabilistic modeling, model building, statistical design , analysis of simulation experiments and model coding. Traditionally, the focus on simulation building is on model coding. However, if the other simulation activities are not well conducted, the simulation model will not be reliable. In this chapter however, we shall focus only on data and information collection and statistical design. However, the simulation model of the maintenance float system presented in this chapter has been designed by paying attention to the listed activities. Our goal is to demonstrate how simulation metamodels are developed as a specific example of maintenance float systems.

There are several definitions of metamodels but we prefer the definition provided by Friedman and Pressman [1988]. They define the term "metamodel" as "any auxiliary model which is used to aid in the interpretation of a more detailed model." The aim of a metamodel is to achieve the following: model simplification, enhanced exploration and interpretation of the model, generalization to other models of the same type, sensitivity analysis, optimization, answering inverse questions, and providing the researcher with a better understanding of the behavior of the system under study. These attributes explain why metamodels have gained increased application over the past few years. Such benefits expand the power of simulation. As we noted earlier, one of the major drawbacks of simulation is the lack of the ability to generalize. Now, with metamodels, such drawbacks can be harnessed. Further, the application of metamodels cuts down on the cost of rerunning the simulation program anytime a problem parameter is changed. Gardenier [1990] provides another definition of metamodels as "models of simulation, models which express the input-output relationship in the form of a regression equation". Indeed, metamodels depend largely on the application of statistical tools specifically in the experimental design stage, output analysis, and model formulation. Various forms of regression analysis have been used in building metamodels. Notable applications and use can be found in Kleijnen [1987] and Madu and Kuei [1993]. Further, there are two major review papers on metamodeling. Friedman and Friedman [1984] present literature reviews on metamodeling for articles published on the topic prior to 1984. Madu and Kuei [1994] cover articles that were published up until 1994. However, before we go further, we shall introduce a maintenance float system and then show how metamodels can be applied to analyze the problem.

2. MAINTENANCE FLOAT SYSTEMS

A maintenance float system is a network with N independent and identical units (i.e. machines) that are required to be simultaneously in operation. Standby units denoted as F machines often support these units. The network operates like a closed system where there is no arrival or departure of a machine from the system once the system is in operation. When a unit within the system fails, it is sent in for repair and at the completion of repair, the repaired unit awaits a further call for service. Clearly, the assumption is that all failed units are repairable. Thus, the system starts with a total of N+F units and ends with the same number. However, at each time, only the N units are in operation. Furthermore, the units in operation and those in standby status are indistinguishable. Thus, they are equally substitutable. There are many applications of this type of problem i.e., airline companies that maintain a fleet of aircraft and standbys. The decision to acquire a new aircraft is not easily made; electric utility companies often maintain standby generators. Other potential application areas are public transportation systems, health care delivery services and military operations.

The critical question that is addressed in modeling maintenance float systems is to determine the optimum combination of standby units and repairpersons needed to minimize the expected operating cost of the maintenance float system.

The maintenance float problem can be expressed as a closed queuing network problem where there are N number of machines required in operation and F number of machines on standby. If a machine in operation fails, it goes in for repair and while under repair, it is replaced by a standby machine. At the completion of the repair, the failed machine is either put back in operation, that is if there are less than N machines in operation, or it is maintained in standby that is if there are exactly N machines in operation. There are no replacement decisions. Thus, no new machine is acquired or brought into the system when the operation

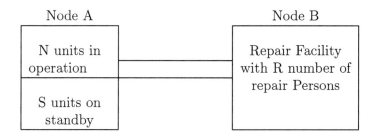

Figure 6.1   Maintenance float network.

is underway. Closed queuing problems often do not have closed form solutions. As a result, they are best solved through simulation rather than mathematical modeling.

Figure 6.1 shows a basic maintenance float network. This basic network can be extended to more complex cases. However, we shall restrict our consideration to the basic model in the hope that the reader will later be able to deal with more difficult cases.

2.1   The State of Maintenance Float Research.   The traditional approach to model the maintenance float problem is to treat it as a closed queuing network problem. Earlier models that treated the maintenance float problem as such were presented by Koeninsberg [1958,1960] and then later by Buzen [1973]. Buzen's [1973] developed an algorithm to solve the original problem formulated by Gordon and Newell [1967]. Buzen's algorithm has since been applied to a wide range of closed queuing problems of the Jacksonian type [Gross, Miller, and Soland, 1983]. Madu [1988a, 1988b, and 1990] extended the Buzen's algorithm to the application of maintenance float problems and applied it to obtain measures of performance for maintenance float systems. However, he noted that this algorithm becomes cumbersome and difficult to use when the number of nodes in the network is increased. In fact, the computational time increases exponentially.

Levine [1965] showed that maintenance floats can be computed using a reliability-based method for finding maintenance float factors. He illustrated this approach by assuming an exponential failure rate and exponential service rate. This method was further extended by Lowe and Lewis [1983] to the case of Weibull failure distribution and subsequently by Madu [1987] to consider gamma failure distributions. Several extensions of Levine's paper have been presented in the literature [Georgantzas and Madu, 1988 and Madu and Georgantzas, 1988].

Another line of modeling of maintenance float systems is based on metamodeling. The application of this approach to solve maintenance float system problems was introduced by Madu [1990] and it has been extended since then to solve more complex maintenance float problems. Madu [1990] addressed the basic maintenance float system identical to the one discussed in this chapter. Madu and Kuei [1992] applied a group screening technique to develop metamodels for maintenance float systems with more than two nodes. Kuei and Madu [1994] applied both polynomial metamodeling and Taguchi design to develop maintenance metamodels for closed queuing networks with more than two nodes. A recent article by Madu and Madu [1999] showed how design optimization could be achieved in conducting metamodeling by using signal-to-noise ratio. We shall now present a maintenance float problem and analyze it using the procedures for simulation metamodeling.

2.2 Maintenance Float Problem.

Consider a job shop that has 10 independent and identical machines. All 10 units are required to be simultaneously in operation in order to meet the system's availability requirements. The operating units need to be supported by standby units since the operating machines are subject to random failures, and any breakdown will hamper the overall system availability. However, due to budgetary constraints, there can be no more than 5 standby units, but there must be at least 1 standby unit. The operating and standby units are assumed to be equally interchangeable thus, there is no need to keep track of which machines are in operation or on standby status at the beginning of the operation. It has also been decided that the number of repairpersons at the repair center can not be more than the number of available standby machines. Thus, there can be at most 5 repairpersons in the system. From historical data on the performance of these machines, it has been concluded that both machine failure times and repair times are independent and identically distributed and follow the exponential distribution. The rate of machine failure is 6/hr. Management is interested in evaluating the effect of repair rates on the average equipment utilization. It has decided that repair rates can vary in the range of 12/hr to 60/hr.

Due to the great need to improve customer satisfaction and remain competitive, management has also decided that an average equipment utilization of at least 95% must be maintained by the facility. At a current rate of repair of 18/hr and if the cost of lost production is $250,000/hr, the cost of standby units is $10,000/hr, and the cost of repair is $35/hr/repairperson, how many standby units and repair persons should be maintained by this facility? These two questions form the basis for our further discussions. We shall demonstrate how to address them using both simulation and metamodeling.

2.3　Experimental Design of the Simulation Model.

This particular problem is used for illustrative purposes only. Notice that both the failure and repair times are exponentially distributed. As such, this closed queuing network problem can be solved to obtain closed form solutions. Koeningsberg [1958, 1960] solved this type of problem. Madu [1988a and b] solved this problem specifically for maintenance floats with two and multiple nodes using Buzen's algorithm. However, when the failure time can not be expressed using exponential distribution, there is no closed form solution. Simulation becomes a better way to solve the problem. The simulation metamodeling procedure illustrated here is applicable in such cases. A framework for the application of simulation metamodeling is presented on the following page (Figure 6.2).

Going through this procedure will enable us to be thorough in our analysis of the problem and ensure that the model is designed efficiently. We shall therefore, analyze this problem by following the procedure outline in Figure 6.2.

2.3.1　Step 1: Specify Metamodel Form.

This step involves the postulation of a metamodel form. Normally, this will require the model builder to be knowledgeable of the problem. There are many different models that could be built. In this particular problem, we can start by identifying the system variables as follows:

N = the number of machines in operation at any given time;
S = the number of standby units
R = the number of repairpersons;
r = the ratio of mean time to repair to mean time between failure or rather, the server utilization factor when expressed as rates;
α= level of significance.
AEU = average equipment utilization
β_I = the regression coefficient for the ith input factor
β_O = constant
ε= is the approximation error

Let us now use this information to postulate a regression metamodel

$$AEU = \beta_0 + \beta_1 S + \beta_2 R + \beta_3 r + \varepsilon \qquad (6.1)$$

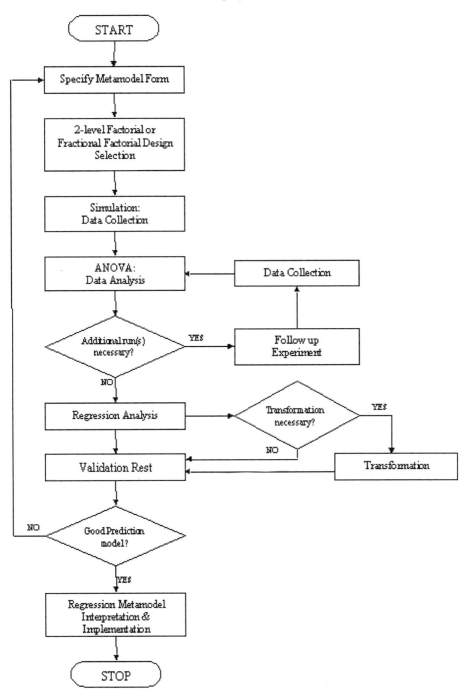

Figure 6.2   Metamodel construction procedure.

This metamodel for AEU is simply a multiple linear regression model that makes the assumption that only main effects are important to explain the variations in AEU. This model is known as the first-order metamodel. It can easily be expanded to include two-factor or more interactions. However, in most practical applications, attention is limited to main effects and two-factor interactions because it is very difficult to give practical interpretation to 3 or more factors interactions. Thus, we can comfortably ignore them for this problem. In equation 1, we assume a linear model but it is possible to incorporate nonlinear effects. As Schmidt and Meile [1989] point out, many system responses are not truly linear. However, within a narrow design space, linear approximations of a nonlinear response can result in a fruitful description of the input-output relationship. This is the case with the problem described here. Once we identify our design plan, it will be clear that a linear approximation may very well explain the model. However, even when a model is nonlinear, it may be possible to achieve a linear transformation. As Box et al. [1978] note, it is important to seek transformations that satisfy and synthesize information from three desiderata simultaneously: (a) the models are structurally adequate ($E(\varepsilon) = 0$) and the errors are independent, (b) the model responses have a constant variance ($Var(\varepsilon) = \sigma^2$), and (c) the model responses are normally distributed. For models that looked at interaction effects, the reader should refer to Madu and Kuei [1994] and to Kuei and Madu [1994] for nonlinear regression metamodels .

2.3.2   Experimental Design and Data Dollection.

Once the type of regression metamodel has been decided on, a decision is made on the type of experimental design. There are different types of experimental designs but the common choice is normally between full factorial and fractional factorial designs. Fractional factorial designs are generally, very efficient especially when there are several input factors. Full factorial designs are generally very expensive and may often be impracticable [Dey [1985], and Hunter and Naylor [1970]. Furthermore, when the interest is in estimating lower-order effects, fractional factorial designs should be used. The model builder can save a considerable amount of time by using them. In our example, we used full factorial design because we have a very small number of input factors to work with. Our full factorial design plan dictates the mode of our simulation experiment. In our example, we set N = 10 as a fixed variable. There are three variables S, R and r and they are expressed from the problem in the following ranges: $1 \leq S \leq 5; 5 \leq R \leq 10$; and $0.1 \leq r \leq 0.5$. These ranges are quite narrow so that we can postulate a linear model. Furthermore, using a two-level full factorial design, we will have a total of 2^3 design points. Thus, we will need a total of 8 combinations of input variables to completely study this problem. Each of these design points will represent a simulation run in our simulation model.

Run No.	S	R	r	AEU
1	5	10	0.5	0.944
2	5	10	0.1	0.999
3	5	5	0.5	0.874
4	5	5	0.1	0.999
5	1	10	0.1	0.744
6	1	10	0.5	0.984
7	1	5	0.5	0.728
8	1	5	0.1	0.984

Table 6.1   Full factorial experimental design plan.

With the design plan given in Table 6.1, a steady state simulation model was developed using GPSS V on an IBM 360 mainframe computer. The simulation model was designed to eliminate warm-up effect prior to generating the steady state outputs of AEUs. The AEU values presented in the table represent the averages of ten replications of each simulation run. The hypothesized model given as equation (6.1) is now used to develop a metamodel for the maintenance float problem.

3.   THE METAMODEL

There are many ways to develop and test a metamodel. Many of the techniques are discussed in major books dedicated to metamodeling [Kleijnen [1987], Madu and Kuei [1993]]. We shall however, apply a method known as stability analysis on the maintenance float problem presented here. A regression analysis of the eight sample observations presented in Table 6.1 is given in Table 6.2.

A level of significance of 0.3 is used to check the significance of both the main effects and the two-factor interaction effects if they are considered. Based on the results of Table 6.2, it is shown that the number of standby units is not statistically significant in explaining the variations observed in AEU (p = 0.47296). However, this result shows that the two-factor interaction of standby units and the ratio of mean time to repair to mean time between failure (r) are statistically significant at p = 0.01762 and have a very high partial r^2 of 0.8830. Likewise, the ratio (r) by itself

VAR	Reg. coef.	Std. error coef.	T(d.f.=3)	P	Partial r	Std. error model	R^2 (Multiple R^2)
Float	-0.0061	0.0075	-0.819	0.47296	0.1862	0.0235	0.9816 (0.9908)
Repair	0.0043	0.0033	1.295	0.28594	0.3586		
Ratio	-0.7187	0.0748	-9.607	0.00239	0.9685		
F x R	0.0988	0.0208	4.759	0.01762	0.8830		
Const.	1.0199						

F=40.006; P=6.172E-03; DW=1.8868; Adjusted R=0.9571
R x R: float x ratio

Table 6.2 Regression analysis of the maintenance flaot problem with no runs deleted.

is highly significant with a partial r^2 of 0.9685. The regression model obtained here is also statistically significant (F = 40.006, p = 6.172 E-03). Further, the Durbin-Watson statistics with a value of 1.8898 shows the lack of autocorrelation in the data.

Initially, we postulated a regression metamodel that considered only main effects. However, this initial model showed that S and R are not statistically significant in explaining the variations in AEU. However, with $\alpha = 0.3$, R was observed to be significant. Kleinjnen and Strandrige [1988] have used such α levels to develop metamodels for a FMS. However, the use of *stability analysis* provides a more sophisticated approach for testing the significance or non-significance of the coefficients of the input variables. Basically, this procedure involves the deletion of one run at a time and then running the regression analysis without the deleted run. For example, the eight different runs in Table 6.1 will have to be deleted one at a time and then the regression analysis is run. The values of the non-significant effects will change showing instability or the lack of stability in those values. Thus, input variables with unstable coefficients can be considered "insignificant" in explaining the variations in AEU and can be dropped. Table 6.3 presents the stability analysis .

Notice from this Table, that β_0 represents the coefficient of the constant term, β_1 represents the coefficient of the standby units (S), β_2 represents the coefficient of the number of repairpersons (R), β_3 represents the coefficient of the ratio of mean time to repair to mean time between failure (r) and β_4 represents the coefficient of the two-factor interaction between S and r. R^2 is the coefficient of determination, F is the F statistic, and p is the attained level of significance. A more detailed result is given in Table 6.4.

RD	β_0	β_1	β_2	β_3	β_4	R^2	F	P
1	1.0421	-0.0041	0.0011	-0.6985	0.0785	0.9990	516.821	1.093 E-03
2	1.0046	-0.0106	0.0057	-0.7098	0.0898	0.9830	29.3	0.233
3	1.0461	-0.0081	0.0011	-0.7390	0.1190	0.9990	518.696	1.925 E-03
4	1.0136	-0.0106	0.0057	-0.7277	0.1077	0.9916	29.300	0.333
5	1.0160	-0.0059	0.0047	-0.7073	0.0965	0.9725	17.711	0.0542
6	1.0315	-0.0106	0.0057	-0.7635	0.1077	1.0315	30.376	-0.0321
7	1.0183	-0.0064	0.0047	-0.7302	0.1010	0.9693	15.797	0.0604
8	0.9867	-0.0106	0.0057	-0.6740	0.0898	0.9838	30.376	0.0321
None	1.0199	-0.0061	0.0043	-0.7187	0.9816	0.9816	40.006	6.172 E-03

RD : Runs deleted

Table 6.3 Stability analysis of the maintenance float problem.

It can be observed from Tables 6.3 and 6.4 that the coefficients of the number of standby units are quite unstable. In some run deletions for example, the coefficients switch from significance to non-significance (see Table 6.4). Further, from Table 6.3, we notice wide swings in the coefficients, as runs are deleted one at a time. It is also observed that the coefficients of the number of repairpersons tend to show signs of instability although they are relatively more stable than the coefficients of the number of standby units. This instability in the coefficient of number of repairpersons is more pronounced when we look at the coefficients when the first and third runs were respectively deleted. Since R is a major decision variable and since its two-factor interaction with other variables is insignificant, we have decided to retain it in the regression metamodel even though it is apparently unstable. However, it is important to explore other functional relationships that might exist between R and AEU. From this analysis therefore, we have deleted the main effect of S and using the regression analysis presented in Table 6.2, we have derived a regression metamodel for AEU as given in equation (6.2) below

$$AEU = 1.0199 + 0.0043R - 0.7187r + 0.0988S \text{ x } r \qquad (6.2)$$

From this model, we can conduct sensitivity analysis. Some of the potential relationships between AEU and the variables are apparent from looking at the signs of the coefficients. It is clear that AEU will drop by 0.7187 for each unit increase in r, but we know that a unit value for r is not possible otherwise the maintenance system will be incapable of providing satisfactory services. In fact, for better services, the value of r should be decreasing not increasing. At low values of r, average equipment utilization will be higher. AEU is however, not very sensitive to changes in r. For example, when r changes from 0.1 to 0.3, only a 2.5% decrease in AEU is observed. Similarly, when it changes from 0.3 to 0.5, only a 4.62% change in AEU is observed.

VAR	Reg. coef.	Std error coef.	T(d.f.=3)	P	Partial r	Std error model	R^2 (Multiple R^2)
	Run 1	deleted					
Float	-0.0041	0.0021	-1.947	0.19094	0.6546	0.0065	0.999
							(0.995)
Repair	0.0011	0.0011	1.000	0.42265	0.3333		
Ratio	-0.6895	0.0211	-33.135	0.00091	0.9982		
F x R	0.0785	0.0067	11.781	0.00713	0.9858		
Const.	1.0421						
F=516.821; P=1.932E-03; DW=1.8333; Adjusted R^2=0.9971							
	Run 2	deleted					
Float	-0.0106	0.0106	-1.55	0.89075	0.0119	0.0259	0.9832
Repair	0.0057	0.0042	1.354	0.30852	0.4781		(.9916)
Ratio	-0.7098	0.0837	-9.479	0.01362	0.9729		
F x R	0.0898	0.0265	3.392	0.077	0.8519		
Const.	1.0046						
F=29.3; P=0.0333E-03; DW=1.9879; Adjusted R^2=0.9479							
	Run 3	deleted					
Float	-0.0081	0.0021	-3.864	0.6092	0.8819	0.0065	0.999
Repair	0.0011	0.0011	1.000	0.42265	0.3333		(0.9995)
Ratio	-0.739	0.0211	-35.052	0.00081	0.9984		
F x R	0.1190	0.0067	17.844	0.0031	0.9938		
Const.	1.0461						
F=518.696; P=1.925E-03; DW=2.0833; Adjusted R^2=0.9971							
	Run 4	deleted					
Float	-0.0106	0.0106	-1.002	0.42207	0.3340	0.0259	0.9832
Repair	0.0057	0.0042	1.354	0.30852	0.4781		(0.9916)
Ratio	-0.7277	0.0837	-8.693	0.01297	0.9742		
F x R	0.1077	0.0265	4.069	0.05543	0.8922		
Const.	1.0136						
F=29.3; P=0.0333E-03; DW=1.9879; Adjusted R^2=0.9479							
	Run 5	deleted					
Float	-0.0059	0.0092	-0.639	0.58812	0.1696	0.0286	0.9725
Repair	0.0047	0.0047	1.000	0.42265	0.3333		(0.9862)
Ratio	-0.7073	0.1167	-6.063	0.02615	0.9484		
F x R	0.0965	0.0292	3.307	0.08054	0.8454		
Const.	1.016						
F=17.7111; P=0.0542E-03; DW=2.0833; Adjusted R^2=0.9176							

	Run 6	deleted					
Float	-0.0106	0.0106	-1.002	0.42207	0.3340	0.0259	0.9838
Repair	0.0057	0.0042	1.354	0.30852	0.4781		(0.9919)
Ratio	-0.7635	0.1059	7.211	0.01869	0.9630		
F x R	0.1077	0.0265	4.069	0.05543	0.8922		
Const.	1.0315						

F=30.376; P=0.0321E-03; DW=2.3352; Adjusted R^2=0.9514

	Run 7	deleted					
Float	-0.0064	0.0092	-0.689	0.56206	0.1918	0.0286	0.9693
Repair	0.0047	0.0047	1.000	0.42265	0.3333		(0.9845)
Ratio	-0.7302	0.1167	-6.259	0.02459	0.9514		
F x R	0.1010	0.0292	3.464	0.7417	0.8572		
Const.	1.0183						

F=15.797; P=0.0604E-03; DW=2.0833; Adjusted R^2=0.9080

	Run 8	deleted					
Float	-0.0106	0.0106	-0.155	0.89075	0.0119	0.9838	
Repair	0.0057	0.0042	1.354	0.30852	0.4781		(0.9919)
Ratio	-0.6740	0.0748	6.365	0.02380	0.953		
F x R	0.0898	0.0265	3.464	0.0770	0.8519		
Const.	0.9867						

F=30.376; P=0.0321E-03; DW=2.228; Adjusted R^2=0.9514
Std. Error est. = 0.0259

Table 6.4 Statistical analysis of the maintenance float problem with statistical significance tests.

3.1 Model Validation. It is important to judge how well the model predicts reality. To validate the metamodel presented as equation (6.2), we randomly selected a new set of input values for the independent variables that fall within the domain of definition of equation (6.2) but were not used in developing the model. These new input values for S and R are given in Table 6.5 while r is set to 0.3. New simulations were run and validated while the metamodel was applied in computing AEU for these input values. The results of both the simulation and the regression metamodels are shown in Table 6.5. As results show, the percentage deviation of the regression metamodel predicted AEU values from the simulation values is within 4%. Thus, this metamodel offers a good prediction of the AEU values.

S	R	AEU Simulation	AEU Metamodel	% Error deviation
3	7	0.958	0.923	3.65
4	9	0.981	0.962	1.94
5	5	0.984	0.974	1.02
5	7	0.991	0.983	0.81
5	10	0.992	0.995	-0.40

Table 6.5   Validation of the maintenance float regression metamodel.

4.   COST OPTIMIZATION

In this section, we use the regression metamodel presented as equation (6.2) to determine the optimal number of standby units and the number of repairpersons needed to meet the system availability requirement of at least 95%. The cost optimization model is therefore, stated as follows:

$$Min. \ TC = (1 - AEU)C_d + C_S S + C_R R \qquad (6.3)$$

Subject to

$$AEU \geq 0.95$$

Where the terms in the model are defined as follows: Cd is the cost of lost production per hour, C_S is the cost of standby units per hour and C_R is the cost of repairpersons per hour. For the intervals $5 \leq R \leq 10$ and $1 \leq S \leq 5$, given for the problem, we use the model of equation (6.2) to compute the AEU for the different combinations of R and S. These results are presented in Table 6.6.

Using the cost information provided for this problem and the constraint on AEU, we determine that the optimal number of standby units and number of repairpersons are 4 and 10 respectively. This combination of S and R will yield an AEU = 0.966 and a minimum total cost of $48,850. These results are presented in Table 6.7.

Although the problem presented here is not as complex as when using simulation and regression metamodeling to analyze it, it is however difficult to develop an analytical solution to it. Some of the analytical solutions make the assumption of ample services or no waiting time for repair [Levine (1965), Lowe and Lewis (1983), and Madu (1987)]. Madu and Georgantzas [1988] studied the implications of such assumptions. Others have also solved this problem by assuming exponential service times, which implies that the coefficient of variation is 1 [Levine (1965),

S	R=5	R=6	AEU values R=7	R=8	R=9	R=10
1	0.855	0.860	0.864	0.868	0.873	0.877
2	0.885	0.889	0.894	0.898	0.902	0.907
3	0.915	0.919	0.923	0.928	0.932	0.936
4	0.944	0.949	0.953	0.957	0.962	0.966
5	0.974	0.978	0.983	0.987	0.991	0.955

Table 6.6   AEU values for defined intervals of S and R obtained through the metamodel.

S	AEU	R	C_5	C_r	C_d	TC
4	0.953	7	40000	245	11750	51995
4	0.957	8	40000	280	10750	51031
4	0.962	9	40000	315	9500	49815
4	0.966	10	40000	350	8500	48850
5	0.974	5	40000	175	6500	56675
5	0.978	6	40000	210	5500	55710
5	0.983	7	40000	245	4250	54495
5	0.987	8	40000	280	3250	53530
5	0.991	9	40000	315	2250	52565
5	0.995	10	40000	350	1250	51600

Table 6.7   Cost table for the "Optimum" R and S.

Madu (1988b,c), and Koeningsberg (1958, 1960)]. Madu et al. [1990] showed that coefficient of variation influences AEU so it is important to be able to consider a wider range of distributions and different coefficient of variations. Such can be achieved using the approach presented in this chapter.

5. STEPS FOR DEVELOPING A METAMODEL

It is important to define the important steps to developing metamodels since this tool is efficient in analyzing complex problems. Our steps are outlined below:

1. Define the problem - Problem definition should be the first step. If the wrong problem is defined, the wrong solution will be derived. It is important that at this stage, the objectives and the constraints of the problem are identified as well as the relationships between the variables. It is important to have an idea of the main effects and the interaction effects that are of importance in the problem.

2. Define the ranges for the input variables - It is important to know the boundaries and limits of the input variables. This requires the experience and judgment of the model builder. Pilot studies may be conducted through simulation or the rule of thumb may be used to determine the appropriate range of values for the controllable variables.

3. Develop the experimental design - This requires sound statistical knowledge to decide what type of information is needed and which design plan will be more efficient in generating such information. Full or fractional factorial design may be used depending on the type of information that is needed. Traditionally, the focus has been on controllable factors. However, there are studies now that show how noise factors can be integrated in metamodeling [Madu and Madu, 1999].

4. Build a simulation model - A simulation is developed and derives its input values from the design plan given as step 3. The simulation itself has to be tested for adequacy and true representation of the actual system that is being modeled. Thus, it must be verified and validated.

5. Develop the metamodel - The output generated from simulation after it has been statistically tested and deemed appropriate, can

be used in developing the regression metamodel . The model derived depends again on what the model builder intends to estimate and knowledge of the type of relationships that exist between the variables.

6. Validate the metamodel - It important to show that the metamodel is adequate and will lead to satisfactory estimation of the dependent variable. The metamodel must also satisfy the statistical properties say of regression if regression is the form of the model. New metamodels should be generated when significant changes in the system are anticipated.

6. CONCLUSION

This chapter shows how to develop a simulation metamodel for a maintenance float system. We went through the necessary steps and showed how the regression metamodel is built, validated, and used in cost optimization. A summary of the steps taken is also presented. The metamodel approach is recommended in situations where it is difficult to develop an analytical solution. We shall now outline some of the benefits of the metamodel:

1. Simulation metamodels are flexible and enable us to solve real life problems in their original forms.

2. By using experimental design techniques, we can efficiently build a simulation metamodel. This makes the application of simulation techniques attractive.

3. The use of metamodels allows us to generalize simulation outputs. However, these generalizations must fall within the bounds of the model.

4. Sensitivity analysis can equally be conducted within the bounds of the model without re-running costly simulation programs.

5. When the regression metamodel is valid, its estimates are satisfactory and comparable to that of simulation.

6. Regression metamodels are easier and simpler to use.

7. The ease of use of these models can bridge the gap between practice and theory and encourage practitioners to actually use metamodels in decision making.

References

[1] Law, M., and McComas, M. G. Pitfalls to Avoid in the Simulation of Manufacturing Systems. *Ind. Engng.* **31**(5)(1989) 28-31.

[2] Friedman, L.W. and Pressman, I. The Metamodel in Simulation Analysis: Can it be Trusted? *J. Opt.Res. Soc.* **39**(10)(1968) 939-948.

[3] Kleijnen, J.P.C. *Statistical Tools for Simulation Practitioners.* Marcel Dekket, New York 1987.

[4] Kleijnen, J.P.C., and Stanridge, C. R. Experimental Design and Regression Analysis in Simulation: an FMS Case Study. *European Journal of Operations Research.***33**(1988) 257-261.

[5] Georgantzas, N.C., and Madu, C.N. Maintenance Float Policy Estimation with Imperfect Information. *Computers Ind. En-grg.***16**(2)(1988) 257-268.

[6] Levine, B. Estimating Maintenance Float Factors on the Basis of Reliability Theory. *Ind. Qual. Control,***21**(2)(1965) 401-405.

[7] Lowe, P.H., and Lewis, W. Reliability Analysis Based on the Weibull Distribution: an Application to Maintenance Floar Factors. *International Journal of Production Research,* **21**(4)(1983) 461-470.

[8] C.N. Madu. Determination of Maintenance Floats Using Buzens Algorithm. *International Journal of Production Research,***26**(1988) 1385-1394.

[9] Madu, C.N., and Georgantzas, N.C. Waiting Line Effects in Analytic Maintenance Float Policy. *Decision Sci.,* **19**(3)(1988) 521-534.

[10] Madu, C.N. The Study of a Maintenance Float Model with Gamma Failure Distribution. *International Journal of Production Research,* **25**(9), (1987) 1305-1323.

[11] Koeningsberg, E. Finite Queues and Cyclic Queues. *Operations Research,***8**(1960) 246-253.

[12] Koeningsberg, E. Cyclic Queues. *Operations Research,* **9**(1958) 22-35.

[13] Madu, C.N., Chanin, M.N., Georgantzas, N.C., and Kuei, C.H. Co-efficient of Variation: a Critical Factor in Maintenance Float Policy. *Computers Operations Research,* **17**(2)(1990) 177-185.

[14] Madu, C.N. A Closed Queueing Maintenance Network with Two Repair Centers. *Journal of Operations Research Society,***39**(10)(1988) 959-967.

[15] Gross, D., Miller, D.R., and Soland, T.M. A Closed Queuing Network Model for Multiechelon Repairable Item Provisioning. *IIE Transactions*, **15**(4)(1983) 344-352.

[16] Madu, C.N., and Chanin, M.N. Maintenance Float Analysis: a Regression Metamodel A roach. Working Paper, Pace University, New York, 1989.

[17] Gordon, W.J., and Newell, G.F. Closed Queueing Systems with Exponential Servers. *Operations Research*, **15**(2)(1967) 254-265.

[18] Buzen, J.P. Computational Algorithms for Closed Queuing Networks with Exponintial Servers. *Commun. ACM* **16**(9)(1973) 527-531.

[19] Box, G.E.P., Hunter, W.G., and Hunter, J.S. *Statistics for Experimenters*. Wiley, New York, 1978.

[20] Dey, A. *Orthogonal Fractional Factorial Designs*. John Wiley and sons, New York, 1985.

[21] Friedman, L.W., and Friedman, H.H. Statistical Considerations in Computer Simulation: the State of the Art. *J. Statistic. Comput. Simul.* **19**(1984) 237-263.

[22] Gardenier,T. K. PRE_PRIM as a Pre-processor to Simulations: a Cohesive Unit. *Simulation.* (1990) 65-70.

[23] Hunter, J.S., and Naylor, T.H. Experimental Designs for Computer Simulation Experiments. *Management Sci.* **16**(7) (1970) 422-434.

[24] Kuei, C.H., and Madu, C.N. Polynomial Decomposition and Taguchi Design for Maintenance Float System. *European Journal of Operations Research*, **72**(1994) 364-375.

[25] Madu, C. N. Simulation in Manufacturing: a Regression Metamodel approach. *Computers and Industrial Engineering*, **18**(3)(1990) 381-389.

[26] Madu, C.N., and Kuei, C.H. Group Screening and Taguchi Design Applications in a Multiechelon Maintenance Float Policy. *Computers Operations Research*, **19**(2)(1992) 95-105.

[27] Schmidt, M.S., and Meile, L.C. Taguchi Designs and Linear Programming Speed New Product Formulation. *Interfaces*, **19**(5)(1989) 49-56.

[28] Madu, I.E., and Madu, C.N. Design Optimization using Signal-to-Noise Ratio. *Simulation*, 1999.

[29] Madu, C.N., and Kuei, C.H. *Experimental Statistical Designs and Analysis in Simulation Modeling*. Quorum Books, Westport, Connecticut, 1993.

III

PREVENTIVE MAINTENANCE

Chapter 7

BASIC PREVENTIVE MAINTENANCE POLICIES AND THEIR VARIATIONS

T. Dohi, N. Kaio, and S. Osaki

Department of Industrial and Systems Engineering,
Hiroshima University

dohi@gal.sys.hiroshima-u.ac.jp

kaio@shudo-u.ac.jp

osaki@gal.sys.hiroshima-u.ac.jp

Abstract   This chapter is concerned with the basic preventive maintenance policies arising in the context of the mathematical maintenance theory . Simple but practically important preventive maintenance optimization models , which involve age replacement and block replacement, are reviewed in the framework of the well-known renewal reward argument. Some variations to these basic models as well as the corresponding discrete time models are also introduced with the aim of the application of the theory to the practice.

Keywords: preventive maintenance, age replacement , block replacement, order replacement, renewal reward policies, continuous and discrete models, cost models

1.   INTRODUCTION

The mathematical maintenance policies have been developed mainly in the research area of operations research/ management science, to generate an effective preventive maintenance schedule. The most important problem facing those developing mathematical maintenance policies is to design a maintenance plan by utilising one of two maintenance options, preventive replacement and corrective replacement . In the preventive replacement, the system or unit is replaced by a new one before it fails. On the other hand, the corrective replacement involves replacing the failed unit. Up to the present time, a huge number of replacement methods have been proposed in the literature. At the same time, some technical books on this problem have been published. For example, Arrow, Karlin and Scarf [4], Barlow and Proschan [10, 11], Jorgenson, McCall and Radner [34], Gnedenko, Belyayev and Solovyev [33], Gertsbakh [31], Ascher and Feingold [5] and Osaki [62], Osaki and Cao [61], Ozekici [67] and Christer, Osaki and Thomas [22]. Recently, Barlow [7] and Aven and Jensen [6] presented excellent text books to review the mathematical maintenance theory . In addition, some survey papers will be useful to review the history of this research context, such as McCall [42], Osaki and Nakagawa [65], Pierskalla and Voelker [69], Sherif and Smith [75] and Valdez-Flores and Feldman [84].

Since the mechanism causing a failure in almost all real complex systems may be considered to be uncertain, the mathematical technique to deal with maintenance problems should be based on the usual probability theory. If we are interested in the dynamic behavior of system failures depending on time, the problems are essentially reduced to the study of the stochastic processes presenting phenomena on both failure and replacement. In fact, since the theory of stochastic processes depends strongly on the mathematical maintenance theory , many text books on the stochastic processes have treated several maintenance problems . See, for example, Feller [29], Karlin and Taylor[40, 41] and Taylor, Karlin [79] and Ross [72]. In other words, in order to design a maintenance schedule effectively, both the underlying stochastic process governing the failure mechanism and the role of maintenance options carried out on the process have to be analyzed carefully. In that sense, the mathematical maintenance theory is one of the most important parts in applied probability modeling.

In this article, we present the basic preventive maintenance policies arising in the context of the maintenance theory. Simple but practically important preventive maintenance optimization models , which involve age replacement and block replacement, are reviewed in the framework of the well-known renewal reward argument. Some variations to these basic models as well as the corresponding discrete time models are also introduced with the aim of the application of the theory to the practice. In almost all text books and technical papers, the discrete-time preventive maintenance models have been paid little attentions. The main reason is that the discrete-time models are ordinarily considered as trivial analogies of the continuous-time ones. However, we often face some maintenance problems modeled in discrete-time setting in practice. If one considers the situation where the number of take-offs from airports influences the damage to airplanes, the parts of the airplane should be replaced at a pre-specified number of take-off rather than the elapsed time. In addition , in the Japanese electric power company which is the subject of this investigation, the failure time data of electric switching devices are recorded as group data (the number of failures per year) as it is not easy to carry out the preventive replacement schedule at the unit of week or month since the service team is engaged in other works, too. From our questionnaire, it is helpful for practitioners that the preventive replacement schedule should be determined roughly at the unit of year. This will motivate discrete-time maintenance models.

2.   MATHEMATICAL PRELIMINARIES

Continuous Probability Theory
    First of all, define the terminologies used in this chapter. In general, "failure" means that the unit or system loses the pre-specified function. The time to failure X is called the lifetime or the failure time and can be defined as a non-negative random variable. Suppose that X (≥ 0) is a continuous random variable having probability density function $f(t)$ $(t \geq 0)$. Then the probability distribution function or the lifetime distribution $F(t)$ is represented as

$$F(t) = \Pr\{X \leq t\} = \int_0^t f(x)dx, \quad t \geq 0, \tag{7.1}$$

where $F(0) = 0$ and $F(\infty) = 1$. The reliability function is the probability that the unit does not fail until time t and is defined by

$$\overline{F}(t) = \Pr\{X > t\} = \int_t^\infty f(x)dx = 1 - F(t), \quad t \geq 0, \tag{7.2}$$

where $\overline{F}(0) = 1$ and $\overline{F}(\infty) = 0$. and this is also called the survivor function . Throughout this chapter, we use the symbol $\overline{\phi}(\cdot) = 1 - \phi(\cdot)$.
    Define the probability that the unit fails at $(t, t + \triangle]$, provided that it is operative at time t, as follows.

$$\Pr\{t < X \leq t + \triangle | X > t\} = \frac{f(t)}{\overline{F}(t)} \cdot \triangle + o(\triangle), \quad t \geq 0, \tag{7.3}$$

where $\Pr\{A_1|A_2\}$ implies the conditioned probability for the event A_1 provided that the event A_2 occurs and the function $o(\triangle)$ satisfies $\lim_{\triangle \to 0} o(\triangle)/\triangle = 0$. Then, the function

$$r(t) = \frac{f(t)}{\overline{F}(t)}, \quad t \geq 0 \tag{7.4}$$

is called the failure rate or the hazard rate . Using the failure rate, the reliability function and the probability density function are represented as

$$\overline{F}(t) = \exp\left[-\int_0^t r(x)dx\right], \quad t \geq 0 \tag{7.5}$$

and

$$f(t) = r(t)\exp\left[-\int_0^t r(x)dx\right], \quad t \geq 0, \tag{7.6}$$

respectively. In particular, the function $H(t) = \int_0^t r(x)dx$ $(t \geq 0)$ is called the cumulative hazard or the mean value function .

In the reliability theory , we concentrate our attention on investigating the behavior of the failure rate with respect to elapsed age. It is well known in general that the failure rate for the displays the so-called bathtub curve . This shows that the failure rate decreases in the initial failure period, subsequently becomes constant in the random failure period and increases in the wear–out failure period. The lifetime distribution is given as IFR (increasing failure rate) and DFR (decreasing failure rate), if $r(t)$ $(t \geq 0)$ increases and decreases, respectively. In particular, when $r(t)$ is constant, $F(t)$ equals CFR (constant failure rate) and is the exponential distribution . The mean lifetime $E[X]$ is defined as

$$E[X] = \int_0^\infty tf(t)dt = \int_0^\infty \overline{F}(t)dt, \tag{7.7}$$

where, in general, $E[\cdot]$ denotes the mathematical expectation . If the unit is repaired or replaced by a new one after the failure occurs, especially, $E[X]$ strictly represents MTBF (mean time between failures) or if not, MTTF (mean time to failure).

Define the convolution between two lifetime distributions $F(t)$ and $G(t)$ $(t \geq 0)$ by

$$F(t) * G(t) = \int_0^t F(t-x)g(x)dx = \int_0^t G(t-x)f(x)dx, \quad t \geq 0, \tag{7.8}$$

where $g(t) = dG(t)/dt$. For identical distributions, the n-fold $(n = 0, 1, 2, \cdots)$ convolution of the lifetime distribution $F^{(n)}(t)$ is recursively defined as follows.

$$F^{(0)}(t) = 1(t), \quad (: \text{unit function}), \quad t \geq 0, \tag{7.9}$$

$$F^{(n)}(t) = F(t) * F^{(n-1)}(t), \quad t \geq 0, \quad n = 1, 2, 3, \cdots. \tag{7.10}$$

Similarly, the n-fold convolution of the probability density function $f^{(n)}(t)$ $(n = 0, 1, 2, \cdots)$ becomes

$$f^{(0)}(t) = \delta(t), \quad (\text{: Dirac's delta function}), \quad t \geq 0, \qquad (7.11)$$

$$
\begin{aligned}
f^{(n)}(t) &= \int_0^t f^{(n-1)}(t-x)f(x)dx \\
&= \frac{dF^{(n)}(t)}{dt}, \quad t \geq 0, \quad n = 1, 2, 3, \cdots. \qquad (7.12)
\end{aligned}
$$

Evidently, the n-fold convolution of the lifetime distribution means the probability that a unit fails at n times up to time t.

The most important stochastic process used in the maintenance theory is the renewal process. Let $N(t)$ $(t \geq 0)$ be the number of failures (renewals) during the time interval $(0, t]$. Then the renewal function $M(t)$ is defined as the expectation of the random variable $N(t)$ for fixed t. This gives us the well-known result, we have

$$
\begin{aligned}
M(t) &= E[N(t)] = \sum_{k=1}^{\infty} F^{(k)}(t) \\
&= F(t) + F(t) * M(t) = \int_0^t m(x)dx, \quad t \geq 0, \qquad (7.13)
\end{aligned}
$$

where $m(t)$ $(t \geq 0)$ is called the renewal density and given by

$$
\begin{aligned}
m(t) &= \frac{dM(t)}{dt} = \sum_{k=1}^{\infty} f^{(k)}(t) \\
&= f(t) + \int_0^t m(t-x)f(x)dx \\
&= f(t) + \int_0^t f(t-x)m(x)dx, \quad t \geq 0. \qquad (7.14)
\end{aligned}
$$

The following theorem is known as the elementary renewal theorem and gives asymptotic properties of $M(t)/t$ and $m(t)$.

Theorem 2.1.

$$\lim_{t \to \infty} \frac{M(t)}{t} = \frac{1}{E[X]}, \qquad (7.15)$$

$$\lim_{t \to \infty} m(t) = \frac{1}{E[X]}. \qquad (7.16)$$

For details on the renewal processes , see Feller [29].

Discrete Probability Theory

Let D $(D = 0, 1, 2, \cdots)$ be the discrete non-negative random variable with period one and denote the system lifetime , having probability mass function $f(d)$ $(d = 0, 1, 2, \cdots, \; f(0) = 0)$. Then the lifetime distribution $F(d)$ is

$$F(d) = \sum_{j=0}^{d} f(j), \quad d = 0, 1, 2, \cdots, \tag{7.17}$$

where $F(0) = 0$ and $F(\infty) = 1$. The reliability function $\overline{F}(d)$ is

$$\overline{F}(d) = \sum_{j=d+1}^{\infty} f(j), \quad d = 0, 1, 2, \cdots, \tag{7.18}$$

where $\overline{F}(0) = 1$ and $\overline{F}(\infty) = 0$. Then the MTTF and the failure rate become

$$E[D] = \sum_{d=0}^{\infty} df(d) = \sum_{d=0}^{\infty} \overline{F}(d) \tag{7.19}$$

and

$$r(d) = \frac{f(d)}{\overline{F}(d-1)} = \frac{f(d)}{\sum_{j=d}^{\infty} f(j)}, \quad d = 0, 1, 2, \cdots, \tag{7.20}$$

respectively.

In a fashion similar to the continuous case, if the failure rate $r(d)$ $(d = 0, 1, 2, \cdots)$ is a non-decreasing (non-increasing) sequence in d, the lifetime $F(d)$ is said to be IFR (DFR). When $r(d)$ is said to be constant, then $F(d)$ is CFR and obeys the geometric distribution . Using the failure rate $r(d)$, the reliability function and the probability mass function can be represented as

$$\overline{F}(d) = \prod_{j=0}^{d} \overline{r}(j), \quad d = 0, 1, 2, \cdots \tag{7.21}$$

and

$$f(d) = r(d) \prod_{j=0}^{d-1} \overline{r}(j), \quad d = 0, 1, 2, \cdots, \tag{7.22}$$

respectively.

For two discrete lifetime distributions $F(d)$ and $G(d)$ ($g(d) = G(d) - G(d-1), d = 0, 1, 2, \cdots$), define the convolution ;

$$F(d) * G(d) = \sum_{j=0}^{d} F(d-j)g(j) = \sum_{j=0}^{d} G(d-j)f(j), \quad d = 0, 1, 2, \cdots.$$

$$(7.23)$$

Similar to Eqs. (7.9) and (7.10), the n-fold convolution $F^{(n)}(d)$ ($n = 0, 1, 2, \cdots$) for identical distributions are recursively defined as

$$F^{(0)}(d) = 1, \quad d = 0, 1, 2, \cdots, \tag{7.24}$$

$$F^{(n)}(d) = F(d) * F^{(n-1)}(d), \quad d = 0, 1, 2, \cdots, \quad n = 1, 2, 3, \cdots. \tag{7.25}$$

Then the corresponding n-fold convolution for identical probability mass functions is

$$f^{(0)}(d) = \begin{cases} 1, & d = 0 \\ 0, & d = 1, 2, 3, \cdots, \end{cases} \tag{7.26}$$

$$\begin{aligned}
f^{(n)}(d) &= \sum_{j=0}^{d} f^{(n-1)}(d-j)f(j) \\
&= F^{(n)}(d) - F^{(n)}(d-1), \quad d = 0, 1, 2, \cdots, \quad n = 1, 2, 3, \cdots.
\end{aligned}$$

$$(7.27)$$

Now, we are in a position to develop the discrete renewal theory . Let $N(d)$ ($d = 0, 1, 2, \cdots$) be the random variable representing the number of failures which occurred during the time period $(0, d]$. Then the (discrete) renewal function $M(d) = \mathrm{E}[N(d)]$ is defined by

$$\begin{aligned}
M(d) &= E[N(d)] = \sum_{k=1}^{\infty} F^{(k)}(d) \\
&= F(d) + F(d) * M(d) = \sum_{j=0}^{d} m(j), \quad d = 0, 1, 2, \cdots, (7.28)
\end{aligned}$$

where $m(d)$ ($d = 0, 1, 2, \cdots$, $m(0) = 0$) is called the renewal probability mass function and means the probability that the failures occur at time d, that is,

$$m(d) = M(d) - M(d-1) = \sum_{k=1}^{\infty} f^{(k)}(d)$$

$$= f(d) + \sum_{j=0}^{d} m(d-j)f(j) = f(d) + \sum_{j=0}^{d} f(d-j)m(j),$$
$$d = 0, 1, 2, \cdots. \tag{7.29}$$

In general, the discrete renewal theory has the one-to-one correspondence to the continuous version. For instance, we have the following elementary renewal theorem.

Theorem 2.2.

$$\lim_{d \to \infty} \frac{M(d)}{d} = \frac{1}{E[D]}, \tag{7.30}$$

$$\lim_{d \to \infty} m(d) = \frac{1}{E[D]}. \tag{7.31}$$

For more details on the discrete renewal theory , see Taylor and Karlin [79], pp. 299-304, Munter [44] and Kaio and Osaki [39].

3.   BLOCK REPLACEMENT

For block replacement models , the preventive replacement is executed periodically at a prespecified time kt_0 ($t_0 \geq 0$) or kN ($N = 0, 1, 2, \cdots$), ($k = 1, 2, 3, \cdots$). If the unit fails during the time interval $((k-1)t_0, kt_0]$ or $((k-1)N, kN]$, then the corrective maintenance is made at the failure time . The main property for the block replacement is to be easier to administer in general, since the preventive replacement time is scheduled in advance and one needs not observe the age of a unit. In this section, we develop the following three variations of block replacement model:

(i) A failed unit is replaced instantaneously at failure (Model I)

(ii) A failed unit remains inoperable until the next scheduled replacement comes (Model II)

(iii) A failed unit undergoes minimal repair (Model III)

The cost components used in this section are the following:

c_p (> 0): unit preventive replacement cost at time kt_0 or kN, ($k = 1, 2, 3, \cdots$).

c_c (> 0): unit corrective replacement cost at failure time .

c_m (> 0): unit minimal repair cost at failure time .

Model I

First, we consider the continuous–time block replacement model (Barlow and Proschan [10]). A failed unit is replaced by a new one during the replacement interval t_0, and the scheduled replacement for the non-failed unit is performed at kt_0 ($k = 1, 2, 3, \cdots$). Let $F(t)$ be the continuous lifetime distribution with finite mean $1/\lambda$ (> 0). From the well-known renewal reward theorem , it is possible to formulate the expected cost per unit time in the steady-state for the block replacement model as follows.

$$B_c(t_0) = \frac{c_c M(t_0) + c_p}{t_0}, \quad t_0 \geq 0, \tag{7.32}$$

where the function $M(t) = \sum_{k=1}^{\infty} F^{(k)}(t)$ denotes the mean number of failures during the time period $(0, t]$ (the renewal function). The problem is, of course, to derive the optimal block replacement time t_0^* which minimizes $B_c(t_0)$.

Define the numerator of the derivative of $B_c(t_0)$ as

$$j_c(t_0) = c_c[t_0 m(t_0) - M(t_0)] - c_p, \tag{7.33}$$

where the function $m(t) = dM(t)/dt$ is the renewal density . Then, we have the optimal block replacement time t_0^* which minimizes the expected cost per unit time in the steady-state $B_c(t_0)$.

Theorem 3.1. (1) Suppose that the function $m(t)$ is strictly steadily increasing with respect to t (> 0).

(i) If $j_c(\infty) > 0$, then there exists one finite optimal block replacement time t_0^* ($0 < t_0^* < \infty$) which satisfies $j_c(t_0^*) = 0$. Then the corresponding minimum expected cost is

$$B_c(t_0^*) = c_c m(t_0^*). \tag{7.34}$$

(ii) If $j_c(\infty) \leq 0$, then $t_0^* \to \infty$, that is, it is optimal to carry out only the corrective replacement , and the corresponding minimum expected cost is

$$B_c(\infty) = \lambda c_c. \tag{7.35}$$

(2) Suppose that the function $m(t)$ is decreasing with respect to t (> 0). Then, the optimal block replacement time is $t_0^* \to \infty$.

Next, we formulate the discrete–time block replacement model (Nakagawa [45]). In the discrete–time setting, the expected cost per unit time in the steady-state is

$$B_d(N) = \frac{c_c M(N) + c_p}{N}, \quad N = 0, 1, 2, \cdots, \tag{7.36}$$

where the function $M(n) = \sum_{k=1}^{\infty} F^{(k)}(n)$ is the discrete renewal function for the discrete lifetime distribution $F(n)$ $(n = 0, 1, 2, \cdots)$. Define the numerator of the difference of $B_d(N)$ as

$$j_d(N) = c_c[Nm(N+1) - M(N)] - c_p, \qquad (7.37)$$

where the function $m(n) = M(n) - M(n-1)$ is the renewal probability mass function . Then, we have the optimal block replacement time N^* which minimizes the expected cost per unit time in the steady-state $B_d(N)$.

Theorem 3.2. (1) Suppose that the $m(n)$ is strictly increasing with respect to n (> 0).

(i) If $j_d(\infty) > 0$, then there exists one finite optimal block replacement time N^* $(0 < N^* < \infty)$ which satisfies $j_d(N-1) < 0$ and $j_d(N) \geq 0$. Then the corresponding minimum expected cost satisfies the inequality

$$c_c m(N^*) < B_d(N^*) \leq c_c m(N^* + 1). \qquad (7.38)$$

(ii) If $j_d(\infty) \leq 0$, then $N^* \to \infty$, that is, it is optimal to carry out only the corrective replacement , and the corresponding minimum expected cost is

$$B_d(\infty) = \lambda c_c. \qquad (7.39)$$

(2) Suppose that the function $m(n)$ is decreasing with respect to n (> 0). Then, the optimal block replacement time is $N^* \to \infty$.

Remarks. A large number of variations on the block replacement model have been studied in the literature. Though we assume in the model above that the cost component is constant, some modifications are possible. Tilquin and Cleroux [82] and Berg and Epstein [14] extended the original model in terms of cost structure.

Model II

For the first model, we have assumed that a failed unit is detected instantaneously just after failure. This implies that the operating unit is monitored by a sensing device. Since such a case is not always general, however, we assume that failure is detected only at kt_0 $(t_0 \geq 0)$ or kN $(N = 0, 1, 2, \cdots)$, $(k = 1, 2, 3, \cdots)$ (see Osaki [60]). Consequently, in Model II, a unit is always replaced at kt_0 or kN, but is not replaced at the time of failure, and the unit remains inoperable for the time duration from the occurrence of failure until its detection.

In the continuous–time model, since the expected duration from the occurrence of failure until its detection per cycle is given by $\int_0^{t_0}(t_0 - t)dF(t) = \int_0^{t_0} F(t)dt$, we have the expected cost per unit time in the steady-state :

$$C_c(t_0) = \frac{c_c \int_0^{t_0} F(t)dt + c_p}{t_0},$$
(7.40)

where c_c is changed to the cost of failure per unit time, that is, the cost occurs per unit time for system down. Define the numerator of the derivative of $C_c(t_0)$ with respect to t_0 as $k_c(t_0)$, i.e.

$$k_c(t_0) = c_c\left\{F(t_0)t_0 - \int_0^{t_0} F(t)dt\right\} - c_p.$$
(7.41)

Theorem 3.3. (i) If $k_c(\infty) > 0$, then there exists one unique optimal block replacement time t_0^* $(0 < t_0^* < \infty)$ which satisfies $k_c(t_0^*) = 0$, and the corresponding minimum expected cost is

$$C_c(t_0^*) = c_c F(t_0^*).$$
(7.42)

(ii) If $k_c(\infty) \leq 0$, then $t_0^* \to \infty$ and $C_c(\infty) = c_c$.

On the other hand, in the discrete–time setting, the expected cost per unit time in the steady-state is

$$C_d(N) = \frac{c_c \sum_{k=1}^{N-1} F(k) + c_p}{N}, \quad N = 0, 1, 2, \cdots,$$
(7.43)

where the function $F(n)$ is the lifetime distribution $(n = 0, 1, 2, \cdots)$. Define the numerator of the difference of $C_d(N)$ as

$$i_d(N) = c_c[NF(N) - \sum_{k=1}^{N-1} F(k)] - c_p.$$
(7.44)

Then, we have the optimal block replacement time N^* which minimizes the expected cost per unit time in the steady-state $C_d(N)$.

Theorem 3.4. (i) If $j_d(\infty) > 0$, then there exists one finite optimal block replacement time N^* $(0 < N^* < \infty)$ which satisfies $i_d(N - 1) < 0$ and $i_d(N) \geq 0$. Then the corresponding minimum expected cost satisfies the inequality

$$c_c F(N^* - 1) < C_d(N^*) \leq c_c F(N^*)$$
(7.45)

(ii) If $i_d(\infty) \leq 0$, then $N^* \to \infty$.

Remarks. It is noticed that Model II has not been studied much in many literature, since this can not detect the failure instantly and is certainly not superior to Model I in terms of cost minimization. However, as described previously, one can see that the continuous monitoring to the operating unit is not always possible for all practical applications.

Model III

In the final model, we assume that minimal repair is performed when a unit fails and the failure rate is not disturbed by each repair. If we consider a stochastic process $\{N(t), t \geq 0\}$ in which $N(t)$ represents the number of minimal repairs up to time t, the process $\{N(t), t \geq 0\}$ is governed by a non-homogeneous Poisson process with mean value function

$$\Lambda(t) = \int_0^t r(x)dx, \tag{7.46}$$

which is also called the hazard function (rate of occurrence of failure). Noting this fact, Barlow and Hunter [8] gave the expected cost per unit time in the steady-state for the continuous–time model:

$$V_c(t_0) = \frac{c_m \Lambda(t_0) + c_p}{t_0}. \tag{7.47}$$

Define the numerator of the derivative of $V_c(t_0)$ as

$$l_c(t_0) = c_m[t_0 r(t_0) - \Lambda(t_0)] - c_p. \tag{7.48}$$

Then, we have the optimal block replacement time (with minimal repair) t_0^* which minimizes the expected cost per unit time in the steady-state $V_c(t_0)$.

Theorem 3.5. (1) Suppose that $F(t)$ is strictly IFR .

(i) If $l_c(\infty) > 0$, then there exists one finite optimal block replacement time with minimal repair t_0^* $(0 < t_0^* < \infty)$ which satisfies $l_c(t_0^*) = 0$. Then the corresponding minimum expected cost is

$$V_c(t_0^*) = c_m r(t_0^*). \tag{7.49}$$

(ii) If $l_c(\infty) \leq 0$, then $t_0^* \to \infty$ and the corresponding minimum expected cost is

$$V_c(\infty) = c_m r(\infty). \tag{7.50}$$

(2) Suppose that $F(t)$ is DFR . Then, the optimal block replacement time with minimal repair is $t_0^* \to \infty$.

Next, we formulate the discrete–time block replacement model with minimal repair (Nakagawa [45]). In the discrete–time setting, the expected cost per unit time in the steady-state is

$$V_d(N) = \frac{c_m \Lambda(N) + c_p}{N}, \quad N = 0, 1, 2, \cdots, \tag{7.51}$$

where the function $\Lambda(n)$ is the mean value function of the discrete non-homogeneous Poisson process . Define the numerator of the difference of $V_d(N)$ as

$$l_d(N) = c_m[Nr(N+1) - \Lambda(N)] - c_p, \tag{7.52}$$

where the function $r(n) = \Lambda(n) - \Lambda(n-1)$ is the failure rate function. Then, we have the optimal block replacement time with minimal repair N^* which minimizes the expected cost per unit time in the steady-state $V_d(N)$.

Theorem 3.6. (1) Suppose that $F(n)$ is strictly IFR .

  (i) If $l_d(\infty) > 0$, then there exists one finite optimal block replacement time with minimal repair N^* $(0 < N^* < \infty)$ which satisfies $l_d(N-1) < 0$ and $l_d(N) \geq 0$. Then the corresponding minimum expected cost satisfies the inequality

$$c_m r(N^*) < V_d(N^*) \leq c_m r(N^* + 1). \tag{7.53}$$

  (ii) If $l_d(\infty) \leq 0$, then $N^* \to \infty$.

  (2) Suppose that $F(n)$ is DFR . Then, the optimal block replacement time with minimal repair is $N^* \to \infty$.

Remarks. So far as we know, a great number of papers on models have been published. Morimura [43], Tilquin and Clerroux [83] and Cleroux, Dubuc and Tilquin [23] involve several interesting modifications. Later, Park [68], Nakagawa [46-50], Nakagawa and Kowada[52], Phelps [70], Berg and Cleroux [13], Boland [17], Boland and Proschan [18], Block, Borges and Savits [16] and Beichelt [12] proposed extended minimal repair models from the standpoint of generalization. One of the most interesting models with minimal repair is the (t, T)-policy. The (t, T)-policy is a combined policy with three kinds of maintenance options:

minimal repair, failure replacement and preventive replacement . That is, the minimal repair is executed for failures during the first period $[0, t)$, but the failure replacement is made for $[t, T]$, where T is the preventive replacement time. First, Tahara and Nishida [78] formulated this model. After being investigated by Phelps [71] and Segawa, Ohnishi and Ibaraki [74], recently, Ohnishi [57] proved the optimality of the (t, T)-policy under average cost criterion, via the dynamic programming approach. This tells us that the (t, T)-policy is optimal if we have only three kinds of maintenance options.

4.   AGE REPLACEMENT

As is well known, in the age replacement model, if the unit does not fail until a pre-specified time t_0 ($t_0 \geq 0$) or N ($N = 0, 1, 2, \ldots$), then it is replaced by a new one preventively. Otherwise, it is replaced at the failure time . The corrective and the preventive replacement costs are denoted by c_c and c_p, respectively, where, without loss of generality, $c_c > c_p$. This model plays a central role in all replacement models, since the optimality of the age replacement model has been proved by Bergman [15] if the replacement by a new unit is only maintenance option (i.e., if no repair is considered as an alternative option). In the rest part of this section, we introduce three kinds of age replacement models.

Basic Age Replacement Model (Barlow and Proschan [10], Barlow and Hunter [9] and Osaki and Nakagawa [64])

From the renewal reward theorem , it can be seen that the expected cost per unit time in the steady-state for the age replacement model is

$$A_c(t_0) = \frac{c_c F(t_0) + c_p \overline{F}(t_0)}{\displaystyle\int_0^{t_0} \overline{F}(t)dt}, \quad t_0 \geq 0. \tag{7.54}$$

If one can assume that the density $f(t)$ for the lifetime distribution $F(t)$ ($t \geq 0$)exists, the failure rate $r(t) = f(t)/\overline{F}(t)$ necessarily also exists. Define the numerator of the derivative of $A_c(t_0)$ with respect to t_0, divided by $\overline{F}(t_0)$ as $h_c(t_0)$, *i.e.*

$$h_c(t_0) = r(t_0) \int_0^{t_0} \overline{F}(t)dt - F(t_0) - \frac{c_p}{c_c - c_p}. \tag{7.55}$$

Then, we have the optimal age replacement time t_0^* which minimizes the expected cost per unit time in the steady-state $A_c(t_0)$.

Theorem 4.1. (1) Suppose that the lifetime distribution $F(t)$ is strictly IFR .

> **(i)** If $r(\infty) > K = \lambda c_c/(c_c - c_p)$, then there exists one finite optimal age replacement time t_0^* $(0 < t_0^* < \infty)$ which satisfies $h_c(t_0^*) = 0$. Then the corresponding minimum expected cost is

$$A_c(t_0^*) = (c_c - c_p)r(t_0^*). \tag{7.56}$$

> **(ii)** If $r(\infty) \le K$, then $t_0^* \to \infty$ and $A_c(\infty) = B_c(\infty) = \lambda c_c$.

(iii) If $F(t)$ is DFR , then $t_0^* \to \infty$.

Next, let us consider the case where the cost is discounted by the discount factor α $(\alpha > 0)$ (see Fox [30]). The present value of a unit cost after t $(t \ge 0)$ period is $\exp(-\alpha t)$. In the continuous-time age replacement problem, the expected total discounted cost over an infinite time horizon is

$$A_c(t_0; \alpha) = \frac{c_c \displaystyle\int_0^{t_0} e^{-\alpha t} f(t)dt + c_p e^{-\alpha t_0}\overline{F}(t_0)}{\alpha \displaystyle\int_0^{t_0} e^{-\alpha t}\overline{F}(t)dt}, \quad t_0 \ge 0. \tag{7.57}$$

Define the numerator of the derivative of $A_c(t_0; \alpha)$ with respect to t_0, divided by $\overline{F}(t_0)\exp(-\alpha t_0)$ as $h_c(t_0; \alpha)$,

$$h_c(t_0; \alpha) = r(t_0)\int_0^{t_0} e^{-\alpha t}\overline{F}(t)dt - \int_0^{t_0} e^{-\alpha t}f(t)dt - \frac{c_p}{c_c - c_p}. \tag{7.58}$$

Then, we have the optimal age replacement time t_0^* which minimizes the expected total discounted cost over an infinite time horizon $A_c(t_0; \alpha)$.

Theorem 4.2. (1) Suppose that the lifetime distribution $F(t)$ is strictly IFR .

> **(i)** If $r(\infty) > K(\alpha)$, then there exists one finite optimal age replacement time t_0^* $(0 < t_0^* < \infty)$ which satisfies $h_c(t_0^*; \alpha) = 0$, where

$$K(\alpha) = \frac{c_c F^*(\alpha) + c_p \overline{F}^*(\alpha)}{(c_c - c_p)\overline{F}^*(\alpha)/\alpha}, \tag{7.59}$$

$$F^*(\alpha) = \int_0^\infty e^{-\alpha t}f(t)dt. \tag{7.60}$$

Then the corresponding minimum expected cost is

$$A_c(t_0^*; \alpha) = \frac{(c_c - c_p)r(t_0^*)}{\alpha} - c_p. \qquad (7.61)$$

(ii) If $r(\infty) \le K(\alpha)$, then $t_0^* \to \infty$ and

$$A_c(\infty; \alpha) = c_c F^*(\alpha) / \overline{F}^*(\alpha). \qquad (7.62)$$

(2) If $F(t)$ is DFR (decreasing failure rate) , then $t_0^* \to \infty$.

Following Nakagawa and Osaki [56], let us consider the discrete age replacement model. Define the discrete lifetime distribution $F(n)$ $(n = 0, 1, 2, \cdots)$, the probability mass function $f(n)$ and the failure rate $r(n) = f(n)/\overline{F}(n-1)$. From the renewal reward theorem , it can be seen that the expected cost per unit time in the steady-state for the age replacement model is

$$A_d(N) = \frac{c_c F(N) + c_p \overline{F}(N)}{\displaystyle\sum_{i=1}^{N} \overline{F}(i-1)}, \quad N = 0, 1, 2, \cdots. \qquad (7.63)$$

Define the numerator of the difference of $A_d(N)$ as

$$h_d(N) = r(N+1) \sum_{i=1}^{N} \overline{F}(i-1) - F(N) - \frac{c_p}{c_c - c_p}. \qquad (7.64)$$

Then, we have the optimal age replacement time N^* which minimizes the expected cost per unit time in the steady-state $A_d(N)$.

Theorem 4.3. (1) Suppose that the lifetime distribution $F(N)$ is strictly IFR .

(i) If $r(\infty) > K$, then there exists one finite optimal age replacement time N^* $(0 < N^* < \infty)$ which satisfies $h_d(N^*-1) < 0$ and $h_d(N^*) \ge 0$. Then the corresponding minimum expected cost satisfies the following inequality

$$(c_c - c_p)r(N^*) < A_d(N^*) \le (c_c - c_p)r(N^* + 1). \quad (7.65)$$

(ii) If $r(\infty) \le K$, then $N^* \to \infty$ and $A_d(\infty) = B_d(\infty) = \lambda c_c$.
(2) If $F(N)$ is DFR , then $N^* \to \infty$.

We introduce the discount factor β $(0 < \beta < 1)$ in the discrete-time age replacement problem. The present value of a unit cost after

n $(n = 0, 1, 2, \ldots)$ period is β^n. In the discrete-time age replacement problem, the expected total discounted cost over an infinite time horizon is

$$A_d(N; \beta) = \frac{c_c \sum_{j=0}^{N} \beta^j f(j) + c_p \beta^N \overline{F}(N)}{\frac{1 - \beta}{\beta} \sum_{i=1}^{N} \beta^i \overline{F}(i - 1)}, \quad N = 0, 1, 2, \cdots. \quad (7.66)$$

Define the numerator of the difference of $A_d(N; \beta)$ as

$$h_d(N; \beta) = r(N + 1) \sum_{i=1}^{N} \beta^i \overline{F}(i - 1) - \sum_{j=0}^{N} \beta^j f(j) - \frac{c_p}{c_c - c_p}. \quad (7.67)$$

Then, we have the optimal age replacement time N^* which minimizes the expected total discounted cost over an infinite time horizon $A_d(N; \beta)$.

Theorem 4.4. (1) Suppose that the lifetime distribution $F(N)$ is strictly IFR.

(i) If $r(\infty) > K(\beta)$, then there exists one finite optimal age replacement time N^* $(0 < N^* < \infty)$ which satisfies $h_d(N^* - 1; \beta) < 0$ and $h_d(N^*; \beta) \geq 0$. Then the corresponding minimum expected cost satisfies the following inequalities.

$$\frac{(c_c - c_p) r(N^*)}{(1 - \beta)/\beta} - c_p < A_d(N^*; \beta) \quad (7.68)$$

and

$$A_d(N^*; \beta) \leq \frac{(c_c - c_p) r(N^* + 1)}{(1 - \beta)/\beta} - c_p. \quad (7.69)$$

where

$$K(\beta) = \frac{c_c \sum_{j=0}^{\infty} \beta^j f(j) + c_p \sum_{j=0}^{\infty} (1 - \beta^j) f(j)}{(c_c - c_p) \sum_{i=1}^{\infty} \beta^i \overline{F}(i - 1)}. \quad (7.70)$$

(ii) If $r(\infty) \leq K(\beta)$, then $N^* \to \infty$ and

$$A_d(\infty; \beta) = \frac{c_c \sum_{j=0}^{\infty} \beta^j f(j)}{\sum_{j=0}^{\infty} (1 - \beta^j) f(j)}. \tag{7.71}$$

(2) If $F(N)$ is DFR , then $N^* \to \infty$.

Theorem 4.5. (1) For the continuous-time age replacement problems, the following relationships hold.

$$A_c(t_0) = \lim_{\alpha \to 0} \alpha A_c(t_0; \alpha), \tag{7.72}$$

$$h_c(t_0) = \lim_{\alpha \to 0} h_c(t_0; \alpha), \tag{7.73}$$

$$K = \lim_{\alpha \to 0} K(\alpha). \tag{7.74}$$

(2) For the discrete-time age replacement problems, the following relationships hold.

$$A_d(N) = \lim_{\beta \to 1} (1 - \beta) A_d(N; \beta), \tag{7.75}$$

$$h_d(N) = \lim_{\beta \to 1} h_d(N; \beta), \tag{7.76}$$

$$K = \lim_{\beta \to 1} K(\beta). \tag{7.77}$$

Remarks. Glasser [32], Schaeffer [73], Cleroux and Hanscom [24], Osaki and Yamada [66] and Nakagawa and Osaki [55] extended the basic age replacement models mentioned above. Here, we introduce an interesting topic on the age replacement policy under the different cost criterion from the expected cost per unit time in the steady-state . Based on the seminal idea by Derman and Sacks [25], Ansell, Bendell and Humble [3] analyzed the age replacement model under alternative cost criterion. In the continuous–time model with no discount, let Y_i and S_i denote the total cost and the time length for i-th cycle ($i = 1, 2, \cdots$), respectively, where $Y_i = c_c I_{\{X_i < t_0\}} + c_p I_{\{X_i \geq t_0\}}$, $S_i = \min\{X_i, t_0\}$, X_i is the lifetime for the i-th cycle and $I_{\{\cdot\}}$ is the indicator function.

In Eq.(7.54), we find that

$$\lim_{t\to\infty} \frac{E[\text{total cost on } (0,t]]}{t} = E[Y_i]/E[S_i] = A_c(t_0). \qquad (7.78)$$

On the other hand, let $NU(t)$ denote the number of cycles up to time t. Then, we define

$$\eta(t) \equiv \frac{1}{NU(N,t)} \sum_{i=1}^{NU(t)} E[Y_i/S_i], \qquad (7.79)$$

where $\eta(t)$ is the mean of the ratio $E[Y_i/S_i]$ during $NU(t)$ cycles. From the independence of each cycle, we have

$$
\begin{aligned}
A_c^*(t_0) &= \lim_{t\to\infty} \eta(t) = E[Y_i/S_i] \\
&= \int_0^{t_0} (c_c/t)dF(t) + \int_{t_0}^{\infty} (c_p/t_0)dF(t). \qquad (7.80)
\end{aligned}
$$

This interesting cost criterion is strictly: the expected cost ratio and is of course different from $E[Y_i]/E[S_i]$. Ansell, Bendell and Humble [3] compared this model with an approximated age replacement policy with finite time horizon by Christer [19, 20], Christer and Jack [21].

5.   ORDER REPLACEMENT

In both block and age replacement problems, a spare unit is available whenever the original unit fails. However, it should be noted that this assumption is questionable in most practical cases. In fact, if a sufficiently large number of spare units are always kept on hand, a large inventory holding cost will be needed. Hence, if the system failure may be considered as a rare event for the operating system, the spare unit will be ordered when it is required. There were seminal contributions by Wiggins[85], Allen and D'Esopo[1, 2], Simon and D'Esopo [77], Nakagawa and Osaki[53] and Osaki[58]. A large number of order replacement models have been analyzed by many authors. For instance, the reader should refer to Thomas and Osaki[80, 81], Kaio and Osaki[35-38] and Osaki, Kaio and Yamada[63]. A comprehensive bibliography in this research area is listed in Dohi, Kaio and Osaki[27].

Continuous-Time Model
    Let us consider a replacement problem for one-unit system where each failed unit is scrapped and each spare is provided, after a lead time , by order. The original unit begins operating at time $t = 0$, and the

planning horizon is infinite. If the original unit does not fail up to a prespecified time $t_0 \in [0, \infty]$, the regular order for a spare is made at the time t_0 and after a lead time L_2 (> 0) the spare is delivered. Then if the original unit has already failed before $t = t_0 + L_2$, the delivered spare takes over its operation immediately. But even if the original unit is still operating, the unit is replaced by the spare preventively. On the other hand, if the original unit fails before the time t_0, the expedited order is made immediately at the failure time and the spare takes over its operation just after it is delivered after a lead time L_1 (> 0). In this situation, it should be noted that the regular order is not made. The same cycle repeats itself continually.

Under this model, define the interval from one replacement to the following replacement as one cycle. Let c_1 (> 0), c_2 (> 0), k_1 (> 0), w (> 0) and s (< 0) be the expedited ordering cost, the regular ordering cost , the system down (shortage) cost per unit time, the operation cost per unit time and the salvage cost per unit time, respectively. Then, the expected cost per unit time in the steady-state is

$$O_c(t_0) = V_c(t_0)/T_c(t_0), \tag{7.81}$$

where

$$
\begin{aligned}
V_c(t_0) &= c_1 \int_0^{t_0} dF(t) + c_2 \int_{t_0}^{\infty} dF(t) + k_1 \bigg\{ \int_0^{t_0} L_1 dF(t) \\
&\quad + \int_{t_0}^{t_0+L_2} (t_0 + L_2 - t) dF(t) \bigg\} + w \bigg\{ \int_0^{t_0+L_2} t dF(t) \\
&\quad + \int_{t_0+L_2}^{\infty} (t_0 + L_2) dF(t) \bigg\} + s \int_{t_0+L_2}^{\infty} (t - t_0 - L_2) dF(t) \\
&= c_1 F(t_0) + c_2 \overline{F}(t_0) + k_1 \bigg\{ (L_1 - L_2) F(t_0) + \int_{t_0}^{t_0+L_2} F(t) dt \bigg\} \\
&\quad + w \int_0^{t_0+L_2} \overline{F}(t) dt + s \int_{t_0+L_2}^{\infty} \overline{F}(t) dt, \quad t_0 \geq 0 \tag{7.82}
\end{aligned}
$$

and

$$
\begin{aligned}
T_c(t_0) &= \int_0^{t_0} (t + L_1) dF(t) + \int_{t_0}^{\infty} (t_0 + L_2) dF(t) \\
&= (L_1 - L_2) F(t_0) + L_2 + \int_0^{t_0} \overline{F}(t) dt. \tag{7.83}
\end{aligned}
$$

Define the numerator of the derivative of $O_c(t_0)$ with respect to t_0, divided by $\overline{F}(t_0)$ as $q_c(t_0)$, i.e.

$$
q_c(t_0) = \Big\{ (k_1 - w + s) R(t_0) + (w - s) + \Big[k_1(L_1 - L_2)
$$

$$+ (c_1 - c_2) \Big] r(t_0) \Big\} \Big\{ (L_1 - L_2) F(t_0) + L_2 + \int_0^{t_0} \overline{F}(t) dt \Big\}$$

$$- \Big\{ w \int_0^{t_0+L_2} \overline{F}(t) dt + c_1 F(t_0) + c_2 \overline{F}(t_0) + k_1 \Big[(L_1 - L_2) F(t_0)$$

$$+ \int_{t_0}^{t_0+L_2} F(t) dt \Big] + s \int_{t_0+L_2}^{\infty} \overline{F}(t) dt \Big\} \Big\{ (L_1 - L_2) r(t_0) + 1 \Big\},$$

$$\text{(7.84)}$$

where the function

$$R(t_0) = \frac{F(t_0 + L_2) - F(t_0)}{\overline{F}(t_0)} \tag{7.85}$$

has the same monotone properties as the failure rate $r(t_0)$, that is, $R(t)$ is increasing (decreasing) if and only if $r(t)$ is increasing (decreasing). Then, we have the optimal order replacement time t_0^* which minimizes the expected cost per unit time in the steady-state $O_c(t_0)$.

Theorem 5.1. (1) Suppose that the lifetime distribution $F(t)$ is strictly IFR.

(i) If $q_c(0) < 0$ and $q_c(\infty) > 0$, then there exists one finite optimal order replacement time t_0^* ($0 < t_0^* < \infty$) which satisfies $q_c(t_0^*) = 0$. Then the corresponding minimum expected cost is

$$O_c(t_0^*) = \frac{(k_1 - w + s) R(t_0^*) + (w - s) + \mu(t_0^*)}{(L_1 - L_2) r(t_0^*) + 1}, \tag{7.86}$$

where

$$\mu(t_0) = \{ k_1(L_1 - L_2) + (c_1 - c_2) \} r(t_0). \tag{7.87}$$

(ii) If $q_c(\infty) \le 0$, then $t_0^* \to \infty$ and

$$O_c(\infty) = \frac{w/\lambda + c_1 + k_1 L_1}{L_1 + 1/\lambda}. \tag{7.88}$$

(iii) If $q_c(0) \ge 0$, then $t_0^* = 0$ and

$$O_c(0) = \frac{1}{L_2} \Big\{ w \int_0^{L_2} \overline{F}(t) dt + c_2 + k_1 \int_0^{L_2} F(t) dt$$

$$+ s \int_{L_2}^{\infty} \overline{F}(t) dt \Big\}. \tag{7.89}$$

(2) Suppose that $F(t)$ is DFR . If the inequality

$$\left\{ w \int_0^{L_2} \overline{F}(t)dt + c_2 + k_1 \int_0^{L_2} F(t)dt \right.$$
$$+ s \int_{L_2}^{\infty} \overline{F}(t)dt \left. \right\}(L_1 + 1/\lambda)$$
$$< (w/\lambda + c_1 + k_1 L_1)L_2 \tag{7.90}$$

holds, then $t_0^* = 0$, otherwise, $t_0^* \to \infty$.

Discrete–Time Model

In the discrete order-replacement model , the function in Eq.(7.85) is given by

$$R(N) = \frac{\sum_{n=1}^{N+L_2} f(n) - \sum_{n=1}^{N} f(n)}{\sum_{n=N+1}^{\infty} f(n)}. \tag{7.91}$$

Then, the expected cost per unit time in the steady-state is

$$O_d(N) = V_d(N)/T_d(N), \tag{7.92}$$

where

$$V_d(N) = w \sum_{i=1}^{N+L_2} \sum_{n=i}^{\infty} f(n) + c_1 \sum_{n=1}^{N-1} f(n) + c_2 \sum_{n=N}^{\infty} f(n)$$
$$+ k_1 \left\{ (L_1 - L_2) \sum_{n=1}^{N-1} f(n) + \sum_{i=N+1}^{N+L_2} \sum_{n=1}^{i-1} f(n) \right.$$
$$+ s \sum_{i=N+L_2+1}^{\infty} \sum_{n=i}^{\infty} f(n) \left. \right\} \tag{7.93}$$

and

$$T_d(N) = (L_1 - L_2) \sum_{n=1}^{N-1} f(n) + L_2 + \sum_{i=1}^{N} \sum_{n=i}^{\infty} f(n). \tag{7.94}$$

Notice in the equations above that L_1 and L_2 are positive integers.

As with eqation(7.84), define the numerator of the difference of $O_d(N)$ with respect to N, divided by $\overline{F}(N)$ as $q_d(N)$, i.e.

$$q_d(N) = \left\{ (k_1 - w + s)R(N) + (w - s) + [k_1(L_1 - L_2) \right.$$
$$+ (c_1 - c_2)]r(N) \left. \right\} \left\{ (L_1 - L_2) \sum_{n=1}^{N-1} f(n) + L_2 \right.$$

$$+ \sum_{i=1}^{N} \sum_{n=i}^{\infty} f(n) \Big\} - \Big\{ w \sum_{i=1}^{N+L_2} \sum_{n=i}^{\infty} f(n) + c_1 \sum_{n=1}^{N-1} f(n)$$

$$+ c_2 \sum_{n=N}^{\infty} f(n) + k_1 \Big[(L_1 - L_2) \sum_{n=1}^{N-1} f(n)$$

$$+ \sum_{i=N+1}^{N+L_2} \sum_{n=1}^{i-1} f(n) \Big] + s \sum_{i=N+L_2+1}^{\infty} \sum_{n=i}^{\infty} f(n) \Big\}$$

$$\times [(L_1 - L_2) r(N) + 1]. \tag{7.95}$$

Then, we have the optimal order replacement time N^* which minimizes the expected cost per unit time in the steady-state $O_d(N)$.

Theorem 5.2. **(1)** Suppose that the lifetime distribution $F(N)$ is strictly IFR.

    (i) If $q_d(0) < 0$ and $q_d(\infty) > 0$, then there exists one finite optimal order replacement time N^* $(0 < N^* < \infty)$ which satisfies $q_d(N^* - 1) < 0$ and $q_d(N^*) \geq 0$.

    (ii) If $q_d(\infty) \leq 0$, then $N^* \to \infty$.

    (iii) If $q_d(0) \geq 0$, then $N^* = 0$.

(2) Suppose that $F(N)$ is DFR. Then $N^* = 0$, otherwise, $N^* \to \infty$.

Remark. This section has dealt with typical order-replacement models in both continuous and discrete–time setting. These models can be extended from the view points. Thomas and Osaki[81] and Dohi, Kaio and Osaki [27] presented continuous models with stochastic lead times . Osaki, Kaio and Yamada [63] proposed a combined model with minimal repair and alternatively introduced the concept of the negative ordering time . Recently, Dohi, Kaio and Osaki [26], Dohi, Shibuya and Osaki[28] and Shibuya, Dohi and Osaki[76] applied the order-replacement model to analyze the special continuous review cyclic inventory control problems

6.   CONCLUSION

    This chapter has been concerned with the basic preventive replacement policies and their variations, in terms of both continuous and discrete–time modeling. For further details on the discrete models, see Nakagawa [45, 51] and Nakagawa and Osaki [54]. Since the mathematical maintenance models are applicable to a variety of real problems, such a modeling technique will be useful for practitioners and researchers. Though we have reviewed only the most basic maintenance models in the limited space available here, a number of earlier models should be re-formulated in discrete–time setting, because, in most cases, the continuous–time models can be regarded as approximated models for actual maintenance problems and the maintenance schedule is often desired in discretized circumstance . These motivations will be evident from the recent development of computer technologies and their related computation abilities.

Acknowledgments

The authors would like to thank the three editors, M. Ben-Daya, A. Raouf and S. O. Duffuaa, for their kind invitation to contribute to this volume. This work was partially supported by a Grant-in-Aid for Scientific Research from the Ministry of Education, Sports, Science and Culture of Japan under Grant No. 09780411 and No. 09680426.

References

[1] Allen, S. G., and D'Esopo, D. A. An ordering policy for repairable stock items. *Ope. Res.*, **16**, (1968) 669-674.

[2] Allen, S. G., and D'Esopo, D. A. An ordering policy for stock items when delivery can be expedited. *Ope. Res.*, **16** (1968) 880-883.

[3] Ansell, J., Bendell, A., and Humble, S. Age replacement under alternative cost criteria. *Management Sci.*, **30** (1984) 358-367.

[4] Arrow, K. J., Karlin, S. and Scarf, H. *Studies in Applied Probability and Management Science*. Stanford University Press (eds.), Stanford, 1962.

[5] Ascher, H. and Feingold, H. *Repairable Systems Reliability*. Marcel Dekker, N.Y., 1984.

[6] Aven, T. and Jensen, U. *Stochastic Models in Reliability*. Springer-Verlag, Berlin, 1999.

[7] Barlow, R. E., *Engineering Reliability*. SIAM, Berlin, 1998.

[8] Barlow, R. E., and Hunter, L. C. Optimum preventive maintenance policies . *Ope. Res.*, **8** (1960) 90-100.

[9] Barlow, R. E., and Hunter, L. C. Reliability analysis of a one-unit system. *Ope. Res.*, **9** (1961) 200-208.

[10] Barlow, R. E., and Proschan, F. *Mathematical Theory of Reliability*. John Wiley & Sons, N.Y., 1965.

[11] Barlow, R. E., and Proschan, F. *Statistical Theory of Reliability and Life Testing: Probability Models*. Holt, Rinehart and Winston, N.Y., 1975.

[12] Beichelt, F. A unifying treatment of replacement policies with minimal repair. *Naval Res. Logist.*, **40** (1993) 51-67.

[13] Berg, M., and Cleroux, R. The block replacement problem with minimal repair and random repair costs. *J. Stat. Comp. Simul.*, **15** (1982) 1-7.

[14] Berg, M., and Epstein, B. A modified block replacement policy. *Naval Res. Logis. Quart.*, **23** (1976) 15-24.

[15] Bergman, B. On the optimality of stationary replacement strategies. *J. Appl. Prob.*, **17** (1980) 178-186.

[16] Block, H. W., Borges, W. S., and Savits, T. H. Age-dependent minimal repair. *J. Appl. Prob.*, **22** (1985) 370-385.

[17] Boland, P. J. Periodic replacement when minimal repair costs vary with time. *Naval Res. Logist. Quart.*, **29** (1982) 541-546.

[18] Boland, P. J., and Proschan, F. Periodic replacement with increasing minimal repair costs at failure. *Ope. Res.*, **30** (1982) 1183-1189.

[19] Christer, A. H. Refined asymptotic costs for renewal reward process. *J. Opl. Res. Soc.*, **29** (1978) 577-583.

[20] Christer, A. H. Comments on finite-period applications of age-based replacement models. *IMA J. Math. Appl. Busin. Indust*, **1**, (1987) 111-124.

[21] Christer, A. H. and Jack, N. An integral-equation a roach for replacement modelling over finite time horizons. *IMA J. Math. Appl. Busin. Indust.*, **3**, (1991) 31-44.

[22] Christer, A. H., Osaki, S., and Thomas, L. C. *Stochastic Modelling in Innovative Manufacturing. Lecture Notes in Economics and Mathematical Systems*, **445**, Springer-Verlag (eds.), Berlin, 1997.

[23] Cleroux, R., Dubuc, S., and Tilquin, C. The age replacement problem with minimal repair and random repair costs. *Ope. Res.*, **27** (1979) 1158-1167.

[24] Cleroux, R., and Hanscom, M. Age replacement with adjustment and depreciation costs and interest charges. *Technometrics*, **16** (1974) 235-239.

[25] Derman, C., and Sacks, J. Replacement of periodically inspected equipment. *Naval Res. Logist. Quart.*, **7** (1960) 597-607.

[26] Dohi, T., Kaio, N. and Osaki, S. Continuous review cyclic inventory models with emergency order. *J. Ope. Res. Soc. Japan*, **38**, 212-229 (1995).

[27] Dohi, T., Kaio, N. and Osaki, S. On the optimal ordering policies in maintenance theory - survey and applications. *Applied Stochastic Models and Data Analysis*, **14**, 309-321 (1998).

[28] Dohi, T., Shibuya, T. and Osaki, S. Models for 1-out-of-Q systems with stochastic lead times and expedited ordering options for spares inventory. *Euro. J. Opl. Res.*, **103**, 255-272 (1997).

[29] Feller, W. An Introduction to Probability Theory and Its Applications. NY: John Wiley & Sons 1957.

[30] Fox, B. L. Age replacement with discounting. *Ope. Res.* **14** (1966) 533–537 .

[31] Gertsbakh, I. B. *Models of Preventive Maintenance.* North-Holland, Amsterdam, 1977.

[32] Glasser, G. J. The age replacement problem. *Technometrics*, **9** (1967) 83-91.

[33] Gnedenko, B. V., Belyayev, Y. K. and Solovyev, A. D. *Mathematical Methods of Reliability Theory.* Academic Press, N.Y., 1969.

[34] Jorgenson, D. W., McCall, J. J. and Radner, R. *Optimal Replacement Policy.* North-Holland, Amsterdam, 1967.

[35] Kaio, N. and Osaki, S. Optimum ordering policies with lead time for an operating unit in preventive maintenance. *IEEE Trans. Reliab.*, **R-27**, (1978) 270-271.

[36] Kaio, N. and Osaki, S. Optimum planned maintenance with salvage costs . *Int. J. Prod. Res.*, **16**, 249-257 (1978).

[37] Kaio, N. and Osaki, S. Discrete-time ordering policies. *IEEE Trans. Reliab.*, **R-28**, 405-406 (1979).

[38] Kaio, N. and Osaki, S. Optimum planned maintenance with discounting. *Int. J. Prod. Res.*, **18**, 515-523 (1980).

[39] Kaio, N., and Osaki, S. Review of discrete and continuous distributions in replacement models. *Int. J. Sys. Sci.*, **19** (1988) 171-177.

[40] Karlin, S., and Taylor, H. M. A First Course in Stochastic Processes. NY: Academic Press, 1975.

[41] Karlin, S., and Taylor, H. M. A Second Course in Stochastic Processes . NY: Academic Press, 1981.

[42] McCall, J. J. Maintenance policies for stochastically failing equipment: a survey. *Management Sci.*, **11** (1965) 493-521.

[43] Morimura, H. On some preventive maintenance policies for IFR. *J. Ope. Res. Soc. Japan*, **12** (1970) 94-124.

[44] Munter, M. Discrete renewal processes . *IEEE Trans. Reliab.*, **R-20** (1971) 46-51.

[45] Nakagawa, T. A summary of discrete replacement policies. *Euro. J. Opl. Res.*, **17** (1979) 382-392 .

[46] Nakagawa, T. A summary of periodic replacement with minimal repair at failure. *J. Ope. Res. Soc. Japan*, **24** (1979) 213-228.

[47] Nakagawa, T. Generalized models for determining optimal number of minimal repairs before replacement. *J. Ope. Res. Soc. Japan*, **24** (1981a) 325-357.

[48] Nakagawa, T. Modified periodic replacement with minimal repair at failure. *IEEE Trans. Reliab.*, **R-30** (1981b) 165-168.

[49] Nakagawa, T. Optimal policy of continuous and discrete replacement with minimal repair at failure. *Naval Res. Logist. Quart.*, **31** (1984) 543-550.

[50] Nakagawa, T. Periodic and sequential preventive maintenance policies . *J. Appl. Prob.*, **23** (1986) 536-542.

[51] Nakagawa, T. Modified discrete preventive maintenance policies . *Naval Res. Logist. Quart.*, **33**, 703-715 (1986).

[52] Nakagawa, T., and Kowada, M. Analysis of a system with minimal repair and its Application to replacement policy. *Euro. J. Opl. Res.*, **12** (1983) 176-182.

[53] Nakagawa, T. and Osaki, S. Optimum replacement policies with delay. *J. Appl. Prob.*, **11**, (1974) 102-110.

[54] Nakagawa, T. and Osaki, S. Analysis of a repairable system which operates at discrete times. *IEEE Trans. Reliab.*, **R-25**, 110-112 (1976).

[55] Nakagawa, T., and Osaki, S. Reliability analysis of a one-unit system with unrepairable spare units and its optimization applications. *Opl. Res. Quart.*, **27** (1976) 101-110.

[56] Nakagawa, T., and Osaki, S. Discrete time age replacement policies. *Opl. Res. Quart.*, **28** (1977) 881-885.

[57] Ohnishi, M. Optimal minimal-repair and replacement problem under average cost criterion: optimality of (t, T)-policy. *J. Ope. Res. Soc. Japan*, **40** (1997) 373-389.

[58] Osaki, S. An ordering policy with lead time . *Int. J. Sys. Sci.*, **8**, (1977) 1091-1095.

[59] Osaki, S. *Stochastic System Reliability Modeling*. World Scientific, Singapore, 1985.

[60] Osaki, S. *Applied Stochastic System Modeling*. Springer-Verlag, Berlin, 1992.

[61] Osaki, S., and Cao, J. *Reliability Theory and Applications*. World Scientific (eds.), Berlin, 1987.

[62] Osaki, S., and Hatoyama, Y. *Stochastic Models in Reliability Theory. Lecture Notes in Economics and Mathematical Systems*, **235**, Springer-Verlag (eds.), Berlin, 1984.

[63] Osaki, S., Kaio, N. and Yamada, S. A summary of optimal ordering policies. *IEEE Trans. Reliab.*, **R-30**, 272-277 (1981).

[64] Osaki, S., and Nakagawa, T. A note on age replacement . *IEEE Trans. Reliab.*, **R-24** (1975) 92-94.

[65] Osaki, S., and Nakagawa, T. Bibliography for reliability and availability of stochastic systems'. *IEEE Trans. Reliab.* **R-25** (1976) 284-287.

[66] Osaki, S., and Yamada, S. Age replacement with lead time . *IEEE Trans. Reliab.*, **R-25** (1976) 344-345.

[67] Ozekici, S. *Reliability and Maintenance of Complex Systems*. NATO ASI Series, Springer (eds.), Berlin, 1996.

[68] Park, K. S. Optimal number of minimal repairs before replacement. *IEEE Trans. Reliab.*, **R-28** (1979) 137-140.

[69] Pierskalla, W. P., and Voelker, J. A. A survey of maintenance models: the control and surveillance of deteriorating systems. *Naval Res. Logist. Quart.*, **23** (1976) 353-388.

[70] Phelps, R. I. Replacement policies under minimal repair. *J. Opl. Res. Soc.*, **32** (1981) 549-554.

[71] Phelps, R. I. Optimal policy for minimal repair. *J. Opl. Res. Soc.*, **34** (1983) 425-427.

[72] Ross, S. M. Applied Probability Models with Optimization Applications, San Francisco: Holden–Day, 1970.

[73] Schaeffer, R. L. Optimum age replacement policies with an increasing cost factor. *Technometrix*, **13** (1971) 139-144.

[74] Segawa, Y., Ohnishi, M., and Ibaraki, T. Optimal minimal-repair and replacement problem with average dependent cost structure. *Computers & Math. Appl.*, **24** (1992) 91-101.

[75] Sherif, Y. S., and Smith, M. L. Optimal maintenance models for systems subject to failure–a review. *Naval Res. Logist. Quart.*, **28** (1981) 47-74.

[76] Shibuya, T., Dohi, T. and Osaki, S. Spare part inventory models with stochastic lead times . *Int. J. Prod. Econ.*, **55**, 257-271 (1998).

[77] Simon, R. M., and D'Esopo, D. A. Comments on a paper by S. G. Allen and D. A. D'Esopo: An ordering policy for repairable stock items. *Ope. Res.*, **19**, (1971) 986-989 .

[78] Tahara, A., and Nishida, T. Optimal replacement policy for minimal repair model. *J. Ope. Res. Soc. Japan*, **18** (1975) 113-124.

[79] Taylor, H. M., and Karlin, S. An Introduction to Stochastic Modeling. NY: Academic Press, 1984.

[80] Thomas, L. C. and Osaki, S. A note on ordering policy. *IEEE Trans. Reliab.*, **R-27**, (1978) 380-381.

[81] Thomas, L. C. and Osaki, S. An optimal ordering policy for a spare unit with lead time . *Euro. J. Opl. Res.*, **2**, (1978) 409-419.

[82] Tilquin, C., and Cleroux, R. Block replacement with general cost structure. *Technometrics*, **17** (1975) 291-298.

[83] Tilquin, C., and Clerroux, R. Periodic replacement with minimal repair at failure and adjustment costs. *Naval Res. Logist. Quart.*, **22** (1975) 243-254.

[84] Valdez-Flores, C., and Feldman, R. M. A survey of preventive maintenance models for stochastically deteriorating single-unit systems. *Naval Res. Logist.*, **36**, 419-446 (1989).

[85] Wiggins, A. D. A minimum cost model of spare parts inventory control. *Technometrics*, **9** (1967), 661-665.

Chapter 8

A GENERAL FRAMEWORK FOR ANALYZING MAINTENANCE POLICIES

V. Makis, X Jiang, A.K.S. Jardine
Department of Mechanical and Industrial Engineering,
University of Toronto, 5 King's College Road, Toronto,
Ontario, Canada M5S 3G8

and

K. Cheng
Institute of Applied Mathematics,
Academia Sinica,
Beijing, 100080, China

Abstract

In this article, we will show how to use the optimal stopping framework for the formulation and analysis of a wide class of maintenance decision problems. The necessary mathematical background from martingale dynamics and the optimal stopping theory is summarized and the approach is illustrated by analyzing a general repair/replacement model. The model includes many models considered previously in the research literature as special cases. The structure of the optimal policy is obtained under both the average and the discounted cost criteria. An algorithm for finding the optimal policy is presented for the average cost case and a numerical example is given to illustrate the algorithm.

Keywords: optimal stopping , repairable system , minimal repair , random repair cost

1.   INTRODUCTION

Maintenance models of repairable systems have been investigated in the research literature for more than three decades (see for example the recent reviews by Aven (1996) and Jensen (1996)).

In most papers, a particular class of policies determined by one or several parameters was considered and the optimal policy in that class was found by using traditional optimization methods.

Structural results have been obtained mainly by formulating and analyzing the maintenance problems in the Markov or semi-Markov decision framework (see e.g. Hastings (1969), White (1989), Makis and Jardine (1992,1993), Stadje and Zuckerman (1991), L'Ecuyer and Haurie(1987)).

In this paper, we present a different approach to solving repair/replacement problems based on optimal stopping.

The optimal stopping theory has been applied in preventive maintenance mainly to analyze replacement problems with quite general deteriorating processes (see e.g. Bergman (1978), Aven (1983), Aven and Bergman (1986) and Jensen (1989)).

In our model presented in Section 3, although the deteriorating process is simpler, we consider jointly a repair/replacement problem at failure times, and a preventive replacement in continuous time. The former is a discrete time and the latter a continuous time stopping problem.

In Section 2, we summarize the relevant results from the optimal stopping. A general repair/replacement problem can be formulated in the optimal stopping framework as a fractional optimization problem. We will show that the original fractional optimization problem can be transformed to a parametric optimization problem with an additive objective function using λ-maximization technique (e.g. Aven and Bergman (1986)), and the objective function can be further simplified by removing its martingale part without the loss of optimality. The methodology presented in Section 2 is applied in Section 3 to a general repair/replacement model. The form of the average cost optimal policy is found and a computational algorithm is presented. In Section 4, we obtained the corresponding results under the expected discounted cost criterion using the same methodology.

2.   BACKGROUND

In this section we provide the necessary results from the optimal stopping theory and introduce the $\lambda-$maximization technique and the semi-martingale decomposition for the discrete time case.

Denote $N = \{1, 2, ...\}$. Let $(\Omega, \mathcal{F}, P)$ be a probability space and $\{\mathcal{F}_n, \ n \in N\}$, be a complete filtration, i.e., $\{\mathcal{F}_n\}$ is a nondecreasing sequence of sub-σ algebras of $\mathcal{F}$ such that $\mathcal{F}_n$ contains all P-null sets of the complete $\sigma-$algebra $\mathcal{F}$. Note that $\{\mathcal{F}_n\}$ can be considered as describing the history of some stochastic process in discrete time and $\mathcal{F}_n$ then represents the $\sigma-$ field of events prior to time n.

A stopping time τ is a random variable τ: $\Omega \to N \cup \{+\infty\}$ such that $\{\tau = n\} \in \mathcal{F}_n$ for all n. Let $\{Y_n, n \in N\}$ be a sequence of random variables such that $\{Y_n\}$ is adapted to $\{\mathcal{F}_n\}$, i.e., Y_n is $\mathcal{F}_n-$measurable for each n. Define

$$Y_\tau = \sum_{n=1}^{+\infty} I_{\{\tau=n\}}Y_n = \begin{cases} Y_n & on \ \{\tau = n\} \\ limsup_{n\to+\infty}Y_n & on \{\tau = +\infty\} \end{cases} \tag{8.1}$$

where I is the set indicator function, $I_A(\omega) = 1$ if $\omega \in A$ and zero otherwise.

The optimal stopping problem is formulated as follows. Find a stopping time τ^*, if it exists, such that

$$EY_{\tau^*} = sup\{EY_\tau : \tau \in C'\}, \tag{8.2}$$

where C' is the class of stopping times such that EY_τ exists.

It can be shown that it is sufficient to consider the following class of stopping times:

$$C = \{\tau : \tau \ is \ an \ (\mathcal{F}_n) - stopping \ time, EY_\tau^- < +\infty\}$$

i.e., $sup_{C'} EY_\tau = sup_C EY_\tau$.

Let $C_n = \{max(\tau, n) : \tau \in C\}$ and define

$$\gamma_n = esssup_{C_n} E(Y_\tau|\mathcal{F}_n), \tag{8.3}$$

$$\sigma_n = \begin{cases} first \ i \geq n & such \ that \ Y_i = \gamma_i \\ \infty, & if \ no \ such \ n \ exists, \end{cases} \tag{8.4}$$

$$\sigma \equiv \sigma_1.$$

Note that γ_n is the essential supremum of $\{E(Y_\tau|\mathcal{F}_n), \tau \in C_n\}$, if $P(\gamma_n \geq E(Y_\tau|\mathcal{F}_n)) = 1$ for every $\tau \in C_n$ and if γ_n' is any random variable such that $P(\gamma_n' \geq E(Y_\tau|\mathcal{F}_n)) = 1$ for every $\tau \in C_n$, then $P(\gamma_n' \geq \gamma_n) = 1$.

We have the following result.

Theorem 1. *[Chow et al.(1971), Theorem 4.5', p.82]* *If $E(supY_n^+) < \infty$, then σ is optimal in C.*

For the Markov case, the optimal stopping time has a more specific form.

Theorem 2 *[Corollary 2.4, Jiang et al. (1998)]   Let $\{X_n\}$ be a homogeneous Markov chain and $Y_n = \sum_{k=1}^{n-1} \theta_k(X_k) + \varphi_n(X_n)$. Define*

$$V_n(x) = sup_\tau E^x \left(\sum_{k=1}^{\tau-1} \theta_{n+k}(X_k) + \varphi_{n+\tau}(X_\tau) \right), n = 1, 2, ... \qquad (8.5)$$

Then the optimal stopping time σ for sequence $\{Y_n\}$ has the form:

$$\sigma = \begin{cases} first\ n \geq 1 & \varphi_n(X_n) \geq V_n(X_n) + \theta_n(X_n). \\ \infty & if\ no\ such\ n\ exists. \end{cases} \qquad (8.6)$$

Next, we will introduce the λ-maximization technique (see e.g. Aven and Bergman (1986)) to solve the following minimization problem.

Let $A_n = \sum_{i=0}^{n} a_i \geq 0$, $B_n = \sum_{i=0}^{n} b_i \geq 0$. To find

$$\lambda^* = inf_\tau \frac{EA_\tau}{EB_\tau}, \qquad (8.7)$$

we can solve the following parametric optimal stopping problem (with parameter λ).

Find

$$V(\lambda) = sup_{\tau \in D} E(\lambda B_\tau - A_\tau), \qquad (8.8)$$

where D is the class of $(\mathcal{F}_n)$-stopping times such that $E(A_\tau) < \infty$ and $E(B_\tau) < \infty$. One can show that it is sufficient to consider class D of stopping times.

The optimal value λ can be obtained as

$$\lambda^* = sup\{\lambda : V(\lambda) < 0\}. \qquad (8.9)$$

If there is a λ such that $V(\lambda) = 0$, then $\lambda = \lambda^*$.

Assume that $V(\lambda^*) = 0$ and let τ^* be the stopping time maximizing the right-hand side of (8.8) for $\lambda = \lambda^*$.

Then, we have from (8.8),

$$\begin{aligned} 0 &= E(\lambda^* B_{\tau^*} - A_{\tau^*}) \\ &= \lambda^* E(B_{\tau^*}) - E(A_{\tau^*}), \end{aligned}$$

so that $\lambda^* = E(A_{\tau^*})/E(B_{\tau^*})$.

On the other hand, since $V(\lambda^*)$ is the supremum, we have from (8.8) that for any $\tau \in D$,

$$0 \geq E(\lambda^* B_\tau - A_\tau)$$

or, $\lambda^* \leq E(A_\tau)/E(B_\tau)$ so that λ^* is the infimum defined by (8.7) and τ^* is the optimal stopping time minimizing $E(A_\tau)/E(B_\tau)$.

As noted earlier, this technique transforms the original fractional optimization problem (8.7) (which is difficult to analyze) into a parametric optimization problem with an additive objective function.

The problem can be further simplified by applying the semi-martingale decomposition that will remove the martingale part. Semi-martingale decomposition in discrete time can be described as follows.

Assume that $Y_n = \sum_{i=0}^n f_i$ is $(\mathcal{F}_n)$-adapted and integrable for each n. Then it can be decomposed into two parts: $Y_n = \overline{Y}_n + M_n$, where M_n is $(\mathcal{F}_n)$−martingale , and $(\overline{Y}_n)$ is $(\mathcal{F}_{n-1})$ adapted. This decomposition is unique, $\overline{Y}_n = \sum_{i=0}^n E(f_i|\mathcal{F}_{i-1})$, and $M_n = \sum_{i=0}^n (f_i - E(f_i|\mathcal{F}_{i-1}))$, where $\mathcal{F}_{-1}$ is the trivial σ−algebra and $\mathcal{F}_0 \subset \mathcal{F}_1$. Note that (M_n) is $(\mathcal{F}_n)$−martingale if $E(M_m|\mathcal{F}_n) = M_n$ for $m > n$.

Obviously, $\overline{Y}_n$ is $\mathcal{F}_{n-1}$−measurable so that $(\overline{Y}_n)$ is $(\mathcal{F}_{n-1})$− adapted. Next, we show that (M_n) is $(\mathcal{F}_n)$-martingale .

We have for any $m > n$,

$$E(M_m|\mathcal{F}_n) = E(\sum_{i=0}^m (f_i - E(f_i|\mathcal{F}_{i-1}))|\mathcal{F}_n)$$

$$= E(\sum_{i=0}^m f_i|\mathcal{F}_n) - \sum_{i=0}^m E(f_i|\mathcal{F}_{i-1}) - \sum_{i=n+1}^m E(f_i|\mathcal{F}_n)$$

(because $E(f_i|\mathcal{F}_{i-1})$ is $\mathcal{F}_n$- measurable for $i \le n$)

$$= E(\sum_{i=0}^n f_i|\mathcal{F}_n) - \sum_{i=0}^n E(f_i|\mathcal{F}_{i-1}) = M_n$$

(because $\sum_{i=0}^n f_i$ is $\mathcal{F}_n$- measurable).

Denote $Y_n(\lambda) = \sum_{i=0}^n (\lambda b_i - a_i) \equiv \sum_{i=0}^n f_i$ and $\overline{Y}_n(\lambda) = \sum_{i=0}^n E(f_i|\mathcal{F}_{i-1})$, where a_i, b_i are non-negative.

It can be shown that the optimal stopping problems for $(Y_n(\lambda))$ and $(\overline{Y}_n(\lambda))$ are equivalent, i.e.,

$$sup_{\tau \in D} E(Y_\tau(\lambda)) = sup_{\tau \in D} E(\overline{Y}_\tau(\lambda)). \tag{8.10}$$

The methodology presented in this section will be applied in the next section to a general repair/replacement model.

3. THE AVERAGE COST OPTIMAL POLICY FOR A REPAIR/REPLACEMENT PROBLEM

In this section, we consider a general model of a repairable system subject to random failure. First, we provide a detailed description of the model and then formulate a repair/replacement problem in the optimal stopping framework.

We will make the following assumptions.

1. The time to system failure is a generally distributed random variable with distribution function $F(t)$, density $f(t)$ and the failure rate $h(t) = f(t)/(1 - F(t))$, which is a continuous and non-decreasing function of t.

2. At failure time, the system can be either minimally repaired at cost $C(n,t)$, where n is the number of repairs since the last replacement and t is the age of the unit, or replaced at cost C_r. The preventive replacement can occur at any time prior to failure and the preventive replacement cost $C_p < C_r$. We assume that C_p and $C_r = C_p + C_f$ are given constants and $\{C(n,t)\}$ are random variables stochastically increasing in n and t. Further, we assume that $EC(n,t)$ is continuous in t, and that $C_f \leq C(n,t) \leq C_r$ (without loss of generality).

3. All maintenance actions take negligible time.

The objective is to find the repair/replacement policy minimizing the long-run expected average cost per unit time.

Let t be the time of the first replacement if a new system is installed at time zero. Then, the total cost incurred up to time t is given by the following formula:

$$llTC(t) \;=\; C_p + \left(\sum_{i=1}^{N(t)-1} C(i,S_i) + C_f\right) I_{\{t=S_{N(t)}\}}$$

$$+ \left(\sum_{i=1}^{N(t)} C(i,S_i)\right) I_{\{t>S_{N(t)}\}}, \tag{8.11}$$

where S_i is the i–th failure time, $S_0 = 0$, $N(t)$ is the number of failures before t, $N(t) = \sum_{i=1}^{\infty} I_{\{S_i \leq t\}}$, and I is the set indicator function.

Let $(\mathcal{F}_t)$ be the (completed) natural filtration of process $\{TC(t), t \geq 0\}$.

For the average cost criterion, the maintenance decision problem can be formulated as follows. Find an $(\mathcal{F}_t)$–stopping time τ^*, if it exists, minimizing the long-run expected average cost per unit time given by

$$\frac{E(TC(\tau))}{E\tau}. \tag{8.12}$$

We proved in Makis et al. (1999) that this continuous time stopping problem can be reduced to the following discrete time stopping problem.

Let $\eta_i = (S_i, C(i,S_i))$ and $\mathcal{H}_n = \sigma(\eta_i, i \leq n)$, i.e., $\mathcal{H}_n$ is the σ–algebra generated by $\{\eta_i, i \leq n\}$.

Any $(\mathcal{F}_t)$–stopping time τ has a representation

$$\tau = \sum_{i=0}^{\sigma-1}(T_i \wedge S_{i+1} - S_i)\prod_{j=0}^{i-1}I_{\{S_{j+1}\leq T_j\}} \equiv \tau(\sigma, \{T_i\}), \qquad (8.13)$$

where σ is an $(\mathcal{H}_n)$-stopping time, T_0 is a constant and T_n is $\mathcal{H}_n$-measurable for $n \geq 1$. (see Lemma 3 in Makis et al. (1999)). We define $\prod_0^{-1} \equiv 1$.

The total cost $TC(\tau)$ given by (11) can then be written as

$$\begin{aligned} lllTC(\tau) &= \sum_{i=0}^{\sigma-1}(C(i,S_i) - C_f I_{\{S_{i+1}>T_i\}})I_{\{S_i \neq T_i\}}\prod_{j=0}^{i-1}I_{\{S_{j+1}\leq T_j\}} \\ &\equiv TC(\sigma, \{T_i\}), \end{aligned} \qquad (8.14)$$

where $C(0,0) \equiv C_r$ (see Lemma 4 in Makis et al. (1999)).

The stopping time σ is the first failure time when the system is replaced and $\{T_i, i \geq 0\}$ is the sequence of preventive replacement times, T_i is the preventive replacement time planned at failure time S_i (after performing minimal repair).

The equivalent optimal stopping problem in the discrete time is then formulated as follows. Find

$$\lambda^* = \inf_{\sigma}\left(\inf_{\{T_i\}} \frac{E(TC(\sigma, \{T_i\}))}{E(\tau(\sigma, \{T_i\}))}\right) \qquad (8.15)$$

and $(\sigma^*, \{T_i^*\})$ minimizing the right-hand side of (8.15) (if they exist), where the first infimum is through $(\mathcal{H}_n)$–stopping times σ and the second infimum is through $\{T_i\}$, T_i is $\mathcal{H}_i$–measurable, $T_i \geq S_i$ for $i \geq 1$, $T_0 \geq 0$ is a constant.

We can now apply the λ–maximization technique described in Section 2. For $\lambda > 0$, find

$$V(\lambda) = \sup_{\sigma}(\sup_{\{T_i\}} E(Y_\lambda(\sigma, \{T_i\}))). \qquad (8.16)$$

where

$$\begin{aligned} Y_\lambda(\sigma, \{T_i\}) &= \textstyle\sum_{i=0}^{\sigma-1}(\lambda(T_i \wedge S_{i+1} - S_i) - C(i, S_i) + I_{\{S_{i+1}>T_i\}}C_f) \\ &\quad I_{\{S_i \neq T_i\}}\prod_{j=0}^{i-1}I_{\{S_{j+1}\leq T_j\}}. \end{aligned} \qquad (8.17)$$

Then,

$$\lambda^* = \sup\{\lambda : V(\lambda) < 0\} \qquad (8.18)$$

and $(\sigma^*, \{T_i^*\})$ maximize the right-hand side of (16) for $\lambda = \lambda^*$.

Next, the maximization problem in (8.16) can be simplified by removing the martingale part from $Y_\lambda(\sigma, \{T_i\})$ defined by (8.17) through conditioning (the semi-martingale decomposition), i.e., we can consider the following problem.

For $\lambda > 0$, find

$$V(\lambda) = \sup_\sigma(\sup_{\{T_i\}} E(\overline{Y}_\lambda(\sigma, \{T_i\}))), \tag{8.19}$$

where

$$
\begin{aligned}
&\overline{Y}_\lambda(\sigma, \{T_i\}) \\
&= \textstyle\sum_{i=0}^{\sigma-1}(\lambda(E(T_i \wedge S_{i+1} \mid \mathcal{F}_{S_i}) - S_i) - C(i, S_i) \\
&\quad + P\{S_{i+1} > T_i \mid \mathcal{F}_{S_i}\}C_f)I_{\{S_i \neq T_i\}} \textstyle\prod_{j=0}^{i-1} I_{\{S_{j+1} \leq T_j\}} \\
&= \textstyle\sum_{i=0}^{\sigma-1}(\lambda \int_{S_i}^{T_i} u\,dF_{S_i}(u) + \lambda T_i \overline{F}_{S_i}(T_i) - \lambda S_i - C(i, S_i) + C_f \\
&\quad - C_f \int_{S_i}^{T_i} dF_{S_i}(u))I_{\{S_i \neq T_i\}} \textstyle\prod_{j=0}^{i-1} I_{\{S_{j+1} \leq T_j\}} \\
&= \textstyle\sum_{i=0}^{\sigma-1}(\lambda \int_{S_i}^{T_i} \overline{F}_{S_i}(u)du - C(i, S_i) + C_f - C_f \int_{S_i}^{T_i} h(u)\overline{F}_{S_i}(u)du) \\
&\quad \times I_{\{S_i \neq T_i\}} \textstyle\prod_{j=0}^{i-1} I_{\{S_{j+1} \leq T_j\}} \\
&= \textstyle\sum_{i=0}^{\sigma-1}(-C(i, S_i) + C_f \\
&\quad + \int_{S_i}^{T_i}(\lambda - h(u)C_f)\overline{F}_{S_i}(u)du)I_{\{S_i \neq T_i\}} \textstyle\prod_{j=0}^{i-1} I_{\{S_{j+1} \leq T_j\}},
\end{aligned}
$$
$$\tag{8.20}$$

$\overline{F}_s(t) = P(\xi > t \mid \xi > s)$, ξ is the time to failure of the unit. Then, λ^* defined by (8.15) can be obtained from (8.18) and the optimal stopping time $\tau(\sigma^*, \{T_i^*\})$ for the problem in (8.15) is the stopping time maximizing $E(\overline{Y}_\lambda(\sigma, \{T_i\}))$ for $\lambda = \lambda^*$.

Define for $\lambda > 0$,

$$T_\lambda = \inf\{t : \lambda \leq C_f h(t)\}. \tag{8.21}$$

Then, we have

$$V(\lambda) = \sup_\sigma E(\overline{Y}_\lambda(\sigma, \{T_\lambda \vee S_i\})), \tag{8.22}$$

i.e., for a given $\lambda > 0$, T_λ is the optimal preventive replacement time for problem (8.19).

To finish solving the maintenance decision problem, it is now sufficient to find the optimal stopping time for the sequence $\{W_n(\lambda)\}$, where

$$W_n(\lambda) \equiv \textstyle\sum_{i=0}^{n-1}(-C(i, S_i) + C_f + \int_{S_i}^{T_\lambda}(\lambda - h(t)C_f)\overline{F}_{S_i}(t)dt)I_{\{S_i \leq T_\lambda\}}, \tag{8.23}$$

which is obviously a discrete time stopping problem.

We only need to consider the values of $\lambda < \lambda_0 = \lim_{n,t\to\infty} h(t)EC(n,t)$, i.e., λ_0 is the average cost when the minimal repair policy is applied at failure times and no preventive replacement is planned. It follows from Theorem 1 that for $\lambda < \lambda_0$ the optimal stopping time σ_λ maximizing $EW_\sigma(\lambda)$ exists. From Theorem 2, σ_λ has the following form:

$$
\begin{aligned}
\sigma_\lambda \; = \; & \inf\{n \geq 1 : -C_p + \int_0^{T_\lambda}(\lambda - h(t)C_f)\overline{F}(t)dt \geq \\
& \overline{V}_n(\lambda, S_n) + (-C(n, S_n) + C_f \\
& + \int_{S_n}^{T_\lambda}(\lambda - h(t)C_f)\overline{F}_{S_n}(t)dt)I_{\{S_n \leq T_\lambda\}}\},
\end{aligned}
\tag{8.24}
$$

where

$$
\begin{aligned}
\overline{V}_n(\lambda, S_n) \; = \; & \sup_\eta E^{S_n}(\textstyle\sum_{i=1}^{\eta-1}(-C(n+i, S_{n+i}) + C_f \\
& + \int_{S_{n+i}}^{T_\lambda}(\lambda - h(t)C_f)\overline{F}_{S_{n+i}}(t)dt)I_{\{S_{n+i} \leq T_\lambda\}}) \\
& -C_p + \int_0^{T_\lambda}(\lambda - h(t)C_f)\overline{F}(t)dt.
\end{aligned}
\tag{8.25}
$$

Denote for $n \geq 0$,

$$
\begin{aligned}
g_n(\lambda, t) \; = \; & \overline{V}_n(\lambda, t) + C_p - \int_0^{T_\lambda}(\lambda - h(x)C_f)\overline{F}(x)dx \\
& + C_f + \int_t^{T_\lambda}(\lambda - h(x)C_f)\overline{F}_t(x)dx I_{\{t \leq T_\lambda\}}.
\end{aligned}
\tag{8.26}
$$

In particular, for $n = 0, t = 0$ and $\lambda = \lambda^*$, where λ^* is defined by (15), $g_0(\lambda, 0) = \overline{V}_0(\lambda, 0) + C_r = V(\lambda) + C_r = C_r$.

Then for $S_n \leq T_\lambda$, the optimal stopping time σ_λ has the form

$$
\sigma_\lambda = \inf\{n \geq 1 : C(n, S_n) \geq g_n(\lambda, S_n)\}.
\tag{8.27}
$$

From (8.25), we have for $S_n > T_\lambda$,

$$
\overline{V}_n(\lambda, S_n) = -C_p + \int_0^{T_\lambda}(\lambda - h(t)C_f)\overline{F}(t)dt,
\tag{8.28}
$$

and from (8.26), we get for $S_n > T_\lambda$,

$$
g_n(\lambda, S_n) = C_f.
\tag{8.29}
$$

It follows from (8.24) that for given $\lambda < \lambda_0$,

$$
E(W_{\sigma_\lambda}(\lambda)) = E(\overline{Y}_\lambda(\sigma_\lambda, \{T_\lambda \vee S_i\})) = V(\lambda),
\tag{8.30}
$$

so that $\tau(\sigma_\lambda, \{T_\lambda \vee S_i\})$ is the optimal stopping time maximizing $E(Y_\lambda(\sigma, \{T_i\}))$.

Hence, if we denote $g_n(t) \equiv g_n(\lambda^*, t)$ and $T \equiv T_{\lambda^*}$, where λ^* is defined by (8.15), the optimal repair/ replacement policy has the following form:

the unit is minimally repaired at the nth failure time S_n if and only if $S_n \leq T$ and $g_n(S_n) > C(n, S_n)$, otherwise it is replaced. If the unit has not been replaced at failure times before T, it is preventively replaced at time T.

Based on the above structural result on the optimal policy, we can now compute $g_n(t)$ for $t \leq T$ by conditioning on S_{n+1} and $C(n+1, S_{n+1})$. We have

$$
g_n(t) - C_f = \int_t^T (\lambda^* - h(x)C_f)\overline{F}_t(x)dx
$$
$$
+ \int_t^T \int_0^{g_{n+1}(u)} (-v + g_{n+1}(u))dP(C(n+1, u) \leq v)dP(S_{n+1} \leq u|S_n = t)
$$
$$
= \frac{1}{\overline{F}(t)} \int_t^T \overline{F}(u)\{\lambda^* + h(u)(\int_0^{g_{n+1}(u)} R_{n+1,u}(v)dv - C_f)\}du,
$$

$$(8.31)$$

where $R_{n,t}(u) = P(C(n, t) \leq u)$, and T is the optimal preventive replacement time determined by the equation $C_f h(T) = \lambda$. If $C_f h(t) < \lambda$ for all t, $T = +\infty$ and the optimal policy is a repair-cost-limit policy .

The results are summarized in the following theorem.

Theorem 3. The optimal control limit functions $\{g_n(t), n \geq 0\}$ and the optimal average cost λ^* satisfy the following system of differential equations with boundary conditions:

$$
g_n'(t) = -\lambda + h(t)(g_n(t) - \int_0^{g_{n+1}(t)} R_{n+1,t}(u)du), \ n \geq 0, \ 0 \leq t \leq T,
$$

$$(8.32)$$

$$
g_0(0) = C_r, \ g_n(T) = C_f, \ C_f h(T) = \lambda.
$$

The differential equations in (8.32) are obtained by differentiating (8.31). The boundary conditions are obtained easily.

Next, we consider the case where the repair cost $C(n, t)$ does not depend on n. In this case, the optimal policy is given by the following theorem (see Makis et al.(1999)).

Next, we consider the case when the repair cost $C(n, t)$ does not depend on n. In this case, the optimal policy is given by the following theorem (see Makis et al.(1999)).

Theorem 4. If $C(n, t) \equiv C(t)$ for all n, then $g_n(t) = g(t)$ and the optimal control function $g(t)$ and the optimal average cost λ^* are uniquely determined by the following differential equation with boundary conditions:

$$
g'(t) = -\lambda + h(t) \int_0^{g(t)} \overline{R}_t(u)du, \ 0 \leq t \leq T, \tag{8.33}
$$

$$g(0) = C_r, \; g(T) = C_f, \; C_f h(T) = \lambda, \tag{8.34}$$

where $\overline{R}_t(u) = P(C(t) > u)$.

Based on Theorem 4, we now present an algorithm to obtain an ϵ-optimal policy, and the corresponding average cost λ^*.

The algorithm.

Step 0. Choose $\epsilon > 0$ and put $\lambda_L = 0, \lambda_U = C_r/ES_1$.

Step 1. Put $\lambda = (\lambda_L + \lambda_U)/2$.

Step2. Put $T = h^{-1}(\lambda/C_f), \; g(\lambda, T) = C_f$.

Step 3. Find the solution $g(\lambda, t)$ to (8.33) for λ defined in step 1.

Step 4. If $g(\lambda, 0) = C_r$, or $\lambda_U - \lambda_L < \epsilon$, go to step 5. Otherwise, compare $g(\lambda, 0)$ and C_r.

If $g(\lambda, 0) < C_r$, put $\lambda_L = \lambda$ and go to step 1.

If $g(\lambda, 0) > C_r$, put $\lambda_U = \lambda$ and go to step 1.

Step 5. Stop. $g(\lambda, t)$ and T determine an ϵ-optimal policy and λ is the corresponding average cost.

To illustrate the computational algorithm, we consider the following example.

Example. Put $\epsilon = 0.01$, $C_r = 4$, $C_f = 1$, $h(t) = t$ (Weibull hazard function with the shape parameter $\alpha = 2$) and let $C(t) = \mathcal{U}$, where $\mathcal{U}$ is uniformly distributed on interval $[1, 3]$. Using the above algorithm and MATLAB, we obtained $\lambda^* = T = 2.94$ and the optimal repair cost limit function depicted in Figure 8.1. The optimal repair cost limit function determines the optimal policy. When a failure occurs at $t < T$, the observed repair cost $C(t)$ is compared with $g(t)$. If $C(t)$ is less than $g(t)$, the unit is repaired, otherwise it is replaced by a new unit. If no replacement has been carried out before T, the unit is preventively replaced at time T.

This plot also provides useful information regarding the residual value of the unit. If a unit has age t, then its residual value is $g(t) - C_f$. In particular, a new unit has residual value $g(t) = C_r - C_f$, which is equal to the preventive replacement cost C_p.

4.   OPTIMAL POLICY IN THE DISCOUNTED COST CASE

In this section, we examine the structure of the optimal policy minimizing the total expected discounted cost associated with the maintenance model described in Section 3.

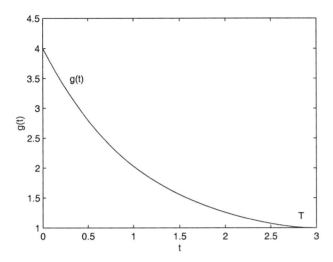

Figure 8.1   The optimal repair cost limit.

Let $0 < \alpha < 1$ be the discount factor and $TC_\alpha(\tau_i)$ be the cost of the i−th unit.

Consider the repair/replacement policy determined by a sequence of stopping times $\pi = \{\tau_i, i = 1, 2, ...\}$, where τ_i is the replacement time of the $i - th$ unit. Then the total discounted cost over an infinite time horizon has the form

$$K_\alpha(\pi) = \sum_{i=1}^{\infty} TC_\alpha(\tau_i) exp(-\alpha \sum_{j=1}^{i-1} \tau_j). \qquad (8.35)$$

It is not difficult to see that it is sufficient to consider the stationary policy, i.e., $\tau_i \equiv \tau$ for all i. The original optimization problem can then be reduced to the following optimal stopping problem. Find

$$inf_\tau \frac{E(TC_\alpha(\tau))}{E(1 - e^{-\alpha\tau})}, \qquad (8.36)$$

and the optimally stopping time τ_α^* (if it exists) minimizing the expression in (8.36).

Using the notation in the Section 3, the total discounted cost $TC_\alpha(\tau)$ has the form

$$
\begin{aligned}
TC_\alpha(\tau) &= \sum_{i=0}^{\sigma-1}((C(i, S_i) - C_f)e^{-\alpha S_i} \\
&\quad + C_f e^{-\alpha S_{i+1}} I_{\{S_{i+1} \leq T_i\}}) I_{\{S_i \neq T_i\}} \prod_{j=0}^{i-1} I_{\{S_{j+1} \leq T_j\}} \qquad (8.37) \\
&\equiv TC_\alpha(\sigma, \{T_i\}),
\end{aligned}
$$

where $C(0,0) \equiv C_r$, and the stopping time τ has the representation given by (8.13), $\tau \equiv \tau(\sigma, \{T_i\})$.

Next,

$$
\begin{aligned}
1/\alpha(1 - e^{-\alpha\tau}) &= \int_0^\infty e^{-\alpha t} I_{(\tau > t)} dt \\
&= \textstyle\sum_{i=0}^{\sigma-1} \int_{S_i}^{T_i} e^{-\alpha t} I_{\{S_{i+1} \geq t\}} dt \prod_{j=0}^{i-1} I_{\{S_{j+1} \leq T_j\}} \equiv \tau_\alpha(\sigma, \{T_i\}),
\end{aligned}
\tag{8.38}
$$

Applying the λ−maximization technique and the semi-martingale decomposition, we obtain the formulas for $Y_\lambda^\alpha(\sigma, \{T_i\})$ and $\overline{Y}_\lambda^\alpha(\sigma, \{T_i\})$, corresponding to $Y_\lambda(\sigma, \{T_i\})$ and $\overline{Y}_\lambda(\sigma, \{T_i\})$ in Section 3. We have,

$$
\begin{aligned}
Y_\lambda^\alpha(\sigma, \{T_i\}) = {} & \textstyle\sum_{i=0}^{\sigma-1} (\lambda \int_{S_i}^{T_i} e^{-\alpha t} I_{\{S_{i+1} \geq t\}} dt \\
& - (C(i, S_i) - C_f) e^{-\alpha S_i} - C_f e^{-\alpha S_{i+1}} I_{\{S_{i+1} \leq T_i\}}) \\
& I_{\{S_i \neq T_i\}} \prod_{j=0}^{i-1} I_{\{S_{j+1} \leq T_j\}},
\end{aligned}
\tag{8.39}
$$

and

$$
\begin{aligned}
\overline{Y}_\lambda^\alpha(\sigma, \{T_i\}) = {} & \textstyle\sum_{i=0}^{\sigma-1} (-(C(i, S_i) - C_f) e^{-\alpha S_i} \\
& + \int_{S_i}^{T_i} e^{-\alpha t} \overline{F}_{S_i}(t)(\lambda - C_f h(t)) dt) \prod_{j=0}^{i-1} I_{\{S_{j+1} \leq T_j\}}.
\end{aligned}
\tag{8.40}
$$

As Section 3,

$$
T_\lambda = \inf\{t \; : \; \lambda \leq C_f h(t)\}. \tag{8.41}
$$

is the optimal preventive replacement time.

We have that

$$
\mathit{inf}_\pi E(K_\alpha(\pi)) = 1/\alpha\lambda_\alpha^*, \tag{8.42}
$$

where

$$
\lambda_\alpha^* = \mathit{sup}\{\lambda : V^\alpha(\lambda) < 0\}, \tag{8.43}
$$

and

$$
V^\alpha(\lambda) = \mathit{sup}_\sigma(\mathit{sup}_{\{T_i\}} E(\overline{Y}_\lambda^\alpha(\sigma, \{T_i\}))) \tag{8.44}
$$

Finally, the maintenance decision problem reduces to the optimal stopping problem for the sequence $\{W_n^\alpha(\lambda)\}$ defined by

$$
\begin{aligned}
W_n^\alpha(\lambda) \equiv {} & \textstyle\sum_{i=0}^{n-1} (-(C(i, S_i) - C_f) e^{-\alpha S_i} \\
& + \int_{S_i}^{T_\lambda} (\lambda - h(t) C_f) e^{-\alpha t} \overline{F}_{S_i}(t) dt) I_{\{S_i \leq T_\lambda\}},
\end{aligned}
\tag{8.45}
$$

which corresponds to $W_n(\lambda)$ in (8.23).

The optimal stopping time $\sigma_\lambda(\alpha)$ has the form

$$
\sigma_\lambda = \inf\{n \geq 1 : C(n, S_n) \geq g_n^\alpha(\lambda, S_n)\}, \tag{8.46}
$$

where $g_n^\alpha(\lambda, t)$ satisfies the boundary conditions $g_0^\alpha(\lambda, 0) = C_r$, $g_n^\alpha(\lambda, S_n) = C_f$ for $S_n \geq T_\lambda$.

Hence, the optimal repair/ replacement policy has the same form as in the average cost case, i.e., the unit is minimally repaired at the nth failure time S_n if and only if $S_n \leq T$ and $g_n^\alpha(\lambda_\alpha^*, S_n) > C(n, S_n)$, otherwise it is replaced. λ_α^* is obtained from (8.43). If the unit has not been replaced at failure times before T, it is preventively replaced at time T, where $T \equiv T_{\lambda_\alpha^*}$ is determined by (8.41).

Similar to the average cost case, we have the following recursive equation for $g_n^\alpha(\lambda_\alpha^*, t)$. For simplicity, denote $g_n^\alpha(\lambda_\alpha^*, t)$ as $g_n(t)$, and λ_α^* as λ.

$$g_n(t) - C_f$$
$$= \frac{1}{e^{-\alpha t}\overline{F}(t)} \int_t^T e^{-\alpha u}\overline{F}(u)\{\lambda + h(u)(\int_0^{g_{n+1}(u)} R_{n+1,u}(v)dv - C_f)\}du.$$
$$(8.47)$$

And the differential form of the above equation is,

$$g_n'(t) = -\lambda + h(t)(g_n(t) -$$
$$\int_0^{g_{n+1}(t)} R_{n+1,t}(u)du) + \alpha(g_n(t) - C_f), \ n \geq 0, \ 0 \leq t \leq T,$$
$$(8.48)$$
$$g_0(0) = C_r, \ g_n(T) = C_f, \ C_f h(T) = \lambda.$$

If the repair cost $C(n, t)$ does not depend on n, the optimal control function $g^\alpha(\lambda_\alpha^*, t)$ and λ_α^* are obtained as the unique solution to the following differential equation with the boundary conditions,

$$g'(t) = -\lambda + h(t) \int_0^{g(t)} \overline{R}_t(u)du + \alpha(g(t) - C_f), \ 0 \leq t \leq T, \quad (8.49)$$

$$g(0) = C_r, \ g(T) = C_f, \ C_f h(T) = \lambda, \quad (8.50)$$

where $\overline{R}_t(u) = P(C(t) > u)$.

Note that when $\alpha \longrightarrow 0$, equation (8.49) has the same form as equation (8.33) for the average cost case. Therefore, λ_α^* decreases to λ^* (where λ^* is the optimal expected average cost), and the repair cost limit function for the discounted cost case $g^\alpha(\lambda_\alpha^*, t)$ increases to $g(\lambda^*, t)$, the repair cost limit function for the average cost case. Therefore, the expected average cost case can be viewed as a special case of the expected discounted cost case.

References

[1] Aven, T. Optimal Replacement under a Minimal Repair Strategy - a General Failure Model. *Advances in Applied Probability* , **15** (1983) 198-211.

[2] Aven, T. and Bergman, B. Optimal Replacement Times - a General Set- up. *Journal of Applied Probability*, **23** (1986) 432-442.

[3] Aven, T. Optimal Replacement of Monotone Repairable Systems . *Reliability and Maintenance of Complex Systems*, S. Ozekici, Ed., Springer,) 224-238, 1996.

[4] Bergman, B. Optimal Replacement under a General Failure Model. *Advances in Applied Probability* , **10**. (1978) 431-451.

[5] Chow, Y. S., Robbins, H. and Siegmund, D.. *Optimal Sto ing Theory*, Dover Publication Inc., New York, 1971.

[6] Hastings, N. A. J. The Repair Limit Replacement Method. *Operational Research Quarterly*, **20(3)** (1969) 337-349.

[7] Jensen, U. Monotone Sto ing Rules for Stochastic Processes in a Semimartingale Representation with Applications. *Optimization*, **20** (1989) 837-852.

[8] Jensen, U. Stochastic Models of Reliability and Maintenance: An Overview, *Reliability and Maintenance of Complex Systems*, S. Ozekici, Ed., Springer, 3-36, 1996.

[9] Jiang, X, K. Cheng and Makis, V. On the Optimality of Repair-Cost-Limit Policy . *Journal of Applied Probability*, **35** (1998) 936-949.

[10] L'Ecuyer, P., and Haurie, A. The Repair vs. Replacement Problem: a Stochastic Approach. *Optimal Control Application and Methods*, **8** (1987) 219-230.

[11] Makis, V. and Jardine, A.K.S. Optimal Replacement Policy for a General Model with Imperfect Repair. *Journal of the Operational Research Society*, **43**(2) (1992) 111-120.

[12] Makis, V. and A.K.S. Jardine, A Note on Optimal Replacement Policy under General Repair. *European Journal of Operational Research*, **69** (1993) 75-82.

[13] Makis, V., Jiang, X. and Cheng, K. Optimal Preventive Replacement under Minimal Repair and Random Repair Cost , to a ear in Mathematics of Operations Research, 1999.

[14] Stadje W. and Zuckerman, D. Optimal Maintenance Strategies for Repairable Systems with General Degree of Repair. *Journal of Applied Probability*, **28** (1991) 384-396.

[15] White, D.J. Repair Limit Replacement. *OR Spektrum*, **11**. (1989) 143-149.

Chapter 9

IMPERFECT PREVENTIVE MAINTENANCE MODELS

Toshio Nakagawa
Department of Industrial Engineering,
Aichi Institute of Technology
nakagawa@ie.aitech.ac.jp

Abstract   This chapter surveys the earlier results of imperfect preventive maintenance (pm) models which could be applied to actual systems; (i) the unit after pm has the same hazard rate as before pm, (ii) the age of the unit becomes x units of time younger at pm, and (iii) the age reduces to at when it was t before pm. The expected cost rates of each model as an objective function are obtained and optimal policies which minimize them are analytically discussed.

Keywords: preventive maintenance, imperfect probability, , age of younger, minimal repair, imperfect inspection, expected cost, optimal policy

1.   INTRODUCTION

Optimal preventive maintenance (pm) policies for redundant models have been summarized in [1, 14, 33] and also in Chapter 7. All models have assumed that "the unit after pm is as good as new". Actually this assumption might not be true. The unit after pm is usually younger after pm, but occasionally, it might be worse than before pm because of faulty procedures, e.g., wrong adjustments, bad parts, or damage done during pm. Generally, the improvement to the unit by pm would depend on the resources used in the pm [16].

Weiss [35] first assumed that the inspection may not be perfect. Similar models indicating that inspection, test and detection of failure are uncertain, were treated in Coleman and Abrams [6], Noonan and Fain [27] and Yun and Bai [37].

Chan and Downs [4], Nakagawa [15] and Murthy and Nguyen [13] assumed that the unit after pm is the same as before pm with a certain probability, and discussed the optimal pm policies. Imperfect maintenances for a computer system first treated by Ingle and Siewiorek [10]. Helvic [9], Yak, et al. [36] and Nakagawa and Yasui [23] considered that while the system is usually renewed after pm, it sometimes remains unchanged, and maintained the MTTF and the availability of the system. Chung [5] studied the imperfect test of intermittent faults occurring in digital systems.

Nakagawa [17,18] considered two imperfect pm models of the unit: (i) The age becomes x units of time younger at each pm and (ii) the hazard rate is reduced in proportion to that before pm or the pm cost. Lie and Chun [12] introduced an improvement factor in hazard rate after maintenance and Canfield [3] considered the system degradation with time where the pm restores the hazard function to the same shape. Nakagawa [24] proposed the sequential policy where the pm is done at suitable times and derived the optimal sequence interval.

Nguyen and Murthy [28] considered the imperfect repair model where the mean life of a repaired unit is less than that of a new unit and discussed the repair limit policy. Brown and Proschan [2] and Fontenot and Proschan [8] assumed that a failed unit is as good as new with a certain probability and investigated some properties of a failure distribution. Similar imperfect repair models were studied by Ebrahimi [7], Natvig [26], Zhao [38], and Wang and Pham [34]. Further, Shaked and Shanthikumar [30] and Shew et al. [31,32] derived multivariate distributions and studied probabilistic quantities of imperfect repair models. Recently, Kijima and Nakagawa [11] studied the replacement policies for a shock model with imperfect pm.

This chapter summarizes our earlier results of imperfect maintenance models which could be applied to real systems in practical fields: An operating unit is inspected and maintained preventively at periodic times jT $(j = 1, 2, \cdots)$.

Policy 1. The unit after pm has the same hazard rate as before pm with probability p, and otherwise, becomes like new.

Policy 2. The age becomes x units of time younger at each pm than its age before pm and failed units undergo minimal repair between pm's.

Policy 3. The age is reduced to at $(0 < a \leq 1)$ when it was t before pm and a failed unit is detected only by inspection.

For example, imperfect maintenance of Policy 1 would be made for the following reasons [23,29]: (i) Hidden faults and failures, and faulty parts which are not detected during maintenance, (ii) human errors such as wrong adjustments and further damage of adjacent parts during mainte- nance, and (iii) replacement with faulty parts or with only partial repair of faulty parts.

Policies 2 & 3 deal generally with systems where failed units are imme- diately detected and failed are detected only by inspection, respectively. Of course, we can easily consider the imperfect pm model where the age becomes x units of time younger at pm and a failed unit is detected only by inspection [21]. Further, we can propose a modified imperfect model where the hazard rate decreases to $ar(t)$ when it was $r(t)$ before pm. Both policies would be applied to most complex systems, such as airplanes, manufacturing equipment, and railroad tracks and wirings, which would be younger after pm. Pham and Wang [29] gave an engine tune-up as one possible example.

The expected cost rates of Policies 1 & 2 and the total expected cost of Policy 3 are derived, and optimal policies which minimize them are analytically discussed. Further, to understand these results easily, simple numerical examples of each model are given.

2.   ANALYSIS OF POLICY 1

2.1   Model and Analysis. Suppose that the unit is main- tained preventively at periodic times jT $(j = 1, 2, \cdots)$, and the hazard rate is the same as before pm with a certain probability. We consider the following imperfect pm policy for a one-unit system which should operate for an infinite time span [15] :

1. The unit has a failure distribution $F(t)$ $(t \geq 0)$ with finite mean $1/\lambda$ and its density $f(t) \equiv dF(t)/dt$, and furthermore, a hazard rate (failure rate) $r(t) \equiv f(t)/\overline{F}(t)$ and a cumulative hazard rate $R(t) \equiv \int_0^t r(u)du$, where $\overline{F}(t) \equiv 1 - F(t)$.

2. The operating unit is repaired at failure or is maintained preventively at time T, whichever occurs first, after its installation or previous pm.

3. The unit after repair is as good as new.

4. The unit after pm has the same hazard rate as it had been before pm with probability p $(0 \leq p < 1)$ and is as good as new with probability $q \equiv 1 - p$.

5. The cost for each pm is c_1 and the cost for each repair is c_2

6. The repair and pm times are negligible.

Consider one cycle from $t = 0$ to the time that the unit is as good as new by either repair or perfect pm. Then, the probability that the unit is repaired at failure is

$$\sum_{j=1}^{\infty} p^{j-1} \int_{(j-1)T}^{jT} dF(t) = 1 - q \sum_{j=1}^{\infty} p^{j-1} \bar{F}(jT), \qquad (9.1)$$

and the expected number of pm's per one cycle is

$$\sum_{j=1}^{\infty} (j-1) p^{j-1} \int_{(j-1)T}^{jT} dF(t) + q \sum_{j=1}^{\infty} j p^{j-1} \bar{F}(jT) = \sum_{j=1}^{\infty} p^{j-1} \bar{F}(jT). \quad (9.2)$$

Further, the mean time duration of one cycle is

$$\sum_{j=1}^{\infty} p^{j-1} \int_{(j-1)T}^{jT} t \, dF(t) + q \sum_{j=1}^{\infty} p^{j-1} (jT) \bar{F}(jT) = \sum_{j=1}^{\infty} p^{j-1} \int_{(j-1)T}^{jT} \bar{F}(t) dt.$$

$$(9.3)$$

Thus, the expected cost per unit of time of one cycle is

$$C(T; p) = \frac{c_1 \sum_{j=1}^{\infty} p^{j-1} \bar{F}(jT) + c_2 \left[1 - q \sum_{j=1}^{\infty} p^{j-1} \bar{F}(jT) \right]}{\sum_{j=1}^{\infty} p^{j-1} \int_{(j-1)T}^{jT} \bar{F}(t) dt}. \qquad (9.4)$$

It is evident that $C(0; p) = \infty$ and $C(\infty; p)$ is λc_2.

2.2   Optimal Policy. We seek an optimal pm time T^* ($0 < T^* \le \infty$) which minimizes $C(T; p)$ in (9.4) . Let

$$H(t; p) \equiv \frac{\sum_{j=1}^{\infty} p^{j-1} j f(jt)}{\sum_{j=1}^{\infty} p^{j-1} j \bar{F}(jt)}. \qquad (9.5)$$

Then, differentiating $C(T; p)$ with respect to T and setting it equal to zero, we have

$$H(T;p)\sum_{j=1}^{\infty}p^{j-1}\int_{(j-1)T}^{jT}\bar{F}(t)dt - q\sum_{j=1}^{\infty}p^{j-1}F(jT) = \frac{c_1}{c_2q - c_1}, \quad (9.6)$$

for $c_2q \neq c_1$.

Let be denoted the left-hand side of (9.6) by $Q(T;p)$. It is easily proved that if $H(T;p)$ is strictly increasing in T then $Q(T;p)$ is also strictly increasing from 0 to

$$Q(\infty;p) = H(\infty;p)(1/\lambda) - 1. \quad (9.7)$$

Thus, we have the following optimal policy when $H(t;p)$ is strictly increasing in t for any p:

(i) If $c_2q > c_1$ and $H(\infty;p) > \lambda c_2q/(c_2q - c_1)$ then there exists a finite and unique T^* which satisfies (9.6), and the resulting expected cost is

$$C(T^*;p) = [c_2 - (c_1/q)]H(T^*;p). \quad (9.8)$$

(ii) If $c_2q > c_1$ and $H(\infty;p) \leq \lambda c_2q/(c_2q-c_1)$, or $c_2q \leq c_1$ then $T^* = \infty$, i.e., no pm should be done, and the expected cost is λc_2.

When $p = 0$, i.e., the pm is perfect, the model corresponds to a usual age replacement policy, and the above results agree with those of [1, p.85].

2.3   Illustrative Example. We give a numerical example when $F(t)$ is a gamma distribution with order 2, i.e., $F(t) = 1-(1+t)e^{-t}$ and $1/\lambda = 2$. Then, $H(t;p)$ in (9.5) is

$$H(t;p) = \frac{t(1 + pe^{-t})}{1 - pe^{-t} + t(1 + pe^{-t})},$$

which is strictly increasing from 0 to 1. Thus, if $c_2q > 2c_1$ then there exists a finite and unique T^* which satisfies (9.6), and otherwise, $T^* = \infty$.

Table 9.1 gives the optimal pm times T^* and the expected cost rates $C(T^*;p)$ for $p = 0.0 \sim 0.4$ when $c_1 = 1$ and $c_2 = 5$. Both T^* and $C(T^*;p)$ are increasing when p is large. The reason is that it is better to repair a failed unit than to perform pm for large p.

Table 9.1: Optimal pm times T^* and expected cost rates

$C(T^*; p)$ for p when $c_1 = 1$ and $c_2 = 5$.

p	T^*	$C(T^*; p)$
0.00	1.31	2.27
0.01	1.32	2.27
0.05	1.36	2.30
0.10	1.43	2.34
0.15	1.52	2.37
0.20	1.64	2.40
0.25	1.80	2.43
0.30	2.02	2.45
0.35	2.33	2.47
0.40	2.79	2.49

3.   ANALYSIS OF POLICY 2

3.1   Model and Analysis. Suppose that the unit is maintained preventively at periodic times jT ($j = 1, 2, \cdots$) and its age becomes x units of time younger at each pm. Further, the unit undergoes only minimal repair [1, 20] at failure between pm's. We consider the following imperfect pm policy for a one-unit system [18]:

2. The operating unit undergoes pm at time T and minimal repair at failure between pm's.

3. The hazard rate is strictly increasing and remains undisturbed by minimal repair.

4. The age of the unit becomes x units of time younger at each pm, where x ($0 \leq x \leq T$) is constant. Further, the unit is replaced if it operates for the time interval NT ($N = 1, 2, \cdots$) for a specified T (> 0).

5. The cost for each pm is c_1, the cost for each minimal repair is c_2, and the cost for replacement at time NT is c_3 ($> c_1$).

1, 6. Same as the assumptions of Sec. 2.

The expected cost per unit of time until repalcement is

$$C(N; T, x) = \frac{(N-1)c_1 + c_2 \sum_{j=0}^{N-1} \int_{j(T-x)}^{T+j(T-x)} r(t)dt + c_3}{NT}$$

$$(N = 1, 2, \cdots). \qquad (9.9)$$

When $x = T$, the unit after pm is as good as new by perfect pm, and when $x = 0$, it has the same age as before pm. It is shown that the expected cost $C(N; T, x)$ in (9.9) is decreasing in x since $r(t)$ is increasing.

3.2   Optimal Policy.

We seek an optimal replacement number N^* ($1 \leq N^* \leq \infty$) which minimizes $C(N; T, x)$ for given T and x ($0 \leq x < T$). This is a modified discrete replacement model [22] with minimal repair at failure. From the inequality $C(N + 1; T, x) \geq C(N; T, x)$, we have

$$L(N; T, x) \geq (c_3 - c_1)/c_2, \tag{9.10}$$

where

$$L(N; T, x) \equiv \sum_{j=0}^{N-1} \int_0^T \{r[t + N(T - x)] - r[t + j(T - x)]\} dt$$

$$(N = 1, 2, \cdots).$$

Further, we have

$$L(N + 1; T, x) - L(N; T, x) =$$
$$(N + 1) \int_0^T \{r[t + (N + 1)(T - x)] - r[t + N(T - x)]\} dt > 0.$$

Therefore, we have the following optimal policy:

(i) If $L(\infty; T, x) > (c_3 - c_1)/c_2$ then there exists a unique minimum N^* which satisfies (9.10) .

(ii) If $L(\infty; T, x) \leq (c_3 - c_1)/c_2$ then $N^* = \infty$, and the expected cost rate is

$$C(\infty; T, x) = \frac{c_1}{T} + c_2 r(\infty). \tag{9.11}$$

It is noted in case (i) that the optimal N^* is increasing with x since $L(N; T, x)$ is a decreasing function of x. Further, it is evident that if $r(t) \to \infty$ as $t \to \infty$ then $N^* < \infty$.

3.3   Illustrative Example.

Consider a numerical example when the failure time has a Weibull distribution , i.e., $\overline{F}(t) = exp(-\lambda t^\alpha)$ ($\lambda > 0$, $\alpha > 1$). Then, the hazard rate is $r(t) = \lambda \alpha t^{\alpha-1}$ which is strictly increasing from 0 to ∞. The expected cost is, from (9.9) ,

$$C(N;T,x) =$$

$$\frac{(N-1)c_1 + c_3 + c_2\lambda\sum_{j=0}^{N-1}\{[T+j(T-x)]^\alpha - [j(T-x)]^\alpha\}}{NT}$$

$$(N = 1, 2, \cdots). \qquad (9.12)$$

From (9.10),

$$\sum_{j=0}^{N-1}\{[T + N(T-x)]^\alpha - [T + j(T-x)]^\alpha$$

$$-[N(T-x)]^\alpha + [j(T-x)]^\alpha\} \geq \frac{c_3 - c_1}{\lambda c_2}, \qquad (9.13)$$

which is strictly increasing in N to ∞. Thus, there exists a unique minimum N^* $(1 \leq N^* < \infty)$ which satisfies (9.13).

In particular, when $\alpha = 2$, equation (9.13) is rewritten as

$$\lambda T(T-x)N(N+1) \geq \frac{c_3 - c_1}{c_2}. \qquad (9.14)$$

Table 9.2 gives the optimal replacement numbers N^* for $x = 0, 0.2T$, $0.4T$, $0.6T$, $0.8T$ and $(c_3 - c_1)/c_2 = 0.5, 1, 3, 5, 10$ when $\lambda T = 0.1, 0.5$ and $T = 1$. It is shown that N^* are increasing slowly with the increases of x and $(c_3 - c_1)/c_2$.

Table 9.2: Optimal replacement numbers N^* for x and $(c_3 - c_1)/c_2$ when $\alpha = 2$ and $T = 1$.

x	$\lambda T = 0.5$ $(c_3-c_1)/c_2$					$\lambda T = 0.1$ $(c_3-c_1)/c_2$				
	0.5	1	3	5	10	0.5	1	3	5	10
0	1	1	2	3	4	2	3	5	7	10
0.2T	1	2	3	4	5	3	4	6	8	11
0.4T	1	2	3	4	6	3	4	7	9	14
0.6T	2	2	4	5	7	4	5	9	11	16
0.8T	2	3	5	7	10	5	7	12	16	22

4.   ANALYSIS OF POLICY 3

4.1   Model and Analysis.

Suppose that the unit is checked and maintained preventively at periodic times jT $(j = 1, 2, \cdots)$, and a failed unit is detected only by inspection and is replaced. We consider the following imperfect inspection policy with pm for a one-unit system [19,21]:

2. The operating unit is checked and maintained preventively at times jT $(j = 1, 2, \cdots)$.

3. The failed unit is detected only by inspection and is replaced.

4. The age of the unit after inspection reduces to at $(0 < a \leq 1)$ when it was t before inspection, i.e., the age becomes $t(1 - a)$ units of time younger at each inspection.

5. The cost for each inspection is c_1 and the cost of time elapsed between failure and its detection per unit of time is c_2.

6. Inspection and replacement times are negligible.

1. Same as the assumption of Sec. 2.

Let $H(t; x)$ be the hazard rate of the unit, i.e., $H(t; x) \equiv [F(t + x) - F(t)]/\overline{F}(t)$ for $x > 0$, $t \geq 0$ and $F(t) < 1$. Then, the probability that the unit does not fail until time t is, from assumption 4,

$$\overline{S}(t; T, a) = \overline{H}(A_k T; t - kT) \prod_{j=0}^{k-1} \overline{H}(A_j T; T)$$

$$\text{for } kT \leq t < (k + 1)T, \tag{9.15}$$

where $\overline{H} \equiv 1 - H$ and $A_k \equiv a + a^2 + \cdots + a^k$ $(k = 1, 2, \cdots)$, $A_0 \equiv 0$. For, $\overline{H}(t; T)$ is the probability that the unit with age t does not fail in an interval of time T.

Using (9.15), we obtain the mean time to failure $\gamma(T; a)$ and the expected number $M(T; a)$ of inspections before failure of the unit. The mean time to failure is

$$\gamma(T; a) = \sum_{k=0}^{\infty} \int_{kT}^{(k+1)T} \overline{S}(t; T, a) dt$$

$$= \sum_{k=0}^{\infty} \left[\prod_{j=0}^{k-1} \overline{H}(A_j T; T) \right] \int_{A_k T}^{(A_k+1)T} \overline{F}(t) dt / \overline{F}(A_k T) \tag{9.16}$$

where $\prod_{j=0}^{-1} \equiv 1$, and the expected number of inspections before failure is

$$M(T;a) = \sum_{k=0}^{\infty} kH(A_kT;T) \prod_{j=0}^{k-1} \overline{H}(A_jT;T)$$

$$= \sum_{k=0}^{\infty} \prod_{j=0}^{k} \overline{H}(A_jT,T). \tag{9.17}$$

When $H(t;x)$ is increasing in t for $x > 0$, we have that $\overline{H}(A_kT;T)$ is decreasing in a for $k \geq 1$ and $T > 0$, and hence, $\overline{S}(t;T,a)$ is decreasing in a. Thus, both $\gamma(T;a)$ and $M(T;a)$ are also decreasing in a. Therefore, we have the following inequalities:

$$\frac{1}{\lambda} \leq \gamma(T;a) \leq \frac{\int_0^T \overline{F}(t)dt}{F(T)}, \tag{9.18}$$

$$\sum_{k=0}^{\infty} \overline{F}[(k+1)T] \leq M(T;a) \leq \frac{\overline{F}(T)}{F(T)}, \tag{9.19}$$

where all equalities hold when F is exponential .

The total expected cost until detection of failure is [1, p.108],

$$C(T;a) = \sum_{k=0}^{\infty} \int_{kT}^{(k+1)T} \{c_1(k+1) + c_2[(k+1)T - t]\} d[-\overline{S}(t;T,a)]$$

$$= (c_1 + c_2T)[M(T;a) + 1] - c_2\gamma(T;a). \tag{9.20}$$

It is evidently seen that $\lim_{T \to 0} C(T;a) = \lim_{T \to \infty} C(T;a) = \infty$. Thus, there exists a positive and finite T^* which minimizes the expected cost $C(T;a)$ in (9.20).

In particular, when $F(t) = 1 - e^{-\lambda t}$,

$$C(T;a) = \frac{c_1 + c_2T}{1 - e^{-\lambda T}} - \frac{c_2}{\lambda}, \tag{9.21}$$

for all a and there exists a unique T^* which satisfies

$$e^{\lambda T} - (1 + \lambda T) = \frac{c_1\lambda}{c_2}. \tag{9.22}$$

which agrees with the result of Weiss [35].

Moreover, assuming that the unit is replaced instantly at detection of failure and its cost is c_3, the expected cost per unit of time is

$$\tilde{C}(T;a) = \frac{(c_1 + c_2 T)[M(T;a) + 1] - c_2 \gamma(T;a) + c_3}{T[M(T;a) + 1]}. \tag{9.23}$$

4.2   Illustrative Example. It is very difficult to discuss an optimal inspection time T^* analytically which minimizes $C(T;a)$ in (9.20). We give a numerical example when the failure time has a Weibull distribution with shape parameter 2, i.e., $\overline{F}(t) = exp[-(t/100)^2]$. Table 9.3 gives the optimal inspection times T^* for a and c_2 when $c_1 = 1$. These times are small when a is large. The reason would be that the hazard rate increases quickly with age when a is large and the unit fails easily.

Table 9.3: Optimal inspection times T^* for a and c_2 when $c_1 = 1$.

a	c_2 0.1	0.5	1.0	2.0	5.0	10.0
0.0	76	47	36	31	23	18
0.2	67	40	32	26	19	15
0.4	60	34	27	22	16	13
0.6	53	29	22	18	13	10
0.8	47	23	18	14	10	8
1.0	42	19	13	9	6	4

5.   CONCLUSIONS

The new notion of imperfect pm arose in 1978, on hearing the news that the new Japanese railway line "Shinkansen" was stopped for one day by construction for rejuvenation, and written on the subject, several papers are contained in this Chapter. Brown and Proschan [2] separately investigated the properties of failure distribution from other points of imperfect repair. Since then, many authors have studied maintenance models from the above two viewpoints of imperfect pm and repair. Recently, Pham and Wang [29] gave good summaries of such models that have arisen over the past 30 years. This paper may be very useful for future research and potential applications of such areas.

Most pm's are imperfect, i.e., systems are not like new, but these become younger after each pm. However, if imperfect policies are applied to real systems, it would be very difficult to estimate rejuvenation parameters and reliability measures. Imperfect pm models considered here are simple, and so would be very useful for application to actual models.

References

[1] Barlow, R. E. and Proschan, F. *Mathematical Theory of Reliability.* John Wiley & Sons, New York, 1965.

[2] Brown, M. and Proschan, F. Imperfect repair. *J. Applied Probability,* **20** (1983) 851-859.

[3] Canfield, R. V. Cost optimization of priodic preventive maintenance. *IEEE Trans. Reliability,* **R-35** (1986) 78-81.

[4] Chan, P. K. W. and Downs, T. Two criteria for preventive maintenance. *IEEE Trans. Reliability,* **R-27** (1978) 272-273.

[5] Chung, K. J. Optimal test-times for intermittent faults. *IEEE Trans. Reliability,* **44** (1995) 645-647.

[6] Coleman, J. J. and Abrams, I. J. Mathematical model for operational readiness. *Operarions Research,* **10** (1962) 126-133.

[7] Ebrahimi, N. Mean time to achieve a failure-free requirement with imperfect repair. *IEEE Trans. Reliability,* **R-34** (1985) 34-37.

[8] Fontenot, R. A. and Proschan, F. Some imperfect maintenance models. in *Reliability Theory and Models,* Academic Press, 1984.

[9] Helvic, B. E. Periodic maintenance on the effect of imperfectness. *10th Int. Symp. Fault-Tolerant Computing,* 204-206, 1980.

[10] Ingle A. D. and Siewiorek, D. P. Reliability models for multiprocess systems with and without periodic maintenance. *7th Int. Symp. Fault-Tolerant Computing,* 3-9, 1977.

[11] Kijima, M. and Nakagawa, T. Replacement policies of a shock model with imperfect preventive maintenance. *European J. of Operational Research,* **57** (1992) 100-110.

[12] Lie C. H. and Chun, Y. H. An algorithm for preventive maintenance policy. *IEEE Trans. Reliabiity,* **R-35** (1986) 71-75.

[13] Murthy, D.N.P. and Nguyen, D.G. Optimal age-policy with imperfect preventive maintenance. *IEEE Trans. Reliabiity,* **R-30** (1981) 80-81.

[14] Nakagawa, T. Optimum preventive maintenance policies for repairable systems. *IEEE Trans. Reliability,* **R-26** (1977) 168-173.

[15] Nakagawa, T. Optimal policies when preventive maintenance is imperfect. *IEEE Trans. Reliability,* **R-28** (1979) 331-332.

[16] Nakagawa, T. Imperfect preventive-maintenance, *IEEE Trans. Reliability.* **R-28** (1979) 402.

[17] Nakagawa, T. Mean time to failure with preventive maintenance. *IEEE Trans. Reliability,* **R-29** (1980) 341.

[18] Nakagawa, T. A summary of imperfect preventive maintenance policies with minimal repair. *R.A.I.R.O. Operations Research*, **14** (1980) 249-255.

[19] Nakagawa, T. Replacement models with inspection and preventive maintenance. *Microelectronics and Reliability*, **20** (1980) 427-433.

[20] Nakagawa, T. A summary of periodic replacement with minimal repair at failure. *J. of Operations Research Soc. of Japan*, **24** (1981) 213-227.

[21] Nakagawa, T. Periodic inspection policy with preventive maintenance. *Naval Research Logistics Quarterly*, **31** (1984) 33-40.

[22] Nakagawa, T. A summary of discrete replacement policies. *European J. of Operational Research*, **17** (1984) 382-392.

[23] Nakagawa, T. and Yasui, K. Optimum policies for a system with imperfect maintenance. *IEEE Trans. Reliability*, **R-36** (1987) 631-633.

[24] Nakagawa, T. Sequential imperfect preventive maintenance policies. *IEEE Trans. Reliability*, **R-37** (1988) 295-298.

[25] Nakagawa, T. A summary of replacement models with changing failure distributions. *R.A.I.R.O. Operations Research*, **23** (1989) 343-353.

[26] Natvig, B. On imformation based minimal repair and the reduction in remaining system lifetime due to the failure of a specific module. *J. Applied Probability*, **27** (1990) 365-375.

[27] Noonan, G. C. and Fain, C. G. Optimum preventive maintenance policies when immediate detection of failure is uncertain. *Operations Research*, **10** (1962) 407-410.

[28] Nguyen, D. C. and Murthy, D. N. P. Optimal repair limit replacement policies with imperfect repair. *J. Operational Research Soc.*, **32** (1981) 409-416.

[29] Pham, H. and Wang, H. Imperfect maintenance. *European J. of Operational Research*, **94** (1996) 425-438.

[30] Shaked, M. and Shanthikumar, J. G. Multivariate imperfect repair. *Operations Research*, **34** (1986) 437-448.

[31] Shew, S. H. and Griffith, W. S. Multivariate age-dependent imperfect repair. *J. of Naval Research*, **38** (1991) 839-850.

[32] Shew, S. H. and Griffith, W. S. Multivariate imperfect repair. *J. of Applied Probability*, **29** (1992) 947-956.

[33] Valdez-Flores, C. and Feldman, R. M. A survey of preventive maintenance models for stochastically deteriorating single-unit systems. *Naval Res. Logist.*, **36** (1989) 419-446.

[34] Wang, H.Z. and Pham, H. Optimal age-dependent preventive maintenance policies with imperfect maintenance. *Int. J. of Reliability, Quality and Safety Engineering*, **3** (1996) 119-135.

[35] Weiss, G. H. A problem in equipment maintenance, *Managemant Science*, **8** (1962) 266-277.

[36] Yak, Y. W., Dillon, T. S. and Forward, K. E. The effect of imperfect periodic maintenance of fault tolerant computer systems. *14th Int. Symp. Fault-Tolerant Computing*, 66-70, 1984.

[37] Yun, W. Y. and Bai, D. S. Repair cost limit replacement policy with imperfect inspection. *Reliability Engineering and System Safety*, **23** (1988) 59-64.

[38] Zhao, M. Availability for repairable components and series systems. *IEEE Trans. Reliability*, **43** (1994) 329-334.

PAGE 183,197,200,202

Chapter 10

DISCOUNTED MODELS FOR THE SINGLE MACHINE INSPECTION PROBLEM

Moncer Hariga
King Saud University
Industrial Engineering Program
College of Engineering
P.O.Box 800, Riyadh 11421
Saudi Arabia
mhariga@ksu.edu.sa

Mohammad A. Al-Fawzan
King Abdulaziz City for Science and Technology
P.O. Box 6086
Riyadh 11442
Saudi Arabia
mfawzan@kacst.edu.sa

Abstract   In this paper, we address the problem of determining optimal inspection schedules for a single machine subject to exponentially distributed failures. We formulated the inspection models using the concept of discounted cash flow analysis to account for the effects of time value of money on the inspection policies. In addition, these models are developed under different assumptions on the planning horizon related to its length (infinite and finite)   and nature (deterministic and stochastic). In the finite horizon problem, we considered both cases where the inspection times are treated as continuous and discrete variables. For the case of continuous inspection times, we prove that the equal inspection interval policy is optimal only under certain conditions. In the case of a random horizon, we show that the problem can be transformed to the infinite horizon problem and, consequently, can be solved using the same procedure. Numerical examples are used to illustrate the models and their solution procedures.

Keywords: Inspection, discounted cash flow approach, finite horizon, infinite horizon, random horizon, dynamic programming

1.   INTRODUCTION

Consider a single unit representing a manufacturing system composed of several components. Under the superposition of the renewal processes related to the failure of the components, it is reasonable to assume that the failure distribution of the single unit is exponentially distributed (Cox and Smith [7] and Drenick [9]). Moreover, many authors, such as Davis [8] and Epstein [10], found strong empirical justification for the exponential distribution . Upon failure, the manufacturing process shifts from an in-control state to an out-of-control state during which an output of sub-standard quality is produced. This shift to the out-of-control state is not apparent and can only be detected through preventive maintenance in the form of inspections.

For some manufacturing systems, a continuous monitoring of their operating states is not economically justifiable. Under such conditions, inspections are useful in monitoring the condition of the system at predetermined times in order to reduce the probability of its malfunctioning. Once an out-of-control state is detected, a repair is carried out to restore the system to its in-control state. During the in-control state, there is an instantaneous profit p, per unit of time, which is equal to the revenue derived from the manufacturing system less the operating costs. On the other hand, a smaller profit, s, per unit of time, is obtained during the out-of-control state due to the extra rework costs of the output with sub-standard quality. In the sequel, we will use the word *machine* to refer to a complex manufacturing system.

Several authors in the reliability and maintenance literature (Goldman and Slattery [11], Jardine [13], and Banerjee and Chuiv [4]) considered different variations of this inspection problem. Baker [3] proposed a simple procedure for finding the optimal inspection frequency that generates maximum profit. The inspection model of Baker is based on the restrictive assumption that a failure completely halts production. Later, Chung [6] and Vaurio [17] developed approximate solution procedures for the same problem addressed by Baker [3]. Hariga [12] extended the single machine inspection problem by allowing the failure time to follow general type of probability distributions. Under the exponential shifting time to the out-of-control state, Ben-Deya and Hariga [5] reformulated

Baker's model with positive inspection and repair times and relaxed the assumption of no production during the out-of-control state. Abdel-Hameed [1] presented inspection models for a deteriorating device for which the deterioration level can only be monitored periodically at discrete times. Simulation was also used by some authors to include realistic factors that cannot be handled by analytical inspection models. Among these authors, Wang and Christer [71] compared simulation with asymptotic formulations of objective functions used for determining optimum inspection intervals . Recently, Alfares [2] presented a simulation model for determining the inspection policy of relief valves in a petrochemical plant.

In the above-cited analytical models, the optimal inspection interval is determined by maximizing the expected net profit per unit of time without considering the effects of discounting the different cash flows derived from the operation of the single machine on the inspection policy. However, because of the time value of money, the cash flows resulting from different inspection policies occur at distinct points in time and, consequently, have different values. Therefore, it is misleading to compare these policies with one another without taking into account the time value of money.

In this paper, we adopt the concept of discounted cash flow analysis to properly account for the effects of time value of money on the inspection policies for a single machine. First of all, we formulate four inspection models under different assumptions on the planning horizon. In the next section, we outline the notation and the problem assumptions and present some preliminary results that will be used in the rest of the paper. In the third section, we discuss the discounted single machine inspection problem under an infinite planning horizon . In the fourth section, we analyze the case of a known and finite horizon during which the inspections are carried out at continuous points in time. The finite horizon problem is reconsidered in the fifth section with discrete inspection times. In the sixth section, we determine the optimal periodic inspection schedule when the finite value of the planning horizon is not known with certainty. Finally, the last section concludes the paper.

2. NOTATION, ASSUMPTIONS, AND PRELIMINARY RESULTS

The following notation will be used throughout the paper. Additional notation will be introduced when needed.

Moreover, all the discounted maintenance inspection models developed in this paper are based on the following assumptions:

C_i Cost of a scheduled inspection;

C_r Cost of repairing a machine if it is found to be in an out-of-control state;

r Continuous discounting rate per unit of time;

$P(T)$ Present worth of the net profit over an inspection interval of length T;

$P(T, \infty)$ Present worth of the net profit of a periodic inspection policy over an infinite planning horizon with an inspection cycle of length T;

$E(T, \infty)$ Equivalent uniform worth per unit of time of $P(T, \infty)$, also known as annuity when one year is taken as the unit of time;

$P(T_j, n, H)$ Present worth of the net profit over a finite horizon, H, consisting of n inspection cycles, where T_j is the length of the jth cycle, $j = 1, 2, ..., n$;

$E(T_j, n, H)$ equivalent uniform worth per unit of time of $P(T_j, n, H)$.

1. The cash flow profiles (profit, inspection cost , and repair cost) are identical in successive inspection intervals .

2. The inspection and repair times are negligible (very small compared to the length of the inspection interval).

3. The inspection is error-free.

4. The elapsed time of a shift from the in-control state to the out-of-control state, t, is exponentially distributed, i.e., the probability density function of t is $f(t) = \lambda e^{-\lambda t}$, where λ is the failure rate.

5. Inspection and necessary repair costs occur at the end of each inspection cycle.

6. Profit and all costs are discounted continuously over time.

Figure 10.1 shows the cash flow structures for the two possible cycles of operation depending on the state of the machine at the inspection time.

For this inspection interval of length T, the discounted values of the different components of the above cash flows are:

Discounted inspection cost $= C_i\, e^{-rT}$

Discounted expected repair cost $= C_r \int_0^T f(t)\, dt\; e^{-rT}$

Discounted expected $=$ Discounted expected profit over a good cycle

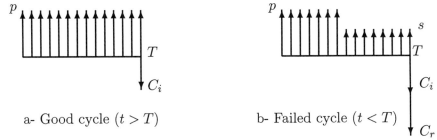

a- Good cycle $(t > T)$ b- Failed cycle $(t < T)$

Figure 10.1   Cash flow structure over one cycle

profit + Discounted expected profit over a failed cycle

$$= \int_T^\infty \left(\int_0^T pe^{-rx} dx \right) f(t) dt$$

$$+ \int_0^T \left(\int_0^t pe^{-rx} dx + \int_t^T se^{-rx} dx \right) f(t) dt$$

After some simplification, the discounted expected profit can be written as

$$p\frac{1 - e^{-rT}}{r} - (p - s) \int_0^T \frac{e^{-rx} - e^{-rT}}{r} f(x) dx$$

Then, the present worth of the net profit over an inspection interval of length T is

$$P(T) = p\frac{1 - e^{-rT}}{r} - \int_0^T \left((p - s)\frac{e^{-rx} - e^{-rT}}{r} + C_r \right) f(x) dx - C_i e^{-rT}$$

$$(10.1)$$

The expression given above developed for the present worth is valid for any inspection cycle of the planning horizon because of the nonrecurring property of the exponential distribution . It is also valid for general time to failure distributions when a perfect maintenance policy is followed instead of performing an inspection just at the end of each cycle. This perfect preventive maintenance will return the machine to as-good-as-new condition when it is found in the in-control state. However, this assumption has been criticized in the maintenance literature as being misleading. In fact, the failure behavior of the machine after preventive maintenance is expected to be improved, but certainly will not be that of a new machine.

For an exponential time to failure distribution, the present worth of the net profit over an inspection interval of length T is obtained from (10.1) by substituting $f(x) = \lambda e^{-\lambda x}$. Doing so, we get

$$P(T) = a_0 - a_1 e^{-rT} - a_2 e^{-(r+\lambda)T} \tag{10.2}$$

where

$$a_0 = \frac{s}{r} + \frac{p-s}{r+\lambda},$$

$$a_1 = \frac{s}{r} + C_i + C_r,$$

$$a_2 = \frac{p-s}{r+\lambda} - C_r.$$

The equivalent uniform worth of the net profit generated during an inspection interval of length T is given by (see White *et al.* [20] for the continuous compounding uniform worth factor of continuous cash flows.)

$$E(T) = \frac{rP(T)}{1 - e^{-rT}} \tag{10.3}$$

3.   INFINITE HORIZON MODEL

In this section, we develop a discounted profit model for the single machine inspection problem under an infinite planning horizon . The present worth of the net profit generated during the first cycle is $P(T)$. Moreover, the present worth of the net profit obtained in the jth cycle is equal to $P(T)$ discounted to time zero. That is, the discounted net profit of the jth cycle is $e^{-jrT}P(T)$. Therefore, the present worth of the total net profit over an infinite planning horizon is

$$P(T, \infty) = \sum_{j=0}^{\infty} e^{-jrT} P(T) = \frac{P(T)}{1 - e^{-rT}} \tag{10.4}$$

For an infinite planning horizon , the equivalent uniform worth of the net profit is

$$E(T, \infty) = rP(T, \infty) = \frac{rP(T)}{1 - e^{-rT}}, \tag{10.5}$$

which is equal to $E(T)$.

The optimal inspection interval , T_∞, is the one that minimizes $P(T, \infty)$ (or $E(T, \infty)$) and is obtained by solving $\frac{dP(T, \infty)}{dT} = 0$. After some simplification, the first derivative of $P(T, \infty)$ is given by

$$\frac{dP(T, \infty)}{dT} = \frac{e^{-rT}}{(1 - e^{-rT})^2} g_\infty(T) \tag{10.6}$$

where

$$g_\infty(T) = a_2(\lambda + r)e^{-\lambda T} - \lambda a_2 e^{-(\lambda + r)T} - (a_0 - a_1)r \tag{10.7}$$

From (10.6), it is clear that T_∞ is the solution to $g_\infty(T) = 0$. In the next proposition, we derive a necessary and sufficient condition for existence of a unique inspection interval maximizing $P(T, \infty)$.

Proposition 1. A finite and unique optimal inspection interval , T_∞, that maximizes $P(T, \infty)$ exists if and only if

$$\frac{p - s}{r + \lambda} - C_i - C_r > 0 \tag{10.8}$$

Proof : First, from (10.7), we have

$$g_\infty(0) = rC_i > 0 \tag{10.9}$$

$$\frac{dg_\infty(T)}{dT} = -\lambda a_2(\lambda + r)e^{-\lambda T}(1 - e^{-rT}) < 0 \tag{10.10}$$

and

$$g_\infty(T \to \infty) = -r(\frac{p - s}{r + \lambda} - C_i - C_r) < 0 \tag{10.11}$$

Moreover,

$$\frac{d^2 P(T,\infty)}{d^2 T}\Big|_{T=T_\infty} = \frac{e^{-rT_\infty}}{(1 - e^{-rT_\infty})^2} \cdot \frac{dg_\infty(T)}{dT}\Big|_{T=\infty} < 0 \qquad (10.12)$$

This last result implies that T_∞ is a maximizer of $P(T,\infty)$, and (10.9)–(10.11) collectively show that T_∞ is the unique finite inspection interval maximizing the present worth of the net profit over an infinite planning horizon.

Note that since $g_\infty(T) > 0$ for any $T < T_\infty$, then

$$\frac{d^2 P(T,\infty)}{d^2 T} = \frac{e^{-rT}}{(1 - e^{-rT})^2}\left[\frac{dg_\infty(T)}{dT} - g_\infty(T)\frac{1 + e^{-rT}}{1 - e^{-rT}}\right] < 0 \ \ \forall T < T_\infty.$$

Therefore, $P(T,\infty)$ is concave in T from $T = 0$ up to $T = \infty$. Moreover, $P(T,\infty)$ is strictly decreasing from $P(T_\infty,\infty)$ at $T = T_\infty$ to a_0 for large T since $g_\infty(T) < 0$ for $T > T_\infty$. Consequently, $P(T,\infty)$ exhibits a quasi-concave form as shown in Figure 10.2.

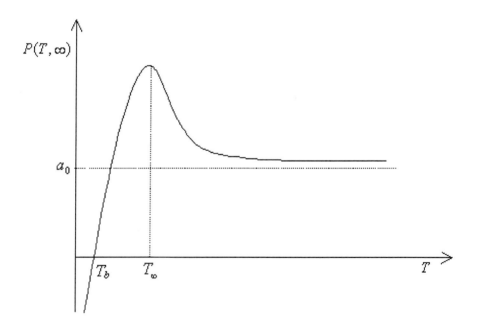

Figure 10.2   Graphical form of $P(T,\infty)$.

¿From this Figure, note that $P(T,\infty)$ is positive when $T > T_b$, where T_b is the break-even inspection interval and is the solution to $P(T,\infty) =$

0, or equivalently to P(T) = 0. After factoring out the term (ra_2) in the first order condition $g_\infty(T) = 0$, we obtain the following expression:

$$e^{-\lambda T}\left(1 + \lambda\frac{1 - e^{-rT}}{r}\right) - \left(1 - \frac{C_i}{a_2}\right) = 0 \qquad (10.13)$$

¿From this equation, it is clear that the optimal inspection interval for non-discounted case $(r \to 0)$ is the solution to

$$e^{-\lambda T}(1 + \lambda T) - \left(1 - \frac{C_i}{\frac{p-s}{\lambda} - C_r}\right) = 0, \qquad (10.14)$$

which is equivalent to the ones developed in Hariga [12] with $s = 0$ and Ben-Daya and Hariga [5] when the inspection and repair times are negligible. Moreover, the net profit per unit of time for the non-discounted case, $Z(T)$, can be obtained from (10.5) by letting r go to zero. Doing so, we obtain the following expression for $Z(T)$

$$Z(T) = \frac{sT - (\frac{p-s}{\lambda} - C_r)e^{-\lambda T} + (\frac{p-s}{\lambda} - C_r - C_i)}{T} \qquad (10.15)$$

In the next proposition, we show that the discounted model has a larger optimal inspection interval than the non-discounted model.

Proposition 2. Let $T_\infty(0)$ be the optimal inspection interval for non-discounted case $(r = 0)$. Then, $T_\infty > T_\infty(0)$.

Proof :
Equation (10.13) defines T_∞ as an implicit function of the discount rate, r. Therefore, after taking the total differential of the left-hand side of (10.13), it is not difficult to prove that

$$\frac{dT_\infty}{dr} = -\frac{\lambda e^{-(r+\lambda)T_\infty} + C_i(p-s)/a_2^2(r+\lambda)^2}{\lambda(r+\lambda)e^{-\lambda T}(e^{-rT} - 1)/r} > 0 \qquad (10.16)$$

where the numerator of the right-hand side of (10.16) is the partial derivative of (10.13) with respect to r and the denominator is the derivative of (10.13) with respect to T. The proof is complete since (10.16) implies that $T_\infty > T_\infty(0)$.

Assuming that the condition in (10.8) is satisfied, we next develop an iterative procedure for finding the optimal inspection interval, T_∞. Substituting the first three terms of the Maclaurin series of $e^{-\lambda T}$ and

$e^{-(\lambda+r)T}$ in (10.13), we obtain the following inspection interval as a starting solution for the iterative procedure.

$$T_0 = \sqrt{\frac{2C_i}{a_2\lambda(\lambda+r)}}$$

The iterative procedure is outlined next in algorithmic form.

Step 0.   Let $T^{(1)} = T_0$ and $k = 1$.
Step 1.   Compute

$$T^{(k+1)} = \frac{1}{\lambda}Ln\left[\frac{1+\lambda\frac{1-e^{-rT^{(k)}}}{r}}{1-\frac{C_i}{a_2}}\right] \tag{10.17}$$

Step 2.   If $\left|T^{(k+1)} - T^{(k)}\right| \le \varepsilon$, then let $T_\infty = T^{(k)}$ and stop (ε is a small number).
Step 3.   If $\left|T^{(k+1)} - T^{(k)}\right| > \varepsilon$, then $k \leftarrow k+1$ and go to Step 1.

The expression given in (10.17) is derived from (10.13). Our experience with this simple iterative procedure revealed that only a few iterations are required before the convergence to the optimal inspection interval . In order to illustrate this iterative procedure, the following example is solved.

Numerical Example
    A machine has a mean time to shift to the out-of-control state equal to 100 weeks, so that $\lambda = 0.01$. Suppose the profit per week during the in-control state is $1000 and the weekly profit under undetected failure is $500 ($p = 1000$ and $s = 500$). The cost of repair is $C_r = \$3000$ and the cost of each inspection is $C_i = \$100$. The weekly discount rate $r = 0.005$. Following the proposed procedure, we start with an inspection interval of 6.629 weeks as an approximate inspection interval . Then, steps 1 to 3 are used repeatedly until convergence to 6.82 weeks as an optimal inspection interval. The present worth of the net profit for the policy of inspecting the machine every 6.82 weeks over an infinite planning horizon is $188100.81. The equivalent uniform weekly worth is $940.504. For the non-discounted case ($r = 0$), the optimal inspection interval is 6.67 weeks and the weekly net profit is $939.7.

Sensitivity Analysis
    Using the above numerical example, a sensitivity analysis is carried out to examine the discounting effects on the inspection policy. The

s/p	C_i/C_r	r			
		0.001	0.005	0.01	0.014
0	0.0333	1.0024	1.0119	1.0241	1.0342
	0.1667	1.0038	1.0181	1.0362	1.0507
	0.3333	1.0040	1.0212	1.0444	1.0632
	0.5	1.0048	1.0251	1.0518	1.0743
0.2	0.0333	1.0029	1.0147	1.0297	1.0421
	0.1667	1.0042	1.0211	1.0430	1.0611
	0.3333	1.0053	1.0258	1.0534	1.0765
	0.5	1.0057	1.0298	1.0619	1.0893
0.4	0.0333	1.0037	1.0190	1.0389	1.0556
	0.1667	1.0052	1.0265	1.0548	1.0787
	0.3333	1.0064	1.0328	1.0681	1.0983
	0.5	1.0073	1.0380	1.0793	1.1150
0.6	0.0333	1.0055	1.0282	1.0585	1.0845
	0.1667	1.0074	1.0383	1.0803	1.1169
	0.3333	1.0090	1.0469	1.0992	1.1456
	0.5	1.0103	1.0543	1.1156	1.1709
0.8	0.0333	1.0115	1.0595	1.1306	1.1969
	0.1667	1.0147	1.0790	1.1754	1.2692
	0.3333	1.0177	1.0965	1.2184	1.3423
	0.5	1.0203	1.1123	1.2596	1.4162

Table 10.1 Effect of discounting on the inspection interval for (λ=0.01).

values of p and C_r are kept equal to 1000 and 3000 respectively, whereas each of the remaining parameters is changed one at a time during this analysis. For each combination of the parameter values, the optimal inspection interval and maximum uniform weekly net profit of the discounted policy are compared with those of the non-discounted one. The results of these numerical comparisons are reported in Tables 10.1–10.6 below. Tables 10.1–10.3 show the effects of the various parameters on the inspection interval ratio, $T_\infty/T_\infty(0)$, whereas Tables 10.4–10.6 report the weekly net profit ratio, $A(T, \infty)/Z(T)$, for the different combinations of the parameter values.

Tables 10.1–10.6 show that the discounting rate, r, is not the only factor that makes the discounted and non-discounted inspection policies different. Tables 10.1–10.3 indicate that the discounted effects on the inspection interval ratio are more pronounced for large values of the different parameters. The same tables show that the T ratio increases as the ratios s/p and C_i/C_r increase. Likewise, the T ratio has positive rela-

s/p	C_i/C_r	r 0.001	0.005	0.01	0.014
	0.0333	1.0021	1.0111	1.0225	1.0319
	0.1667	1.0030	1.0152	1.0311	1.0443
0	0.3333	1.0036	1.0153	1.0186	1.0161
	0.5	1.0043	1.0217	1.0441	1.0633
	0.0333	1.0029	1.0138	1.0284	1.0401
	0.1667	1.0037	1.0187	1.0382	1.0546
0.2	0.3333	1.0040	1.0222	1.0467	1.0668
	0.5	1.0051	1.0259	1.0538	1.0776
	0.0333	1.0036	1.0185	1.0380	1.0543
	0.1667	1.0048	1.0246	1.0508	1.0729
0.4	0.3333	1.0058	1.0299	1.0620	1.0894
	0.5	1.0067	1.0344	1.0717	1.1039
	0.0333	1.0056	1.0288	1.0599	1.0867
	0.1667	1.0073	1.0376	1.0789	1.1150
0.6	0.3333	1.0087	1.0455	1.0963	1.1414
	0.5	1.0100	1.0526	1.1123	1.1661
	0.0333	1.0132	1.0711	1.1586	1.2434
	0.1667	1.0170	1.0927	1.2106	1.3323
0.8	0.3333	1.0205	1.1150	1.2699	1.4409
	0.5	1.0243	1.1389	1.3387	1.5811

Table 10.2   Effect of discounting on the inspection interval for (λ=0.02).

s/p	C_i/C_r	r			
		0.001	0.005	0.01	0.014
	0.0333	1.0022	1.0108	1.0222	1.0315
	0.1667	1.0029	1.0147	1.0300	1.0426
0	0.3333	1.0034	1.0178	1.0364	1.0518
	0.5	1.0042	1.0204	1.0421	1.0601
	0.0333	1.0028	1.0139	1.0283	1.0402
	0.1667	1.0035	1.0183	1.0374	1.0533
0.2	0.3333	1.0037	1.0223	1.0451	1.0646
	0.5	1.0052	1.0251	1.0524	1.0752
	0.0333	1.0037	1.0190	1.0389	1.0556
	0.1667	1.0048	1.0247	1.0509	1.0731
0.4	0.3333	1.0058	1.0298	1.0618	1.0892
	0.5	1.0067	1.0344	1.0717	1.1039
	0.0333	1.0060	1.0309	1.0645	1.0936
	0.1667	1.0076	1.0398	1.0838	1.1225
0.6	0.3333	1.0092	1.0483	1.1027	1.1515
	0.5	1.0107	1.0564	1.1212	1.1805
	0.0333	1.0168	1.0925	1.2133	1.3419
	0.1667	1.0219	1.1241	1.2986	1.5022
0.8	0.3333	1.0284	1.1660	1.4291	1.8054
	0.5	1.0361	1.2250	1.6778	3.2706

Table 10.3   Effect of discounting on the inspection interval for (λ=0.03).

s/p	C_i/C_r	r			
		0.001	0.005	0.01	0.014
0	0.0333	1.00011	1.00076	1.00140	1.00203
	0.1667	1.00057	1.00275	1.00550	1.00776
	0.3333	1.00108	1.00528	1.01055	1.01470
	0.5	1.00162	1.00783	1.01566	1.02178
0.2	0.0333	1.00012	1.00076	1.00151	1.00217
	0.1667	1.00057	1.00294	1.00577	1.00801
	0.3333	1.00106	1.00530	1.01060	1.01487
	0.5	1.00158	1.00790	1.01567	1.02184
0.4	0.0333	1.00017	1.00083	1.00167	1.00235
	0.1667	1.00060	1.00298	1.00595	1.00831
	0.3333	1.00110	1.00548	1.01092	1.01523
	0.5	1.00160	1.00798	1.01585	1.02207
0.6	0.0333	1.00019	1.00094	1.00190	1.00267
	0.1667	1.00064	1.00320	1.00640	1.00896
	0.3333	1.00115	1.00575	1.01145	1.01598
	0.5	1.00165	1.00824	1.01636	1.02277
0.8	0.0333	1.00021	1.00126	1.00252	1.00353
	0.1667	1.00075	1.00376	1.00763	1.01073
	0.3333	1.00131	1.00656	1.01311	1.01832
	0.5	1.00177	1.00907	1.01815	1.02526

Table 10.4   Effect of discounting on the equivalent uniform worth (λ=0.01).

s/p	C_i/C_r	r			
		0.001	0.005	0.01	0.014
	0.0333	1.00011	1.00091	1.00182	1.00258
	0.1667	1.00062	1.00335	1.00682	1.00957
0	0.3333	1.00133	1.00651	1.01302	1.01815
	0.5	1.00196	1.00982	1.01950	1.02715
	0.0333	1.00011	1.00102	1.00192	1.00278
	0.1667	1.00073	1.00353	1.00718	1.00996
0.2	0.3333	1.00129	1.00658	1.01316	1.01834
	0.5	1.00203	1.00974	1.01947	1.02711
	0.0333	1.00022	1.00110	1.00221	1.00311
	0.1667	1.00075	1.00373	1.00746	1.01046
0.4	0.3333	1.00136	1.00677	1.01352	1.01891
	0.5	1.00197	1.00982	1.01959	1.02736
	0.0333	1.00024	1.00128	1.00258	1.00363
	0.1667	1.00081	1.00409	1.00820	1.01151
0.6	0.3333	1.00143	1.00719	1.01439	1.02015
	0.5	1.00204	1.01022	1.02041	1.02853
	0.0333	1.00033	1.00185	1.00371	1.00524
	0.1667	1.00101	1.00517	1.01056	1.01498
0.8	0.3333	1.00172	1.00860	1.01743	1.02469
	0.5	1.00244	1.01187	1.02398	1.03377

Table 10.5 Effect of discounting on the equivalent uniform worth (λ=0.02).

s/p	C_i/C_r	\multicolumn{4}{c}{r}
		0.001	0.005	0.01	0.014
	0.0333	1.00024	1.00119	1.00227	1.00319
	0.1667	1.00080	1.00400	1.00800	1.01131
0	0.3333	1.00146	1.00772	1.01529	1.02151
	0.5	1.00235	1.01173	1.02331	1.03249
	0.0333	1.00024	1.00130	1.00248	1.00347
	0.1667	1.00091	1.00429	1.00845	1.01183
0.2	0.3333	1.00154	1.00770	1.01555	1.02174
	0.5	1.00238	1.01160	1.02304	1.03218
	0.0333	1.00027	1.00135	1.00283	1.00381
	0.1667	1.00089	1.00444	1.00890	1.01248
0.4	0.3333	1.00160	1.00801	1.01602	1.02244
	0.5	1.00232	1.01161	1.02320	1.03245
	0.0333	1.00031	1.00160	1.00324	1.00457
	0.1667	1.00098	1.00495	1.00995	1.01399
0.6	0.3333	1.00171	1.00860	1.01725	1.02420
	0.5	1.00243	1.01215	1.02434	1.03410
	0.0333	1.00056	1.00249	1.00508	1.00730
	0.1667	1.00128	1.00676	1.01387	1.01991
0.8	0.3333	1.00226	1.01108	1.02263	1.03243
	0.5	1.00290	1.01499	1.03083	1.04424

Table 10.6 Effect of discounting on the equivalent uniform worth ($\lambda=0.03$).

tionships with the discounting and failure rates. From Tables 10.4–10.6, it can be observed that the discounted effects on the weekly net profit are minimal since a maximum of 4.44% for the ratio $A\left(T,\infty\right)/Z\left(T\right)$ is obtained over all the examples solved.

4.   FINITE HORIZON MODEL WITH CONTINUOUS INSPECTION TIMES

The assumption of an infinite horizon may not be realistic when the manufacturing technology that is related to the machine currently in use is undergoing rapid development. In this sort of situation, the machine may have to be replaced after a certain period of time with a technologically improved one having for example, a lower operating cost, a larger throughput, and/or a better quality of output. In this case, considering a finite-planning horizon is more appropriate when developing a scheduling inspection model for such a machine.

Suppose that n inspections are being carried out during a finite time span of length H. The timing of the jth inspection is t_j, $j = 1, 2, ..., n$. Using the results obtained in the second section, the expected net profit derived from operating the machine during the jth inspection interval $(j = 1, 2, ..., n)$ and discounted as to time t_j is simply $P\left(T_j\right)$. Therefore, the present worth of the total net profit over a planning horizon H is:

$$P\left(T_j, n, H\right) = \sum_{j=1}^{n} e^{-r t_{j-1}} P\left(T_j\right) \qquad (10.18)$$

and the uniform worth per unit of time is:

$$E\left(T, n, H\right) = \frac{r\,P\left(T_j, n, H\right)}{1 - e^{-rH}} \qquad (10.19)$$

The optimization problem to be solved in order to determine an optimal inspection schedule over a period of length H is:

$$Max \quad P\left(T_j, n, H\right) = \sum_{j=1}^{n} e^{-r t_{j-1}} P\left(T_j\right) \qquad (10.20)$$

Subject to

$$t_j = \sum_{k=1}^{j} T_k \quad , j = 1, 2, ..., n - 1 \qquad (10.21)$$

$$t_n = \sum_{k=1}^{n} T_k = H \tag{10.22}$$

$$t_0 = 0, \quad T_j \geq 0, j = 1, 2, ..., n$$

After substituting t_j given in (10.21) $t_0 = 0$ and $T_n = H - \sum_{k=1}^{n-1} T_k$ into (10.20) and ignoring the non-negativity constraints, the optimization problem is transformed to the following unconstrained mixed-integer non-linear program with $(n - 1)$ continuous decision variables $(T_j, j = 1, 2, ..., n - 1)$ and one discrete variable, n:

$$Max \quad P(T_j, n, H) = \sum_{j=1}^{n-1} e^{-r \sum_{k=1}^{j-1} T_k} P(T_j) + e^{-r \sum_{k=1}^{n-1} T_k} P\left(H - \sum_{k=1}^{n-1} T_k\right). \tag{10.23}$$

The approach for maximizing this problem is to follow a two-stage optimization procedure. In the first stage, the problem is maximized over the $T_j's$ for a given number of inspections, n. Then, in the second stage the problem is maximized over n.

By partially differentiating (10.23) with respect to T_j $(j = 1, 2, ..., n - 1)$ and manipulating the resulting equations, we obtain the following necessary conditions for optimality

$$e^{-rT_j} \left[a_2 (r + \lambda) e^{-\lambda T_j} - \lambda a_2 e^{-(r+\lambda)T_{j+1}} - r(a_0 - a_1) \right] = 0, j = 1, .., n-1. \tag{10.24}$$

Note that since $e^{-rT_j} > 0$ and using (10.7), the above equation in (10.24) can be rewritten as

$$(r + \lambda) \left[e^{-\lambda T_j} - e^{-\lambda T_\infty} \right] = \lambda \left[e^{-(r+\lambda)T_{j+1}} - e^{-(r+\lambda)T_\infty} \right], j = 1, ..., n - 1, \tag{10.25}$$

where T_∞ is the optimal inspection interval for an infinite planning horizon . Moreover, observe that the equations in (10.25) reduce the $(n-1)$ dimensional problem of determining an optimal inspection schedule to a one-dimensional problem. In fact it is easy to see that once T_1 is fixed, all other T_j's can be determined consecutively using the following relationship obtained from equation (10.25)

$$T_{j+1} = T_\infty - \frac{1}{r+\lambda} Ln \left[1 - \frac{r+\lambda}{\lambda} e^{rT_\infty} (1 - e^{-\lambda(T_j - T_\infty)}) \right], j = 2, ..., n-1.$$

$$(10.26)$$

If $\sum_{j=1}^{n} T_j = H$, then T_1 has been selected correctly. Therefore, we need only to search for the optimal T_1 and then determine the rest of the optimal inspection timings from (10.26). This optimal T_1, can be found by any standard one-dimensional search technique such as the bisection method as suggested in the algorithm proposed below. Before presenting this algorithm, we will next discuss some useful properties of the optimal inspection schedule.

Property 1. As $r \to 0$ (non-discounted case), the optimal inspection intervals are equal; that is, $T_j^* = \frac{H}{n}, j = 1, 2, ..., n$.

Proof : Letting $r \to 0$ in (10.25) we obtain $T_j = T_{j+1}T$,$j = 1, 2, ...$, $n-1$. Therefore, since the $T_j's$ must sum to H, we get $T_j = \frac{H}{n}$, $j = 1, 2, ..$., n.

Property 2. The optimum present worth (uniform worth) of the finite horizon model is bounded from above by the optimum present worth (uniform worth) of the infinite horizon model; i.e.,

$$P(T_j, n, H) \le P(T, \infty) \qquad (E(T_j, n, H) \le E(T, \infty)) \qquad (10.27)$$

Proof : Suppose that the optimal inspection policy for the finite horizon problem is repeated indefinitely every H units of time. The present worth of this new inspection policy is given by $\sum_{k=0}^{\infty} e^{-krH} P(T_j, n, H)$ which is larger or equal to $P(T_j, n, H)$. Moreover, by the definition of the optimal inspection policy for the infinite horizon problem, we have

$$P(T, \infty) \ge \sum_{k=0}^{\infty} e^{-irH} P(T, n, H).$$

Therefore,

$$P(T, \infty) \ge P(T_j, n, H).$$

The proof for the uniform worth can be carried out in a similar manner.

Property 3. The bound in (10.27) for the uniform worth is tight when H is a multiple integer of the optimal inspection interval of the infinite horizon problem.

Proof : Suppose $H = nT_\infty$. Then, let $T_j = T_\infty$, $j = 1, 2, ..., n$. In this case

$$P(T_j, n, H) = \frac{1 - e^{-rH}}{1 - e^{-rT_\infty}} P(T_\infty)$$

and

$$E(T_j, n, H) = \frac{r P(T_j, n, H)}{1 - e^{-rH}} = \frac{rP(T_\infty)}{1 - e^{-rT_\infty}} = E(T, \infty).$$

As a result of Property 3, the optimal inspection policy for the finite horizon problem when $H = nT_\infty$ is to inspect every T_∞ units of time. We next present more properties for the optimal inspection policy when the finite horizon is not an integer multiple of T_∞. These properties will be useful for the search of the optimal inspection schedule as outlined in the algorithm below. Let $n_l = \left\lfloor \frac{H}{T_\infty} \right\rfloor$ and $n^u = \left\lceil \frac{H}{T_\infty} \right\rceil$, where $\lfloor x \rfloor$ is the largest integer smaller than x and $\lceil x \rceil$ is the smallest integer larger than x.

Property 4. If $T_1 > T_\infty$, then

1. $T_\infty < T_1 < T_2 < ... < T_n$

2. $n \le n_l$

Proof : If $T_1 > T_\infty$, then using (10.25) we also have $T_j > T_\infty, j = 1, 2, ..., n$. Next, since $T_j > T_\infty$ for all $j's$, (10.25) can be modified as

$$\int_0^{T_j - T_\infty} e^{-\lambda x} dx = e^{-rT_\infty} \int_0^{T_{j+1} - T_\infty} e^{-(r+\lambda)x} dx.$$

Moreover, given that $e^{-\lambda x} > e^{-\lambda x} e^{-r(T_\infty + x)}$ and using the last equality we have

$$\int_0^{T_{j+1} - T_\infty} e^{-\lambda x} dx > \int_0^{T_j - T_\infty} e^{-\lambda x} e^{-r(T_\infty + x)} dx = \int_0^{T_j - T_\infty} e^{-\lambda x} dx$$

Therefore, $T_{j+1} - T_\infty > T_j - T_\infty$ implying that $T_{j+1} > T_j$.

2. Using the result of the first part, i.e., $T_j > T_\infty \quad \forall j$, we have

$$H = \sum_{j=1}^{n} T_j > nT_\infty$$

thus, $n < \frac{H}{T_\infty}$ or $n \le n_l$ since n must be an integer.

Property 5. If $T_1 < T_\infty$, then

1. $T_\infty > T_1 > T_2 > ... > T_n$
2. $n \le n^u$

Proof : The proof is similar to the one carried out in Property 4.

The next property provides an interval of search for the optimal T_1. Recall that once T_1 is found, all subsequent $T_j's$ can be determined from (10.26).

Property 6.

1. For $n \le n_l$, $T_\infty < T_1 < \frac{H}{n}$
2. For $n \ge n^u$, $\frac{H}{n} < T_1 < T_\infty$

Proof : Let T_j, $j = 1, 2, ..., n$ be a feasible inspection schedule ($\sum_{j=1}^{n} T_j = H$). According to Property 4, we have $T_\infty < T_1$ for $n \le n_l$. Next suppose that $T_1 > \frac{H}{n}$. Then, from (10.25) we also have $T_j > \frac{H}{n}$ for $j = 1, 2, ..., n$, leading to $\sum_{j=1}^{n} T_j > H$ and, consequently, to a contradiction of the feasibility assumption for the $T_j's$.

2. Similar to part (1).

Now let

$$SUM (T_1) = T_1 + T_2 (T_1) + ... + T_n (T_1) - H, \qquad (10.28)$$

where $T_j (T_1)$ is determined from (10.26) for a given T_1. Using (10.25) it is easy to prove that $SUM (T_1)$ is an increasing function of T_1. In the next proposition, we will show the existence of a unique solution to $SUM (T_1) = 0$ for any number of inspections carried out during the planning horizon H.

Proposition 4. For both cases of $n \le n_l$ and $n \ge n^u$ the equation $SUM(T_1) = 0$ has a unique solution.

Proof :

When $T_1 = T_\infty$, we get from (10.25) $T_j = T_\infty \; \forall j$.

1. In the case $n \le n_l$ we have $T_\infty < T_1 < \frac{H}{n}$. From (10.25) when $T_1 = T_\infty$,

we get $T_j = T_\infty \quad \forall j$. Therefore, $SUM(T_\infty) = nT_\infty - H = (n - \frac{H}{T_\infty})T_\infty <$ 0 since $n \leq n_l$. Moreover, $T_n\left(\frac{H}{n}\right) > T_{n-1}\left(\frac{H}{n}\right) > ... > T_2\left(\frac{H}{n}\right) > T_1 = \frac{H}{n}$. Therefore, $SUM\left(\frac{H}{n}\right) > 0$. Consequently, $SUM(T_1) = 0$ has a unique solution since $SUM(T_1)$ is strictly increasing from a negative value to a positive one.

2. The proof for the case $n \geq n^u$ is similar to the first part.

Using the above results, we next propose the following algorithm to find the optimal inspection schedule for a given n.

Step 0.   If $n \leq n_l$, let $L = T_\infty$ and $U = \frac{H}{n}$.
           If $n \geq n^u$, let $L = \frac{H}{n}$ and $U = T_\infty$.
Step 1. $T_1 = \frac{L+U}{2}$.
Step 2.   Compute T_j using (10.26) for $j = 2, 3, ..., n$.
Step 3.   If $|SUM(T_1)| < \epsilon$ (a very small number), stop; otherwise go to Step 4.
Step 4.   If $SUM(T_1) > 0$, set $U = T_1$ and go to step 1.
Step 5.   If $SUM(T_1) < 0$, set $L = T_1$ and go to step 1.

Having developed an algorithm to find an optimal inspection schedule for a given n, we next discuss the second optimization stage of searching for the optimal number of inspections. In the following proposition we show that the search is limited to the values n_l and n^u.

Proposition 5. The optimal number of inspections is n_l or n^u; i.e. $n^* \in \{n_l, n^u\}$.

Proof : Let $\{T_j, j = 1, 2, ..., n\}$ be an optimum inspection schedule with $n < n_l$. We will construct a feasible schedule with an extra inspection and having larger present worth which contradicts the optimality of $\{T_j, j = 1, 2, ..., n\}$. Since $n < n_l$ and $T_j > T_\infty$ for $j = 1, 2, ..., n$ (see Property 4), there exists an integer m such that

$$\sum_{j=k}^{n}(T_j - T_\infty) < T_\infty \quad k = m+1, ..., n,$$

and

$$\sum_{j=m}^{n}(T_j - T_\infty) \geq T_\infty.$$

Now, let $\left\{T_j', j = 1, 2, ..., n+1\right\}$ be a feasible schedule defined as follows,

$$T'_j = T_\infty \quad k = m+1, ..., n+1$$

$$T'_m = \sum_{j=m}^{n} (T_j - T_\infty)$$

$$T'_j = T_j \quad j = 1, 2, ..., m-1$$

It is easy to verify that this new schedule is feasible and $T_\infty \leq T'_j \leq T_j$ for $j = 1, 2, ..., n$. Next let $E_j(T_j)$ $(E'_j(T'_j))$ be the uniform worth during an inspection interval of length T_j (T'_j) of the initial(new) schedule (new schedule). Using the quasi concavity shape of the uniform worth function (See Figure 10.1 and recall that $E_j(T_j) = \frac{rP(T_j)}{1-e^{-rT_j}}$), it is not hard to prove that

$$E'_j\left(T'_j\right) > E_j\left(T_j\right) \quad j = m+1, ..., n$$

since $T_\infty \leq T'_j \leq T_j$. Consequently, the new inspection schedule has a larger present worth than the original one. The proof for the case $n > n^u$ is similar.

Note that a similar proof was used by Rachamadugu and Ramasesh [16] to show the sub-optimality of equal lot sizes for the inventory problem with a finite horizon. As a result of Proposition 5, the optimal inspection policy can be found by evaluating the present worth for n equal to n_l and n^u and selecting the one with the larger present worth .

To illustrate the above analysis, we consider the same example as the previous section with $H=25$ weeks. For this example, recall that the optimal inspection interval for the infinite horizon problem, T_∞, was 6.82 weeks. Therefore, $n_l = 3$ and $n^u = 4$. The corresponding inspection schedules and present worth values are shown in Table 10.7 below. It is clear from this table that the optimal inspection policy is to perform 4 inspections over the planning horizon of length 25 weeks.

5.   FINITE HORIZON MODEL WITH DISCRETE INSPECTION TIME

In the previous section, we treated the inspection times, $T_j(j, 1, 2, ..., n)$, as continuous variables. We now assume that inspections are carried out at discrete points in time. Although this assumption seems to be restrictive, there are managerial advantages associated with the discretization

n	3			4			
$P(T_j, n, H)$	\$22088.51			\$22099.91			
$A(T_j, n, H)$	\$939.91			\$940.397			
T_j	8.276	8.332	8.392	6.278	6.260	6.241	6.221

Table 10.7 Results for the finite horizon example.

of the inspection times. In fact, from an operational point of view, it is difficult to implement a schedule that calls for inspections at non-integer times. In practice, inspections are performed every day, week or month. Because of the discrete nature of the inspection times we will formulate the single machine inspection problem using the dynamic programming approach. This approach is simple to use and provides ready made sensitivity analysis for the length of the planning horizon.

Consider the $[j, k]$ inspection interval during which two consecutive inspections are carried out at time j and k $(j < k)$, respectively. Note that we assume that the machine is not inspected at time $j = 0$. Let $P(j, k)$ be the present worth of the net profit derived over the interval $[j, k]$ discounted as to time zero. Then,

$$P(j, k) = e^{-rj} P(k - j) \qquad (10.29)$$

where $P(k - j)$ is given by (10.2) for $T = k - j$.

Moreover, let $M(k)$ be the maximum present worth of the net profit generated over a planning horizon of length k. Then the following recursive equation can be used in a forward fashion to determine the maximum present worth over the time horizon H.

$$M(k) = \max_{0 \le j < k \le H} \{M(j) + P(j, k)\}, \qquad (10.30)$$

with $M(0) = 0$. At the termination of the dynamic programming procedure, $M(H)$ is the maximum present worth of the net profit over the planning horizon H. The optimal inspection schedule can be determined by working backward from time H to time zero.

For the purpose of illustration, we use the dynamic programming procedure to solve the same example as the one used in the previous two sections. The detailed results for this example are reported in Table 10.8.

In the third column of this table, j^* refers to the index where the maximum $M(k)$ is obtained. It can be seen that the maximum present worth for discrete inspection times is \$22098.45 compared to \$22099.91 for the continuous inspection times. The optimal number of inspections

Time horizon, k	$M(k)$	j^*
1	865.83	0
2	1822.36	0
3	2769.71	0
4	3708.0	0
5	4637.34	0
6	5557.83	0
7	6469.58	0
8	7372.70	0
9	8267.29	0
10	9160.18	5
11	10058.11	6
12	10951.40	6
13	11836.25	7
14	12716.64	7
15	13588.70	7
16	14456.31	8
17	15318.68	12
18	16185.57	12
19	17044.31	13
20	17898.73	14
21	18748.84	14
22	19580.90	14
23	20428.67	15
24	21265.04	18
25	22098.45	19

Table 10.8 Results of the dynamic programming procedure for the example problem.

over the planning horizon of 25 weeks is also equal to 4 for the discrete case. The four optimal inspection intervals are of length 7, 6, 6, and 6, respectively. Note that if the planning horizon is changed to 20 weeks, then the corresponding optimal inspection policy can also be obtained from Table 10.8. This policy has a maximum present worth of $17898.73 and performs inspections at times 7, 14, and 20.

6. RANDOM HORIZON MODEL

In this section, we assume that the machine is inspected over a horizon of random duration. The stochastic nature of the horizon may be due to technological obsolescence (McCall [21]) or to uncertainty concerning the mission of the machine [Wells and Bryant [19]). In the inventory context, Moon and Yun [15] studied the economic order quantity model with random horizon using the discounted cash flow approach. Since the machine is being operated over a random horizon, we will develop an optimal periodic inspection policy that calls for inspection every T_r units of time.

Suppose that a given planning horizon H fully accommodates n inspection intervals and terminates during the $(n + 1)th$ interval. The conditional present worth of the net profit over H is

$$\frac{1 - e^{-nrT}}{1 - e^{-rT}} P(T) + e^{-nrT} P(H - nT)$$

The expected present worth is then given by

$$P(T, H) = \sum_{n=0}^{\infty} \int_{nT}^{(n+1)T} \left[\frac{1 - e^{-nrT}}{1 - e^{-rT}} P(T) + e^{-nrT} P(H - nT) \right] g(H) \, dH$$

$$(10.31)$$

where $g(H)$ is the probability density function of the random horizon, which is assumed to be an exponential distribution with mean μ. Substituting $P(\cdot)$ and manipulating, we get

$$
\begin{aligned}
P(T, H) &= \frac{a_0(r + \lambda + \mu) - a_2\mu - a_1(r + \lambda + \mu)e^{-(r+\mu)T}}{(r + \lambda + \mu)(1 - e^{-(r+\mu)T})} \\
&\quad - \frac{a_2(r + \lambda)e^{-(r+\lambda+\mu)T}}{(r + \lambda + \mu)(1 - e^{-(r+\mu)T})} \\
&\quad - \frac{a_1\mu}{r + \mu}
\end{aligned}
\qquad (10.32)
$$

Note that as $\mu \to 0$ (the mean time horizon tends to infinity), $P(T, H)$ converges to $P(T, \infty)$. Letting

$$r' = r + \mu$$

$$
\begin{aligned}
a_0' &= a_0 - \frac{a_2\mu}{(r + \lambda + \mu)} - \frac{a_1\mu}{r + \mu} \\
&= \frac{s}{r + \mu} + \frac{p - s}{r + \lambda + \mu} - \frac{C_r\mu\lambda}{(r + \mu)(r + \lambda + \mu)} - \frac{C_i\mu}{r + \mu}
\end{aligned}
$$

$$a_1' = \frac{a_1 r}{r + \mu}$$

$$a_2' = a_2 \frac{r + \lambda}{(r + \lambda + \mu)}$$

$P(T, H)$ can be rewritten as

$$P(T, H) = \frac{a_0' - a_1' e^{-r'T} - a_2' e^{-(r' + \lambda)T}}{(1 - e^{-r'T})} \qquad (10.33)$$

which is similar to (10.4). Therefore, the same analysis of the third section can be carried out to determine the optimal inspection interval, T_r. In particular, the existence of a unique finite T_r is guaranteed when

$$\begin{aligned}
\left(a_0' - a_1'\right) r' &= \left(a_0 - \frac{a_2 \mu}{r + \lambda + \mu} - a_1\right) \\
&= \frac{\left[(r + \lambda)\left(\frac{p-s}{r+\lambda} - C_i - C_r\right) - \mu C_i\right]}{(r + \lambda + \mu)} > 0
\end{aligned}$$

Moreover the same algorithm of the third section can be used to find the optimal T_r after replacing r with r' and a_2 with a_2'. Note that when $a_0' < 0$, the machine is not worth operating under an uncertain planning horizon. Using the same reasoning as in the proof of Proposition 2, it can be shown that $T_r < T_\infty$.

To illustrate the analysis of this section, we consider the same example with $\mu = 0.04$ (a mean planning horizon of 25 weeks). The optimal inspection interval is 7.15 weeks with an expected present worth of the net profit equal to \$20868.9. Recall that the present worth of the net profit derived over a time horizon of 25 weeks was \$22099.91. Therefore, the difference $P(T_j, n, H) - P(T, H) = \$1231.01$ can be regarded as the maximum cost of randomness. In other words, this difference is the loss due to the lack of certain information about the planning horizon. Table 10.9 below indicates the effect of the mean planning horizon $(1/\mu)$ on the inspection interval and the expected present worth . It can be observed from this table that the inspection interval, T_r, decreases as the mean planning horizon increases. On the other hand, both the expected present profit and the cost of randomness increase as $1/\mu$ increases.

7.   CONCLUSIONS

In this paper, we developed optimal inspection policies for a single machine with the consideration of time value of money by applying the concept of the discounted cash flow analysis. We presented and discussed different models for the case of an infinite planning horizon

$1/\mu$	T_r	$P(T, H)$	$P(T_j, n, H) - P(T, H)$
10	7.71	8921.3	239.05
20	7.24	17067.0	842.69
25	7.15	20868.9	1231.01
50	6.98	37592.7	4014.03
100	6.9	62677.6	11334.1
200	6.86	94033.4	24869.0

Table 10.9 Effect of the mean planning horizon on T_r and $P(T, H)$.

as well as for the situations of deterministic and stochastic finite time span. We solved several numerical examples and carried out sensitivity analyses to illustrate the formulated models and their solution procedures. Possible extension work to the current research is to study the discounted inspection problem for a single machine with a general failure distribution.

References

[1] Abdel-Hameed, M. Inspection, Maintenance and Replacement Models. *Computers & Operations Research*, **22-4** (1995) 435-441.

[2] Alfares, H. A Simulation Model for Determining Inspection Frequency. *Computers & Industrial Engineering*, **36** (1999) 685-696.

[3] Baker, M.J.C. How often should a machine be inspected? *International Journal of Quality and Reliability Management*, **7** (1990) 14-18.

[4] Banerjee, P.K. and Chuiv, N.N. Inspection Policies for Repairable Systems, *IIE Transactions*, **28** (1996) 1003-1010.

[5] Ben-Daya, M. and Hariga, M. A Maintenance Inspection Model : Optimal and Heuristic Solutions. *International Journal of Quality and Reliability Management*, **15** (1998) 481-488.

[6] Chung, K.J. A Note on the Inspection Interval of Machine. *International Journal of Quality and Reliability Management*, **10** (1993) 71-73.

[7] Cox ,D.R. and Smith, W. L. On the Super-Position of Renewal Processes. *Biometrica*, **41** (1954) 91-99.

[8] Davis, D.J. An Analysis of Some Failure Data. *Journal of American Statistical Association*, **47** (1952) 113-150.

[9] Drenick, R.F. The Failure Law of Complex Equipment. *Journal of Society of Industrial Apllied Mathematics*, **8** (1960) 680-690.

[10] Epstein, B. The Exponential Distribution and its Role in Life-Testing. *Industrial Quality Control*, **15** (1958) 2-7.

[11] Goldman, A.S. and Slattery, T.B. *Maintenability: A Major Element of System Effectiveness*. Krieger, Huntington, New York, 1977.

[12] Hariga, M. A Maintenance Inspection Model for a Single Machine with General Failure Distribution. *Journal of Microelectronics and Reliability*, **36** (1995) 353-358.

[13] Jardine, A.K.S. *Maintenance, Replacement and Reliability*. Pitman, London, 1973.

[14] McCall, J. Maintenance Policies for Stochastically Failing Equipment: A Survey. *Management Science*, **11** (1965) 493-524.

[15] Moon, I. and Yun, W. An Economic Order Quantity Model with a Random Planning Horizon. *The Engineering Economist*, **39** (1993) 77-86.

[16] Rachamadugu, R. and Ramasesh, R. Suboptimality of Equal Lot Sizes for Finite Horizon Problems. *Naval Research Logistics*, **41** (1994) 1019-1028.

[17] Vaurio, J.K. A note on optimal inspection intervals , *International Journal of Quality and Reliability Management*, **11** (1994) 65-68.

[18] Wang, W. and Christer, A.H. Modeling Procedure to Optimize Component Safety Inspection Over a Finite Time Horizon. *Quality and Reliability Engineering International*, **13** (1997) 217-224.

[19] Wells, C.E. and Bryant, J.L. Optimal Preventive Maintenance Policies for Systems with Missions of Random Duration. *IIE Transactions*, **17** (1985) 338-345.

[20] White, J.A., Case, K.E., Pratt, D.B., and Agee, M.H. *Principles of Engineering Economic Analysis*. 4th ed., John Wiley & Sons, Inc., New York, 1998.

Chapter 11

A GENERAL APPROACH FOR THE COORDINATION OF MAINTENANCE FREQUENCIES IN CASES WITH A SINGLE SET-UP

R. Dekker, R.E. Wildeman, J.B.G. Frenk and R. Van Egmond
Econometric Institute, Erasmus University Rotterdam, P.O. Box 1738, NL-3000 DR Rotterdam, The Netherlands

Abstract   A maintenance activity carried out on a technical system often involves a system-dependent set-up cost that is the same for all maintenance activities carried out on that system. Grouping activities thus saves costs since execution of a group of activities requires only one set-up. By now, there are already several multi-component maintenance models available in the literature, but most of them suffer from intractability when the number of components grows, unless a special structure is assumed. An approach that can handle many components was introduced in the literature by Goyal et al. However, this approach requires a specific deterioration structure for components. Moreover, the authors present an algorithm that is not optimal and there is no information of how good the obtained solutions are. In this paper we present an approach that solves the model of Goyal et al. to optimality. Furthermore, we extend the approach to deal with more general maintenance models like minimal repair and inspection that can be solved to optimality as well. Even block replacement can be incorporated, in which case our approach is a good heuristic.

Keywords: Maintenance of multicomponent systems, grouping of maintenance frequencies, joint overhaul scheduling problem, block replacement model, inspection model, minimal repair, global optimization techniques.

1.   INTRODUCTION

The cost of maintaining a component of a technical system (a transport fleet, a machine, a road, a traffic light system, etc.) often consists of a cost that depends on the specific component involved and of a fixed cost that only depends on the system. In such a case the system-dependent cost, the so-called set-up cost, is the same for all maintenance activities carried out on the system. For example, the set-up cost can consist of the down-time cost due to production loss if the system cannot be used during maintenance, or of the preparation cost associated with erecting a scaffolding or opening up a machine. Set-up costs can be saved when maintenance activities are executed simultaneously, since execution of a group of activities requires only one set-up. A nice practical application is given by Van der Duyn Schouten et al. (1998).

For literature overviews of the field of maintenance of multi-component systems, we refer to the review article by Cho and Parlar (1991) and the article by Dekker, Van der Duyn Schouten and Wildeman (1997). Currently, there are several methods that can handle multiple components. However, most of them suffer from intractability when the number of components grows, unless a special structure is assumed. For instance, the maintenance of a deteriorating system is frequently described using the Markov decision theory (see, for example Howard 1960, who was the first to use such a problem formulation). Since the state space in such problems grows exponentially with the number of components, the Markov decision modelling of multi-component systems is not tractable for more than three non-identical components (see, for example Bäckert and Rippin 1985). For problems with many components heuristic methods can be applied. For instance, Dekker and Roelvink (1995) present a heuristic replacement criterion for a case in which a fixed group of components is always replaced. Van der Duyn Schouten and Vanneste (1990) study structured strategies, viz. (n, N)-strategies, but provide an algorithm for only two identical components.

An approach that can handle many components was introduced by Goyal and Kusy (1985) and Goyal and Gunasekaran (1992). In this approach a basis interval for maintenance is taken and it is assumed that components can only be maintained at integer multiples of this interval, thereby saving set-up costs. The authors present an algorithm that iteratively determines the basis interval and the integer multiples. The algorithm has two disadvantages. The first is that only components with a very specific deterioration structure can be handled, which makes it impossible to apply it to well-known maintenance models. The second disadvantage is that the algorithm often gives solutions that are not optimal and about which there is no information of how good the solutions are (see Van Egmond, Dekker and Wildeman 1995).

The idea of using a basic cycle time and individual integer multiples was first applied in the definition of the joint replenishment problem in inventory theory, see Goyal (1973). The joint replenishment problem can be considered as a very simple case of the maintenance problem of Goyal and Kusy (1985). A method to solve the joint replenishment problem to optimality was presented by Goyal (1974). However, this method is based on enumeration and is computationally prohibitive. Moreover, it is not clear how this method can be extended to deal with the more general cost functions in the case of maintenance optimization. Many heuristics have appeared in the joint replenishment literature (see Goyal and Satir 1989). But again, it is not clear how these heuristics will perform in the case of the more general maintenance cost functions.

In this contribution we present a general approach for the coordination of maintenance frequencies, thereby pursuing the idea of Goyal et al. (1985, 1992). With this approach we can easily solve the model of Goyal et al. to optimality, but we can also incorporate many other maintenance models like minimal repair, inspection and block replacement. The joint overhaul scheduling problem, in which a large overhaul interval is subdivided into an integer number of subintervals at which minor maintenance is done (see Duffua and Ben-Daya (1994)), can be covered as well. We can also efficiently solve the joint replenishment problem to optimality (see Dekker, Frenk and Wildeman 1995).

Our solution approach is based on global optimization of the problem. We first apply a relaxation and find a corresponding feasible solution. This relaxation yields a lower bound on an optimal solution so that we can decide whether the feasible solution is good enough. If not so, we apply a global-optimization procedure on an interval that is obtained by the relaxation and that contains an optimal solution. For the special cases of Goyal et al., the minimal-repair model and the inspection model it is then possible to apply Lipschitz optimization to find a solution with an arbitrarily small deviation from an optimal solution in little time. For the block replacement model we will apply a good heuristic. The main elements of our approach were published in Dekker et al. (1996).

This paper is structured as follows. In the next section we give the problem formulation. In Section 3. we rewrite the problem so that it may use solution techniques that will be discussed in Section 4.. In Section 5. we present numerical results and in Section 6., we draw conclusions.

2.   PROBLEM DEFINITION

Consider a multi-component system with components i, $i \in \{1, \ldots, n\}$. Creating an occasion for preventive maintenance on one or more of these components involves a set-up cost S, independent of how many components are maintained. The set-up cost can be due to, for example system down-time. Because of this set-up cost S there is an economic dependence between the individual components.

In this paper we consider preventive maintenance activities of the block type, that is, in which the next execution time depends only on the time passed since the latest execution. Otherwise, for example in case of age replacement , execution of maintenance can no longer be coordinated and one has to use opportunity policies or modified block replacement .

On an occasion of maintenance, component i can be preventively maintained at an extra cost of c_i^p. Let $M_i(x)$ be the expected total deterioration costs of component i (due to failures, repairs, etc.), x time units after its latest preventive maintenance . We assume that $M_i(\cdot)$ is continuous. We assume furthermore that after preventive maintenance a component can be considered as good as new. Consequently, the average costs $\Phi_i(x)$ of component i, when component i is preventively maintained on an occasion each x time units, amount to

$$\Phi_i(x) = \frac{c_i^p + M_i(x)}{x}, \quad x > 0. \tag{11.1}$$

Since the function $M_i(\cdot)$ is continuous, the function $\Phi_i(\cdot)$ is also continuous.

To reduce costs by exploiting the economic dependence between components, maintenance on individual components can be combined. We assume that preventive maintenance is carried out at a basis interval of T time units (that is, each T time units an occasion for preventive maintenance is created) and that preventive maintenance on a component can only be carried out at integer multiples of this basis interval T: component i is preventively maintained each $k_i T$ time units, $k_i \in \mathbb{N}$. The objective is the minimization of the total average costs per time unit . The total average costs are the sum of the average set-up cost and the individual average costs of each component . Consequently, we have the following problem:

$$\inf \left\{ \frac{S}{T} + \sum_{i=1}^{n} \Phi_i(k_i T) \ : \ k_i \in \mathbb{N}, \ T > 0 \right\}. \tag{11.2}$$

In the formulation of this problem we assume that components are preventively maintained at fixed intervals that are integer multiples of a basis interval . This strategy is called *indirect grouping*, since the groups are formed indirectly and are not fixed over time. Another grouping strategy is *direct grouping*, where the n components are partitioned into disjoint groups that are fixed over time, that is, a component is then always maintained in the same group.

The disadvantage of direct-grouping models is that they can be solved to optimality for only a small number of components, unless a special structure is assumed (for example, when the frequency of a group of components equals the largest *individual* frequency among the components in that group, see Van Dijkhuizen and Van Harten 1997). Van Eijs, Heuts and Kleijnen (1992) show that in an inventory environment (where the same grouping strategies apply for items that are jointly replenished), the indirect-grouping strategy always outperforms the direct-grouping approach. This is why we will consider indirect grouping in this paper.

The idea of modelling maintenance at fixed intervals that are integer multiples of a basis interval originates from the inventory theory, see Goyal (1973). It was introduced in maintenance by Goyal and Kusy (1985), and Goyal and Gunasekaran (1992). However, these authors allow only a very specific deterioration-cost function $M_i(\cdot)$ for component i, $i \in \{1, \ldots, n\}$. By allowing more general deterioration-cost functions, this modelling approach can also be applied to well-known preventive-maintenance strategies such as minimal repair, inspection and block replacement.

By choosing the appropriate function $M_i(\cdot)$, the following models can be incorporated (see also Dekker 1995, who provides an extensive list of these models; here we mention only some important ones).

Special Case of Goyal and Kusy. Goyal and Kusy (1985) use the following deterioration-cost function: $M_i(x) = \int_0^x (f_i + v_i t^e) dt$, where f_i and v_i are non-negative constants for component i and $e \geq 0$ is the same for all components. Notice that $e = 1$ represents the joint replenishment problem as commonly encountered in inventory literature, see also Dekker, Frenk and Wildeman (1995).

Special Case of Goyal and Gunasekaran. The deterioration-cost function that is used by Goyal and Gunasekaran (1992) is slightly different from that of Goyal and Kusy (1985). They take $M_i(x) = \int_0^{Y_i(x-X_i)} (a_i + b_i t) dt$, where x must of course be larger than X_i, and a_i, b_i, X_i and Y_i are non-negative constants for component i. In this expression, Y_i denotes the average utilization factor of component i and X_i is the time required for maintenance of component i. Consequently, they take $e = 1$ in the deterioration-cost function of Goyal and Kusy, and they take individual down-time and utilization factors into account.

Minimal-Repair Model. According to a standard minimal-repair model (see, for example Dekker 1995), component i is preventively replaced at fixed intervals of length x, with failure repair occurring whenever necessary. A failure repair restores the component into a state as good as before. Consequently, $M_i(x) = c_i^r \int_0^x r_i(t)dt$, where $r_i(\cdot)$ denotes the rate of occurrence of failures, and c_i^r the failure-repair cost. Here $M_i(x)$ expresses the expected repair costs incurred in the interval $[0, x]$ due to failures. Notice that this model incorporates the special case of Goyal and Kusy if we take $c_i^r = 1$ and $r_i(t) = f_i + v_i t^e$.

Inspection Model. In a standard inspection model (see Dekker 1995), component i is inspected at fixed intervals of length x, with a subsequent replacement when at inspection the component turns out to have failed. If a component fails before it is inspected, it stays inoperative until it is inspected. After inspection , a component can be considered as good as new. Here we have $M_i(x) = c_i^u \int_0^x F_i(t)dt$, where c_i^u is the failure cost per unit time and $F_i(\cdot)$ is the cdf of the failure distribution of component i.

Replacement Model. According to a standard block replacement (see Dekker 1995), component i is replaced upon failure and preventively after a fixed interval of length x. Consequently, $M_i(x) = c_i^f N_i(x)$, where $N_i(x)$ denotes the renewal function (expressing the expected number of failures in $[0, x]$), and c_i^f the failure-replacement cost.

In the following section we present a general approach to solve (11.2). This approach allows a different $M_i(\cdot)$ for each component . Thus it is also possible to mix the different models above; it is, for instance, possible to combine the maintenance of a component according to the minimal-repair model with the maintenance of a component according to an inspection model.

3.   ANALYSIS OF THE PROBLEM

Goyal and Kusy (1985) and Goyal and Gunasekaran (1992) apply an iterative algorithm to solve the problem in the previous section for their specific deterioration-cost functions. The authors initialize each $k_i = 1$ and then find the corresponding optimal T by setting the derivative of the cost function of (11.2) as a function of T to zero. Subsequently, the authors find for each i a value of k_i, in two different ways. Goyal and Kusy (1985) find for each i the optimal integer k_i belonging to T by looking in a table that is made in advance for each component and that gives the optimal k_i for disjoint ranges of T. Goyal and Gunasekaran (1992) find for each i the optimal *real-valued* k_i by setting the derivative of the cost function of (11.2) as a function of k_i to zero and by rounding this real-valued k_i to the nearest integer. Once a value for k_i is found, it is compared to the k_i in the previous iteration (in this case the initialization). When for each i the k_i in the two iterations are equal, the algorithm terminates. Otherwise a new optimal T is found for the current values of k_i, and subsequently new values of k_i are determined, and so on, until for all i the k_i in two consecutive iterations are equal.

The advantage of this algorithm is that it is fast. This is primarily due to the special deterioration structure of the components in the cases of Goyal et al., which makes it possible to find an analytical expression for the optimal T given values of k_i, and also to find a value for the k_i in little time.

The specific deterioration structure of the components is at the same time a great disadvantage, since there is little room for differently modelled components. It is possible to extend the algorithm to deal with the more general maintenance models given in the previous section, but in that case a value for an optimal T given values of k_i has to be computed numerically, and the same holds for the corresponding values of k_i. As a result, the algorithm becomes much slower.

The greatest disadvantage of the algorithm of Goyal et al. is, however, that it is often stuck in a local optimal solution (see Van Egmond, Dekker and Wildeman 1995). There is no indication whatsoever of how good the solutions are when this occurs. This implies that even if we extend the algorithm to deal with more general maintenance models (which we will do anyway to study its performance in Section 5.), we do not have any guarantee concerning the quality and optimality of the obtained solutions.

In inventory theory literature many heuristics have appeared for the special cost functions in the joint replenishment problem (see Goyal and Satir 1989). Though some heuristics can be modified to deal with the cost functions of maintenance optimization, the performance of these heuristics cannot be guaranteed. That is why we will focus in this paper on an alternative solution approach that is based on the global optimization of (11.2). We first apply a relaxation and find a corresponding feasible solution to the problem. This relaxation yields a lower bound on an optimal solution so that we can decide whether the feasible solution is good enough. If not so, we apply a global-optimization procedure in an interval that is obtained by the relaxation and that contains an optimal T. For the special cases of Goyal et al., the minimal-repair model and the inspection model it is then possible to find a solution with an arbitrarily small deviation from an optimal solution. For the block replacement model this is not possible, but application of a fast golden-section search heuristic yields a good solution as well. In all cases our approach outperforms that of Goyal et al. (in original form when the special functions are used, and in extended form otherwise). Our approach can also be applied to find an optimal solution to the joint replenishment problem, see Dekker, Frenk and Wildeman (1995). In that case the procedure can be made even more efficient, since the cost functions in the joint replenishment problem have a very simple form.

To make the optimization problem (11.2) mathematically more tractable, we substitute T with $1/T$. Using this transformation, the relaxation that will be discussed in the next section becomes an easily solvable convex-programming problem if each of the individual cost functions $\Phi_i(\cdot)$ is given by one of the special cases of Goyal et al., the minimal-repair model or the inspection model. As will be shown later, one cannot generally derive that the relaxation is a convex-programming problem without this transformation.

Clearly, using the above transformation the optimization problem (11.2) is then equivalent with

$$\inf\left\{ST + \sum_{i=1}^{n} \Phi_i(k_i/T) \; : \; k_i \in \mathbb{N}, \; T > 0\right\} =$$

$$(P) \quad \inf_{T>0}\left\{ST + \sum_{i=1}^{n} \inf\{\Phi_i(k_i/T) \; : \; k_i \in \mathbb{N}\}\right\}.$$

Denote now by $v(P)$ the optimal objective value of (P) and by $T(P)$ an optimal T (if it exists). Notice that if $T(P)$ and certain values of $k_1, k_2, \ldots, k_n$ are optimal for (P), then $T = 1/T(P)$ and the same values of $k_1, k_2, \ldots, k_n$ are optimal for the optimization problem (11.2).

To simplify the optimization problem (P), we introduce the following definition and assumption.

Definition 1 *A function $f(x)$, $x \in (0, \infty)$, is called unimodal with respect to $b \geq 0$ if $f(x)$ is decreasing for $x \leq b$ and increasing for $x \geq b$. That is, $f(y) \geq f(x)$ for every $y \leq x \leq b$, and $f(y) \geq f(x)$ for every $y \geq x \geq b$.*

Observe that by this definition it is immediately clear that any increasing function $f(x)$, $x \in (0, \infty)$, is unimodal with respect to $b = 0$.

Assumption 1 *For each $i = 1, \ldots, n$ the optimization problem (P_i) given by $\inf\{\Phi_i(x) \; : \; x > 0\}$ has a finite optimal solution $x_i^* > 0$. Furthermore, for each i the function $\Phi_i(\cdot)$ is unimodal with respect to x_i^*.*

By Assumption 1 the optimization problem (P) can be simplified considerably. To this end we introduce the interval $I_i^{(k)} := [k/x_i^*, (k+1)/x_i^*]$, $k = 0, 1, \ldots$, and we observe that for $t \in I_i^{(k)}$ it holds that $k/t \le x_i^* \le (k+1)/t$, so that

$$x_i^* \le (k+1)/t \le (k+2)/t \le (k+3)/t \le \cdots$$

and

$$x_i^* \ge k/t \ge (k-1)/t \ge \cdots \ge 1/t.$$

Therefore, as by Assumption 1 the function $\Phi_i(\cdot)$ is unimodal with respect to x_i^*, we have that

$$\Phi_i((k+1)/t) \le \Phi_i((k+2)/t) \le \Phi_i((k+3)/t) \le \cdots$$

and

$$\Phi_i(k/t) \le \Phi_i((k-1)/t) \le \cdots \le \Phi_i(1/t).$$

This implies that for a given t it is easy to determine an optimal integer k_i, since now we have that

$$\inf\{\Phi_i(k_i/t) \ : \ k_i \in \mathbb{N}\} = \begin{cases} \Phi_i(1/t) & \text{if } t \in I_i^{(0)} \\ \min\{\Phi_i(k/t), \Phi_i((k+1)/t)\} & \text{if } t \in I_i^{(k)}, \\ & k = 1, 2, \ldots \end{cases}$$

Consequently, if we define

$$g_i(t) := \inf\{\Phi_i(k_i/t) \ : \ k_i \in \mathbb{N}\},$$

we have

$$g_i(t) = \begin{cases} \Phi_i(1/t) & \text{if } t \in I_i^{(0)} \\ \min\{\Phi_i(k/t), \Phi_i((k+1)/t)\} & \text{if } t \in I_i^{(k)}, \ k = 1, 2, \ldots \end{cases}$$

$$(11.3)$$

It is not difficult to verify that by Assumption 1 and the fact that $\Phi_i(\cdot)$ is continuous, the function $g_i(\cdot)$ is also continuous. In Figure 11.1 an example of the function $g_i(\cdot)$ is given.

Notice that for a given t, the function $g_i(t)$ and the corresponding k can easily be calculated once an optimal solution x_i^* of (P_i) is known. A given t is within the interval $I_i^{(k)}$ for which $k = \lfloor tx_i^* \rfloor$, with $\lfloor \cdot \rfloor$ denoting the lower-entier function. Consequently, if $k = 0$, one function evaluation (viz. of $\Phi_i(1/t)$) is necessary to compute $g_i(t)$. Otherwise, if $k \ge 1$, two function evaluations are necessary. Under Assumption 1 the optimization problem (P) now has the following simplified representation:

$$(Q) \qquad \inf_{T>0} \left\{ ST + \sum_{i=1}^{n} g_i(T) \right\},$$

with $g_i(\cdot)$ given by (11.3). Denote by $v(Q)$ the optimal objective value of (Q) and by $T(Q)$ an optimal T (if it exists). Under Assumption 1 we have that the problems (P) and (Q) are equivalent, which implies that if $T(Q)$ and certain values of $k_1, k_2, \ldots, k_n$ are optimal for (Q), then $T(P) = T(Q)$ and the same values of $k_1, k_2, \ldots, k_n$ are optimal for problem (P) and, as a result, $T = 1/T(Q)$ and $k_1, k_2, \ldots, k_n$ are also optimal for the original optimization problem (11.2). In the remainder we assume that Assumption 1 is satisfied and, consequently, we will consider from now on the optimization of problem (Q).

Below we show that in general the special cases of Goyal et al., the minimal-repair model and the inspection model satisfy Assumption 1 when the optimization problem (P_i) has a finite optimal solution $x_i^* > 0$. To verify this we need the following lemma.

Lemma 1 *If $M_i(\cdot)$ is concave on $(0, b_i)$ and convex on (b_i, ∞), $b_i \geq 0$, and the optimization problem (P_i) given by $\inf\{\Phi_i(x) : x > 0\}$ has a finite optimal solution $x_i^* > 0$, then the function $\Phi_i(\cdot)$ is unimodal with respect to x_i^*.*

PROOF: If $b_i > 0$ and $M_i(\cdot)$ is concave on $(0, b_i)$ then $c_i^p + M_i(\cdot)$ is also concave on $(0, b_i)$. This implies that for every $0 < z_1 < z_2 < b_i$ we have that $c_i^p + M_i(z_1) = c_i^p + M_i((z_1/z_2)z_2) > (z_1/z_2)(c_i^p + M_i(z_2))$. Hence, by equation (11.1), $\Phi_i(z_1) > \Phi_i(z_2)$ and, consequently, $\Phi_i(\cdot)$ is strictly decreasing on $(0, b_i)$. Since x_i^* is a minimum for $\Phi_i(\cdot)$ this implies that $x_i^* \geq b_i$. If $b_i = 0$ then we also have that $x_i^* \geq b_i$, since $x_i^* > 0$.

Since $M_i(\cdot)$ is convex on (b_i, ∞), $c_i^p + M_i(\cdot)$ is also convex on (b_i, ∞), $b_i \geq 0$. By Theorems 3.51 and 3.21 of Martos (1975) this implies that $\Phi_i(t) = (c_i^p + M_i(t))/t$ is unimodal on (b_i, ∞) with respect to x_i^*, and so, if $b_i = 0$, the desired result follows immediately. For the case $b_i > 0$ we showed above that $\Phi_i(\cdot)$ is strictly decreasing on $(0, b_i)$. Together with the result that the continuous function $\Phi_i(\cdot)$ is unimodal on (b_i, ∞) with respect to x_i^*, we then obtain that $\Phi_i(\cdot)$ is unimodal on the interval $(0, \infty)$ with respect to x_i^*, which completes the proof.

Theorem 1 *If each (P_i), $i = 1, \ldots, n$, has a finite solution $x_i^* > 0$ and is formulated according to one of the special cases of Goyal et al., the minimal-repair model with a unimodal rate of occurrence of failures or the inspection model, then Assumption 1 is satisfied.*

PROOF: We will prove that if for a certain $i \in \{1, \ldots, n\}$ the optimization problem (P_i) has a finite solution $x_i^* > 0$ for one of the models mentioned, then the function $\Phi_i(\cdot)$ is unimodal with respect to x_i^*. Consider therefore an arbitrary $i \in \{1, \ldots, n\}$ and distinguish between the different models.

1. Special Case of Goyal and Kusy. It is easy to show (by setting the derivative of $\Phi_i(\cdot)$ to zero) that the optimization problem (P_i) has an optimal solution $x_i^* = \{(c_i^p(e+1))/(v_i e)\}^{1/(e+1)}$. This solution is finite and positive if and only if v_i, e and c_i^p are strictly larger than zero, and by the assumption that $x_i^* > 0$ we can assume that this is the case.

   We have that $M_i(x) = \int_0^x (f_i + v_i t^e) dt = f_i x + (v_i/(e+1)) x^{e+1}$, so that $M_i''(x) = e v_i x^{e-1} > 0$ and, as a result, $M_i(\cdot)$ is (strictly) convex on $(0, \infty)$. By Lemma 1 we then have that $\Phi_i(\cdot)$ is unimodal with respect to x_i^*.

2. Special Case of Goyal and Gunasekaran. It is easy to show (by setting the derivative of $\Phi_i(\cdot)$ to zero) that the optimization problem (P_i) has an optimal solution $x_i^* = \{2(c_i^p - a_i X_i Y_i)/(b_i Y_i^2) + X_i^2\}^{1/2}$. This solution is finite and positive if and only if b_i and Y_i are strictly larger than zero and $c_i^p > X_i Y_i(a_i - b_i X_i Y_i/2)$, and by the assumption that $x_i^* > 0$ we can assume that this is the case.

   We have that $M_i(x) = \int_0^{Y_i(x-X_i)} (a_i + b_i t) dt = a_i Y_i(x - X_i) + b_i Y_i^2(x - X_i)^2/2$, so that $M_i''(x) = b_i Y_i^2 > 0$ and, as a result, $M_i(\cdot)$ is (strictly) convex on $(0, \infty)$. By Lemma 1 we then have that $\Phi_i(\cdot)$ is unimodal with respect to x_i^*.

3. Minimal-Repair Model. If the rate of occurrence of failures $r_i(\cdot)$ is unimodal with respect to a value $b_i \geq 0$, then as $M_i(x) = c_i^r \int_0^x r_i(t) \, dt$ it follows that $M_i'(\cdot)$ is decreasing on $(0, b_i)$ and increasing on (b_i, ∞). Hence $M_i(\cdot)$ is concave on $(0, b_i)$ and convex on (b_i, ∞). Since the optimization problem (P_i) has a finite solution $x_i^* > 0$ we then have by Lemma 1 that $\Phi_i(\cdot)$ is unimodal with respect to x_i^*.

   Notice that if $b_i = 0$, then $r_i(\cdot)$ is increasing on $(0, \infty)$ and $M_i(\cdot)$ is convex on $(0, \infty)$ If $r_i(\cdot)$ is unimodal with respect to a b_i strictly larger than zero, then $\Phi_i(\cdot)$ follows a bathtub pattern. In Lemma 1 we showed that in that case $x_i^* \geq b_i$. As the function $M_i(\cdot)$ is convex on (b_i, ∞), it is a forteriori convex on (x_i^*, ∞), a result that will be used later to prove that the relaxation of (Q) is a convex-programming problem (see Lemma 2).

4. Inspection Model. Since $M_i(x) = c_i^u \int_0^x F_i(t) dt$, we have that $M_i'(x)$ is increasing on $(0, \infty)$, and hence that $M_i(x)$ is convex on $(0, \infty)$. Since the optimization problem (P_i) has a finite solution $x_i^* > 0$ we then have by Lemma 1 that $\Phi_i(\cdot)$ is unimodal with respect to x_i^*.

Consequently, if for each $i = 1, \ldots, n$ one of the above models is used (possibly different models for different i), then $\Phi_i(\cdot)$ is unimodal with respect to x_i^* and so we have verified that Assumption 1 is satisfied.

In Dekker (1995) it is shown that a finite optimal solution x_i^* of the optimization problem (P_i) for the minimal-repair model exists if $\lim_{t \to \infty} r_i(t) = \infty$, or if $\lim_{t \to \infty} r_i(t) = c$ and $\lim_{t \to \infty} \{ct - \int_0^t r(s)ds\} > c_i^p$ for some $c > 0$. A finite optimal solution for the inspection model exists if $c_i^p < c_i^u EX_i$, with EX_i the expected lifetime of component i.

In Figure 11.2 an example of the objective function of problem (Q) is given. In general (Q) has several local minima, even for the simple models described above. This is due to the shape of the functions $g_i(\cdot)$ and it is inherent to the fact that the k_i have to be integer. In the following section we show that when these integrality constraints are relaxed, often a much easier problem is obtained; for the special cases of Goyal et al., the minimal-repair model and the inspection model this relaxed problem turns out to be a single-variable convex-programming problem that is easy to solve. The relaxation can be used to obtain an approximation of the optimal objective value $v(Q)$ and to determine an interval that contains an optimal value $T(Q)$. When the approximation is not good enough, we can apply global-optimization techniques within this interval to find an optimal solution.

4.   SOLVING PROBLEM (Q)

In this section we develop a solution procedure for problem (Q). To this end we first consider a relaxation.

4.1   A Relaxation for (Q).
If we replace in (Q) the constraints $k_i \in \mathbb{N}$ by $k_i \geq 1$ (in the definition of $g_i(\cdot)$) then clearly $v(Q) \geq v(Q_{rel})$ with $v(Q_{rel})$ the optimal objective value of the optimization problem given by

$$(Q_{rel}) \qquad \inf_{T>0} \left\{ ST + \sum_{i=1}^n \inf\{\Phi_i(k_i/T) \ : \ k_i \geq 1\} \right\}.$$

Let $T(Q_{rel})$ be a corresponding optimal value of T (if it exists).

To simplify this problem as we did in the previous section we need a much weaker assumption than Assumption 1.

Assumption 2 *For each* $i = 1, \ldots, n$ *the optimization problem* (P_i) *given by* $\inf\{\Phi_i(x) \ : \ x > 0\}$ *has a finite optimal solution* $x_i^* > 0$. *Furthermore, for each* $i = 1, \ldots, n$ *it holds that* $\Phi_i(\cdot)$ *is increasing on* $[x_i^*, \infty)$.

Theorem 1 showed that for the special cases of Goyal et al., the minimal-repair model with a unimodal rate of occurrence of failures and the inspection model, Assumption 1 is satisfied when (P_i) has a finite solution $x_i^* > 0$. As a result, also Assumption 2 is satisfied for these models.

By Assumption 2 the optimization problem (Q_{rel}) can be simplified. Analogously to equation (11.3) we have

$$g_i^{(R)}(t) := \inf\{\Phi_i(k_i/t) \ : \ k_i \geq 1\} = \begin{cases} \Phi_i(1/t) & \text{if } t < 1/x_i^* \\ \Phi_i(x_i^*) & \text{if } t \geq 1/x_i^*. \end{cases} \quad (11.4)$$

In Figure 11.3 an example of the function $g_i^{(R)}(\cdot)$ is given.

Now (Q_{rel}) has the following simplified representation:

$$(R) \qquad \inf_{T>0} \left\{ ST + \sum_{i=1}^n g_i^{(R)}(T) \right\}.$$

Denote by $v(R)$ the optimal objective value of (R) and by $T(R)$ an optimal T (if it exists). In the remainder we will assume the (R) always has an optimal solution. Notice that by Assumption 2 it follows that $v(R) = v(Q_{rel})$ and $T(R) = T(Q_{rel})$, since (R) and (Q_{rel}) are equivalent. In Figure 11.4 an example of the objective function of problem (R) is given.

Observe that if each $g_i^{(R)}(\cdot)$ is convex, then (R) is a convex-programming problem. We can prove that $g_i^{(R)}(\cdot)$ is convex on $(0, \infty)$ if the function $M_i(\cdot)$ is convex on (x_i^*, ∞), so that (R) is a convex-programming problem if each function $M_i(\cdot)$ is convex on (x_i^*, ∞). This is established by the following lemma.

Lemma 2 *If the function $M_i(\cdot)$ is convex on (x_i^*, ∞), then the function $g_i^{(R)}(\cdot)$ is convex on $(0, \infty)$.*

PROOF: To show that the function $g_i^{(R)}(t)$ is convex it is sufficient to show that $\Phi_i(1/t)$ is convex on $(0, 1/x_i^*)$. If $\Phi_i(1/t)$ is convex on $(0, 1/x_i^*)$ then it follows from the fact that $\Phi_i(1/(1/x_i^*)) = \Phi_i(x_i^*)$ is the minimal value of $\Phi_i(\cdot)$ on $(0, \infty)$, that $\Phi_i(1/t)$ is also decreasing on $(0, 1/x_i^*)$. Then it follows from the definition of $g_i^{(R)}(t)$ (see (11.4)) that $g_i^{(R)}(t)$ is convex on $(0, \infty)$.

So we have to prove that $\Phi_i(1/t)$ is convex on $(0, 1/x_i^*)$. We will prove that $\Phi_i(1/t)$ is convex on $(0, 1/b_i)$ if $M_i(t)$ is convex on (b_i, ∞). So let $M_i(t)$ be convex on (b_i, ∞), then $c_i^p + M_i(t) = t\Phi_i(t)$ is also convex on (b_i, ∞). Define for a function $f(\cdot)$

$$s_f(t, t_0) = \frac{f(t) - f(t_0)}{t - t_0},$$

and let $f(t) := t\Phi_i(t)$ and $g(t) := \Phi_i(1/t)$. Then it is easy to verify that

$$s_f(t, t_0) = \Phi_i(t_0) - (1/t_0)s_g(1/t, 1/t_0). \tag{11.5}$$

Application of the well-known criterion of increasing slopes valid for convex functions (see, e.g., Proposition 1.1.4 in Chapter I of Hiriart-Urruty and Lemaréchal 1993), we have for the convex function $f(t) = t\Phi_i(t)$ on (b_i, ∞) that $s_f(t, t_0)$ is increasing in $t > b_i$ for every $t_0 > b_i$. Using (11.5) this implies that $\Phi_i(t_0) - (1/t_0)s_g(1/t, 1/t_0)$ is increasing in $t > b_i$ for every $t_0 > b_i$. Since $\Phi_i(t_0)$ and $1/t_0$ are constants, the function $-s_g(1/t, 1/t_0)$ is then increasing in $t > b_i$ for every $t_0 > b_i$. Hence, $s_g(1/t, 1/t_0)$ is increasing as a function of $1/t < 1/b_i$ for every $1/t_0 < 1/b_i$, which is equivalent with $s_g(x, x_0)$ is increasing in $x < 1/b_i$ for every $x_0 < 1/b_i$. Using again the criterion of increasing slopes for convex functions it follows that $g(t) = \Phi_i(1/t)$ is convex on $(0, 1/b_i)$.

If $M_i(t)$ is convex on (x_i^*, ∞), that is, if $b_i = x_i^*$, then we have that $\Phi_i(1/t)$ is convex on $(0, 1/x_i^*)$, which completes the proof. (Notice that if $M_i(t)$ is convex on $(0, \infty)$, that is, if $b_i = 0$, then we have that $\Phi_i(1/t)$ is also convex on $(0, \infty)$.)

Theorem 2 *If each (P_i), $i = 1, \ldots, n$, has a finite solution $x_i^* > 0$ and is formulated according to one of the special cases of Goyal et al., the minimal-repair model with a unimodal rate of occurrence of failures or the inspection model, then problem (R) is a convex-programming problem.*

PROOF: In the proof of Theorem 1 we showed that for the minimal-repair model with a unimodal rate of occurrence of failures the function $M_i(\cdot)$ is convex on (x_i^*, ∞). In the case of an increasing rate of occurrence of failures, $M_i(\cdot)$ is even convex on $(0, \infty)$, and thus a forteriori convex on (x_i^*, ∞). We also showed that for the special cases of Goyal et al. and the inspection model the function $M_i(\cdot)$ is convex on $(0, \infty)$, so that $M_i(\cdot)$ is a forteriori convex on (x_i^*, ∞). Consequently, if for each $i = 1, \ldots, n$ one of the above models is used (possibly different models for different i), then by Lemma 2 the corresponding $g_i^{(R)}(\cdot)$ are convex so that problem (R) is a convex-programming problem.

We can now explain why we applied in the previous section the transformation of T into $1/T$ in the original optimization problem (11.2). We saw that (R) is a convex-programming problem if each function $g_i^{(R)}(t)$ is convex on $(0, \infty)$. In the proof of Lemma 2 we showed that this is the case if each function $\Phi_i(1/t)$ is convex on $(0, 1/x_i^*)$. We showed furthermore that the function $\Phi_i(1/t)$ is convex on $(0, 1/x_i^*)$ if $M_i(t)$ is convex on (x_i^*, ∞) (which is generally the case for the models described before). If we did not apply the transformation of T into $1/T$ as we did in the previous section, we would find that the corresponding relaxation is a convex-programming problem only if each function $\Phi_i(t)$ is convex on (x_i^*, ∞), which is a much more restrictive condition and is in general not the case (not even for the models mentioned before). Summarizing, the transformation of T into $1/T$ causes the relaxation to be a convex-programming problem for the models described before, a result that otherwise does not generally hold.

If (R) is a convex-programming problem, it can easily be solved to optimality. When the functions $g_i^{(R)}(\cdot)$ are differentiable (which is the case if the functions $\Phi_i(\cdot)$ are differentiable), we can set the derivative of the cost function in (R) equal to zero and subsequently find an optimal solution with the bisection method. When the functions $g_i^{(R)}(\cdot)$ are not differentiable, we can apply a golden-section search. (For a description of these methods, see Chapter 8 of Bazaraa, Sherali and Shetty 1993).

In all cases it is convenient to have an upper bound on an optimal value $T(R)$. To this end we assume without loss of generality that $1/x_n^* \leq 1/x_{n-1}^* \leq \cdots \leq 1/x_1^*$. The following lemma proves that $T(R) \leq 1/x_1^*$.

Lemma 3 *If we assume without loss of generality that $1/x_n^* \leq 1/x_{n-1}^* \leq \cdots \leq 1/x_1^*$, then for an optimal $T(R)$ of (R) it holds that $T(R) \leq 1/x_1^*$.*

PROOF: For all $T > 1/x_1^*$ it follows that

$$ST + \sum_{i=1}^{n} g_i^{(R)}(T) = ST + \sum_{i=1}^{n} \Phi_i(x_i^*) > \frac{S}{x_1^*} + \sum_{i=1}^{n} \Phi_i(x_i^*)$$

$$= \frac{S}{x_1^*} + \sum_{i=1}^{n} g_i^{(R)}(1/x_1^*).$$

Hence $T(R) \leq 1/x_1^*$.

Once (R) is solved we have an optimal $T(R)$. In some cases this value of $T(R)$ is also an optimal T for (Q) as is established by the following lemma.

Lemma 4 *The value $T < 1/x_n^*$ is a local optimal solution of problem (R) if and only if T is a local optimal solution of problem (Q). Moreover, if for an optimal $T(R)$ of problem (R) it holds that $T(R) \leq 1/x_n^*$, then $T(Q) = T(R)$ is optimal for problem (Q).*

PROOF: Since the functions $g_i(\cdot)$ and $g_i^{(R)}(\cdot)$ (see (11.3) and (11.4)) are identical on $(0, 1/x_n^*]$ (by Assumption 2), $T < 1/x_n^*$ is a local optimal solution of problem (R) if and only if T is a local optimal solution of problem (Q).

To prove the second part, we observe that if $T(R)$ is optimal for problem (R), it holds that

$$ST(R) + \sum_{i=1}^{n} g_i^{(R)}(T(R)) = v(R) \leq v(Q) \leq ST(R) + \sum_{i=1}^{n} g_i(T(R)).$$

Since $T(R) \leq 1/x_n^*$ and hence $g_i(T(R)) = g_i^{(R)}(T(R))$, the upper and lower bound in the above inequality are equal, which implies that $T(R)$ is also optimal for (Q).

An immediate result of this lemma is the following corollary, that yields a lower bound on $T(Q)$.

Corollary 1 *If there does not exist a local optimal solution of (R) within $(0, 1/x_n^*)$, then it holds for the optimal $T(Q)$ of (Q) that $T(Q) \geq 1/x_n^*$.*

Notice that if (R) is a convex-programming problem it is easy to check whether there exists a local optimal solution within $(0, 1/x_n^*)$. When $T(R) \leq 1/x_n^*$ then $T(R)$ is a local (in that case even global) optimal solution and we have that $T(Q) = T(R)$ is optimal for (Q) according to Lemma 4. When $T(R) > 1/x_n^*$, we calculate the objective function of (R) in $1/x_n^*$. When it equals $v(R)$ then $1/x_n^*$ is also optimal for (R), so that $T(Q) = 1/x_n^*$ is also optimal for (Q) again according to Lemma 4. When the objective function of (R) in $1/x_n^*$ is larger than $v(R)$, there does not exist a local optimal solution of (R) within $(0, 1/x_n^*)$ and it holds by Corollary 1 that $T(Q) > 1/x_n^*$. Consequently, we proved the following corollary.

Corollary 2 *Suppose (R) is a convex-programming problem. If the objective function of (R) in $1/x_n^*$ is larger than $v(R)$ and $T(R) > 1/x_n^*$, then $T(Q) > 1/x_n^*$. Otherwise, there is a global optimal solution in $(0, 1/x_n^*]$; an optimal $T(Q)$ is then given by $T(Q) = \min\{1/x_n^*, T(R)\}$.*

Notice that if $T(R) \leq 1/x_n^*$ the corresponding values of k_i are all equal to 1 because of (11.3). This implies that not only $T(R)$ is optimal for (Q) but the corresponding k_i as well, since the functions $g_i(t)$ and $g_i^{(R)}(t)$ are equal for $t \leq 1/x_n^*$. If $T(R) > 1/x_n^*$, this may not be an optimal T for problem (Q). Besides, the values of k_i corresponding with $T(R)$ are not necessarily integer, which implies that the optimal solution of (R) is in general not feasible for (Q) when $T(R) > 1/x_n^*$. Consequently, the first thing to do when $T(R) > 1/x_n^*$ is to find a feasible solution for (Q).

4.2 Feasibility Procedures for (Q).

A straightforward way to find a feasible solution for (Q) is to substitute the value of $T(R)$ in (11.3). We saw in Section 3. that an optimal value of k_i depends on the interval $I_i^{(k)}$ that contains $T(R)$. Using this, we can obtain a feasible solution with the following Feasibility Procedure (FP).

Feasibility Procedure For each $i = 1, \ldots, n$ do the following:

1. Compute $k = \lfloor T(R)x_i^* \rfloor$. This is the value for which $T(R) \in I_i^{(k)}$.

2. If $k = 0$, then $k_i(FP) = 1$ is the optimal k_i-value for (Q) corresponding with $T(R)$ (use (11.3)).

3. If $k \geq 1$ then $k_i(FP) = k$ or $k_i(FP) = k + 1$ is an optimal value, depending on whether $\Phi_i(k/T(R)) \leq \Phi_i((k+1)/T(R))$ or $\Phi_i(k/T(R)) \geq \Phi_i((k+1)/T(R))$ (use (11.3)).

Let $v(FP)$ be the objective-function value of (Q) corresponding with this feasible solution. Then we have

$$v(FP) = ST(R) + \sum_{i=1}^{n} g_i(T(R)) \geq v(Q) \geq v(R).$$

Hence we can check the quality of this solution; if $v(FP)$ is close to $v(R)$ then it is also close to the optimal objective value $v(Q)$.

If it is not close enough we may find a better solution by application of a procedure that is similar to the iterative approach of Goyal et al. We call this the Improved-Feasibility Procedure (IFP).

Improved-Feasibility Procedure

1. Let $k_i(IFP) = k_i(FP)$, $i = 1, \ldots, n$, with $k_i(FP)$ the values given by the feasibility procedure FP.

2. Solve the optimization problem

$$\inf_{T>0} \left\{ ST + \sum_{i=1}^{n} \Phi_i(k_i(IFP)/T) \right\}, \tag{11.6}$$

and let $T(IFP)$ be an optimal value for T.

3. Determine new constants $k_i(IFP)$ by substitution of $T(IFP)$ in (11.3). This implies the application of the feasibility procedure FP to the value $T(IFP)$. Let $v(IFP)$ be the corresponding objective-function value.

For the value $v(IFP)$ generated by the IFP it follows that

$$
\begin{aligned}
v(IFP) &= ST(IFP) + \sum_{i=1}^{n} g_i(T(IFP)) \\
&= ST(IFP) + \sum_{i=1}^{n} \Phi_i(k_i(IFP)/T(IFP)) \\
&\leq ST(IFP) + \sum_{i=1}^{n} \Phi_i(k_i(FP)/T(IFP)) \qquad \text{(Step 3.)} \\
&\leq ST(R) + \sum_{i=1}^{n} \Phi_i(k_i(FP)/T(R)) \qquad \text{(Step 2.)} \\
&= ST(R) + \sum_{i=1}^{n} g_i(T(R)) \\
&= v(FP).
\end{aligned}
$$

Consequently, the solution generated by the IFP is at least as good as the solution obtained with the FP.

The IFP can in principle be repeated with in step 1 the new constants $k_i(IFP)$, and this can be done until no improvement is found. This procedure differs from the iterative algorithm of Goyal et al. in two aspects. The first difference concerns the way integer values of k_i are found given a value of T. We explained in Section 3. that in the algorithm of Goyal and Kusy (1985) optimal k_i are found by searching in a table that is made in advance for each i. This becomes inefficient when the values of k_i are large, since searching in the table then takes much time. Besides, this has to be done in each iteration again. Goyal and Gunasekaran (1992) find for each i an optimal *real-valued* k_i that is rounded to the nearest integer. This may not be optimal. In our procedure we can identify optimal values of k_i for a given value of T *immediately*, by substitution of T in (11.3) (that is, by application of the FP).

The second difference concerns the initialization of the k_i. Goyal et al. initialize each $k_i = 1$ and then find a corresponding optimal T. This often results in a solution that cannot be improved upon by the algorithm but is not optimal, that is, the algorithm is then stuck in a local optimal solution (see Van Egmond, Dekker and Wildeman 1995). In the IFP we start with a value of T that is optimal for (R) and hence might be a good solution for (Q) as well; this may be a much better initialization for the algorithm (we will investigate this in Section 5.).

However, we cannot guarantee that with this alternative initialization the IFP will not suffer from local optimality. If the procedure terminates and the generated solution $v(IFP)$ is not close to $v(R)$, then we cannot guarantee that it is a good solution. In that case we will apply a global-optimization algorithm.

Observe that for the models mentioned before (with an increasing rate of occurrence of failures for the minimal-repair model) the IFP is easily solvable since (11.6) is a convex-programming problem (the functions $\Phi_i(1/t)$ are then convex). Otherwise, the IFP may not be useful since (11.6) can be a difficult problem to solve. In that case we will not apply the IFP but we will use a global-optimization algorithm immediately after application of the FP when $v(FP)$ is not close enough to $v(R)$.

To apply a global optimization we first need an interval that contains an optimal $T(Q)$.

4.3   Upper and Lower Bounds on $T(Q)$.

In this subsection we will derive lower and upper bounds on $T(Q)$. Corollary 1 already provides a lower bound $1/x_n^*$ when there is no local optimal solution of (R) within $(0, 1/x_n^*)$. We saw that if (R) is a convex-programming problem this results in Corollary 2 that states that if $T(R) > 1/x_n^*$ and the objective function of (R) in $1/x_n^*$ is larger than $v(R)$ we have $T(Q) > 1/x_n^*$, while otherwise an optimal $T(Q)$ can be identified immediately by $T(Q) = \min\{T(R), 1/x_n^*\}$.

If the functions $M_i(\cdot)$ are convex and differentiable it is easy to see that the functions $g_i^{(R)}$ are differentiable and that (R) is a differentiable convex-programming problem. Moreover, if at least one of the functions $M_i(\cdot)$ is strictly convex we can prove a better lower bound than $1/x_n^*$ on $T(Q)$. This is established by the following lemma.

Lemma 5 *Consider the optimization problem:*

$$(Q_1) \quad \inf_{T>0} \left\{ ST + \sum_{i=1}^{n} \Phi_i(1/T) \right\},$$

with $v(Q_1)$ the optimal objective value and $T(Q_1)$ an optimal T. If for each $i = 1, \ldots, n$ the function $M_i(\cdot)$ is convex and differentiable on $(0, \infty)$, and for at least one $i \in \{1, \ldots, n\}$ the function $M_i(\cdot)$ is strictly convex on $(0, \infty)$, and the differentiable convex-programming problem (R) has no global optimal solution within $(0, 1/x_n^*)$, then $T(Q) \geq T(Q_1) > 1/x_n^*$.

PROOF: If there does not exist a global optimal solution of (R) within $(0, 1/x_n^*)$, then it is shown analogously to Lemma 4 and Corollary 1 that $T(Q_1) > 1/x_n^*$.

To prove the inequality $T(Q) \geq T(Q_1)$, notice first that (Q_1) equals the optimization problem (Q) when all k_i are fixed to the value 1. Consequently, (Q_1) is a more restricted problem than (Q) and it is easy to verify that $v(Q) \leq v(Q_1)$. Furthermore, if $T(Q)$ and certain values of k_i are optimal for (Q), then it is easy to see that if the functions $\Phi_i(\cdot)$ are differentiable the following holds:

$$S - \sum_{i=1}^{n} \frac{d}{dt}\Phi_i(k_i/t)\Big|_{t=T(Q)} = 0,$$

so that

$$S = \sum_{i=1}^{n} \frac{k_i}{(T(Q))^2}\Phi_i'(k_i/T(Q)).$$

Substitution of this in the optimal objective value of (Q) yields:

$$v(Q) = ST(Q) + \sum_{i=1}^{n}\Phi_i(k_i/T(Q))$$

$$= \sum_{i=1}^{n}\left\{\frac{k_i}{T(Q)}\Phi_i'(k_i/T(Q)) + \Phi_i(k_i/T(Q))\right\}.$$

It is easily verified that

$$\sum_{i=1}^{n}\left\{\frac{k_i}{T(Q)}\Phi_i'(k_i/T(Q)) + \Phi_i(k_i/T(Q))\right\} = \sum_{i=1}^{n}M_i'(k_i/T(Q)),$$

so that

$$v(Q) = \sum_{i=1}^{n}M_i'(k_i/T(Q)). \tag{11.7}$$

Analogously, it can be shown for the optimal objective value of (Q_1) that

$$v(Q_1) = \sum_{i=1}^{n}M_i'(1/T(Q_1)). \tag{11.8}$$

Suppose now that the inequality $T(Q) \geq T(Q_1)$ does not hold, that is, $T(Q) < T(Q_1)$. Since the functions $M_i(\cdot)$ are (strictly) convex and, consequently, the functions $M_i'(\cdot)$ are (strictly) increasing, this implies that (use (11.7) and (11.8))

$$v(Q) = \sum_{i=1}^{n} M_i'(k_i/T(Q)) \geq \sum_{i=1}^{n} M_i'(1/T(Q)) > \sum_{i=1}^{n} M_i'(1/T(Q_1)) = v(Q_1),$$

which is in contradiction with $v(Q) \leq v(Q_1)$. Consequently, $T(Q) \geq T(Q_1)$.

An upper bound on $T(Q)$ is obtained by the following lemma.

Lemma 6 *For an optimal $T(Q)$ of (Q) it holds that*

$$T(Q) \leq (1/S) \left\{ v(FP) - \sum_{i=1}^{n} \Phi_i(x_i^*) \right\},$$

with $v(FP)$ the objective value corresponding with the feasible solution of (Q) generated by the FP.

PROOF: For every $T > 0$ it holds that

$$ST + \sum_{i=1}^{n} \Phi_i(x_i^*) \leq ST + \sum_{i=1}^{n} g_i^{(R)}(T) \leq ST + \sum_{i=1}^{n} g_i(T).$$

Consequently, we have for every T with $ST + \sum_{i=1}^{n} g_i(T) \leq v(FP)$ that

$$ST + \sum_{i=1}^{n} \Phi_i(x_i^*) \leq v(FP),$$

which implies that

$$T \leq (1/S) \left\{ v(FP) - \sum_{i=1}^{n} \Phi_i(x_i^*) \right\}.$$

Since $T(Q)$ is such a T for which $ST + \sum_{i=1}^{n} g_i(T) \leq v(FP)$, the lemma follows.

To use the bound $T(Q_1)$, each $M_i(\cdot)$ must be convex on $(0, \infty)$ and at least one $M_i(\cdot)$ must be strictly convex. We showed in the proof of Theorem 1 that each $M_i(\cdot)$ is convex on $(0, \infty)$ for the special cases of Goyal et al., the minimal-repair model with an increasing rate of occurrence of failures and the inspection model. For the special cases of Goyal et al. each $M_i(\cdot)$ is even strictly convex on $(0, \infty)$, so that the bound $T(Q_1)$ can then always be used. For the minimal-repair and inspection model at least one $M_i(\cdot)$ must be strictly convex.

If the lower bound $T(Q_1)$ can be used, then define $T_l := T(Q_1)$ and otherwise take $T_l := 1/x_n^*$. Let T_u be the upper bound derived in Lemma 6. Then we have that $T(Q) \in [T_l, T_u]$. Consequently, it is sufficient to apply a global-optimization technique on the interval $[T_l, T_u]$ to find a value for $T(Q)$.

4.4   Global-Optimization Techniques.

What remains to be specified is the usage of a global-optimization technique for (Q) on the interval $[T_l, T_u]$ when the feasible solution to (Q) found after application of the FP (or the IFP) is not good enough.

Lipschitz Optimization   Efficient global-optimization techniques exist for the case that the objective function of (Q) is *Lipschitz*. A univariate function is said to be Lipschitz if for each pair x and y the absolute difference of the function values in these points is smaller than or equal to a constant (called the *Lipschitz constant*) multiplied by the absolute distance between x and y. More formally:

Definition 2 *A function $f(x)$ is said to be Lipschitz on the interval $[a, b]$ with Lipschitz constant L, if for all $x, y \in [a, b]$ it holds that $|f(x) - f(y)| \leq L|x - y|$.*

If the objective function of (Q) is Lipschitz on the interval $[T_l, T_u]$, then global-optimization techniques can be applied in this interval to obtain a solution with a corresponding objective value that is arbitrarily close to the optimal objective value $v(Q)$ (see the chapter on Lipschitz optimization in Horst and Pardalos 1995). For the special cases of Goyal et al., the minimal-repair model with an increasing rate of occurrence of failures, and the inspection model, we can prove that the objective function of (Q) is Lipschitz on $[T_l, T_u]$, and we can derive a Lipschitz constant (see appendix).

There are several Lipschitz-optimization algorithms (see Horst and Pardalos 1995), and we implemented some of them. The simplest one, called the *passive algorithm*, evaluates the function to be minimized at the points $a + \epsilon/L$, $a + 3\epsilon/L$, $a + 5\epsilon/L, \ldots$, and takes the point at which the function is minimal. The function value in this point does not differ more than ϵ from the minimal value in $[a, b]$. We implemented the algorithm of Evtushenko that is based on the passive algorithm, but that takes a following step larger than $2\epsilon/L$ if the current function value is larger than 2ϵ above the current *best* known value, which makes the algorithm faster than the passive algorithm. However, this algorithm can still be very time consuming, especially when the Lipschitz constant L is large. The algorithm of Evtushenko and the other algorithms described in Horst and Pardalos (1995) turned out to be too time consuming, and were therefore not of practical use to our problem.

Fortunately, however, the shape of the objective function of problem (Q) is such that the Lipschitz constant is decreasing in T (this is shown in the appendix). Using this, the algorithm of Evtushenko can easily be extended to deal with a *dynamic* Lipschitz constant; after a certain number of steps (going from left to right) the Lipschitz constant is recomputed, such that larger steps can be taken. This is repeated after the same number of steps, and so on, until the interval $[a, b]$ is covered. This approach turned out to work very well for our problem; the increment in speed was sometimes a factor 1000 compared to the version of Evtushenko, and this made Lipschitz optimization of practical use to our problem.

Golden-Section Search Heuristic   We can also apply alternative methods that do not use the notion of Lipschitz optimization. One such a method is a golden-section search. The Golden-section search is usually applied (and is optimal) for functions that are strictly unimodal, which the objective function of (Q) is generally not. However, we will apply an approach in which the interval $[T_l, T_u]$ is divided into a number of subintervals of equal length, on each of which a golden-section search is applied. The best point of these intervals is taken as solution. We then divide the subintervals into intervals that are twice as small and apply on each a golden-section search again. This doubling of intervals is repeated until no improvement is found. We refer to this approach as the *multiple-interval golden-section search heuristic*, the results of which are given in Section 5.

4.5   A Solution Procedure for (Q).   We are now ready to formulate a solution procedure for (Q). We consider first a solution procedure for the special cases of Goyal et al., the minimal-repair model with a unimodal rate of occurrence of failures, and the inspection model, in which cases problem (R) is a convex-programming problem. Subsequently, we indicate the changes when, for example, block replacement is used.

We can summarize the results in this section in the formulation of the following solution procedure for (Q):

1. Solve the convex-programming problem (R) using $T(R) \leq 1/x_1^*$. An optimal value $T(R)$ can be found by application of a bisec-

tion technique if the objective function of (R) is differentiable, but otherwise a golden-section search can be applied.

2. If $T(R) \leq 1/x_n^*$ then $T(Q) = T(R)$ is optimal for (Q); stop.

3. If $T(R) > 1/x_n^*$, check whether the objective function of (R) in $1/x_n^*$ equals $v(R)$. If so, $T(Q) = 1/x_n^*$ is optimal for (Q); stop.

4. Otherwise, we have that $T(Q) > 1/x_n^*$ and we first find a feasible solution for problem (Q) by applying the FP or IFP. If the corresponding objective value is close enough to $v(R)$, then it is also close to $v(Q)$, so that we have a good solution; stop.

5. If the solution is not good enough, apply a global-optimization technique on the interval $[T_l, T_u]$ to find a value for $T(Q)$.

If this solution procedure is applied to the block replacement model, some details have to be modified slightly. The first modification concerns the solution of the relaxation (R) that is in general not a convex-programming problem, but, since it has fewer local minima, is much easier to solve than problem (Q). Therefore, to find a solution to problem (R), we will apply a single iteration of the multiple-interval golden-section search heuristic described in the previous subsection, that is, the number of subintervals is fixed and will not be doubled until no improvement is found.

Though the optimization problem (11.6) is not a convex-programming problem for the block replacement model and is therefore more difficult to solve, we will still use the IFP with a single golden-section search applied to solve problem (11.6); even as such the IFP outperforms the approach of Goyal and Kusy, as will be shown by the experiments in the next section.

Since the nice results that we derived for the special cases of Goyal et al., minimal repair and inspection do not generally hold for the block replacement model, determination of a Lipschitz constant becomes more difficult, if possible at all. Therefore, we will not apply Lipschitz optimization to determine a value of $v(Q)$. Instead, we will use the multiple-interval golden-section search heuristic as described in the previous subsection.

5.   NUMERICAL RESULTS

In this section the solution procedure described in the previous section will be investigated and it will be compared with the iterative approach of Goyal et al. This will first be done for the special case of Goyal and Kusy, the minimal-repair model with an increasing rate of occurrence of failures, and the inspection model, in which cases an optimal solution can be found by Lipschitz optimization. This makes it possible to make a good comparison and also to investigate the performance of the multiple-interval golden-section search heuristic. Subsequently, the performance of the solution procedure for the block replacement model is investigated, using the golden-section search heuristic. Since the iterative algorithm of Goyal and Kusy (1985) performs better than that of Goyal and Gunasekaran (1992), we chose the first one to compare it with our solution procedure. All algorithms are implemented in Borland Pascal version 7.0 on a 66 MHz personal computer.

For all models we have 6 different values for the number n of components and 7 different values for the set-up cost S . This yields 42 different combinations of n and S, and for each of these combinations 100 random problem instances are taken by choosing random values for the remaining parameters. For the minimal-repair, inspection and block replacement model the lifetime distribution for component i is given by a Weibull-(λ_i, β_i) distribution (a Weibull-(λ, β) distributed stochastic variable has a cumulative distribution function $F(t) = 1 - e^{-(t/\lambda)^\beta}$). The data are summarized in Table 11.1.

Results for the special case of Goyal and Kusy, the minimal-repair model and the inspection model   For the special case of Goyal and Kusy, the minimal-repair model and the inspection model, the value $v(Q)$ can be determined by Lipschitz optimization with an arbitrary deviation from the optimal value; we allowed a relative deviation of 10^{-4} (i.e., 0.01%). In Table 11.2 the relevant results of the 4200 problem instances for each model are given.

From the table it can be seen that solving the relaxation takes very little time. Applying the IFP also takes little time. (In all cases the running times for the inspection model are higher than for the special case of Goyal and Kusy and the minimal-repair model, since for the inspection model a numerical routine has to be applied for each function evaluation, whereas for the other two models the cost functions can be computed analytically.) Notice that some deviations are negative. This is due to the relative deviation of 0.01% allowed in the optimal objective value determined by the Lipschitz optimization; a heuristic can give a solution with objective value up to 0.01% smaller than that produced by the Lipschitz-optimization procedure.

In all cases the IFP (that is an intelligent modification of the approach of Goyal et al.) outperforms the iterative algorithm of Goyal and Kusy, while the running times of the IFP are equal or faster. The differences are smallest for the special case of Goyal and Kusy. This can be explained by the fact that in the model of Goyal and Kusy there is little variance possible in the lifetime distributions of the components, mainly because the exponent e has to be the same for all components. In the inspection model, however, there can be large differences in the individual lifetime distributions, and this can cause much larger deviations for the iterative algorithm of Goyal and Kusy; the average deviation for Goyal and Kusy's algorithm is then 1.253% and the maximum deviation even 66.188%, which is much higher than the deviations for the IFP. The IFP performs well for all models.

Since for many examples the algorithm of Goyal and Kusy and the IFP find the optimal solution, the average deviations of these algorithms do not differ so much (in many cases the deviation is zero percent). However, there is a considerable difference in the number of times that large deviations were generated. This is illustrated in Table 11.3 that gives the percentage of the examples in which the deviation was larger than 1% and 5% for the three models discussed in this subsection. From this table it is clear that the IFP performs much better than the algorithm of Goyal and Kusy and that if the algorithm of Goyal and Kusy does not give the optimal solution, the deviation can be large. The conclusion is that solving the relaxation and subsequently the improved feasibility procedure is better than and at least as fast as the iterative algorithm of Goyal and Kusy. This also implies that the algorithm of Goyal and Kusy can be improved considerably if another initialization of the k_i and T is taken, viz. according to the solution of the relaxation.

From the results of Table 11.2 it can further be seen that the multiple-interval golden-section search heuristic performs very well in all cases. The average deviation is almost zero, and the maximum deviation is relatively small. The heuristic is initialized with four subintervals and this number is doubled until no improvement is found. It turned out that four subintervals is mostly sufficient. The running time of the heuristic is also quite moderate: less than a second for the special case of Goyal and Kusy and the minimal-repair model, and almost 12 seconds for the inspection model (where a numerical routine has to be applied for each function evaluation).

Usually, Lipschitz optimization takes a lot of time. For the special cases in this subsection, Lipschitz optimization can be made much faster by the application of a dynamic Lipschitz constant, as was explained in the previous subsection. For the special case of Goyal and Kusy, Lipschitz optimization then took on average 5.83 seconds, for the minimal-repair model only 0.82 seconds, and for the inspection model 23.82 seconds. This is more than the golden-section search heuristic, but still not very much. The running time depends of course on the number of components; the experiments show that the runtime of the Lipschitz-optimization procedure increases approximately linearly in n. The running time also depends on the precision that is required. For less precision Lipschitz optimization becomes much faster. Future generations of computers will make the advantage of the golden-section search heuristic over Lipschitz optimization less important.

We can conclude that if a solution is required in little time, we can solve the relaxation and apply the improved feasibility procedure to obtain a solution for the above problem with a deviation of less than one percent on average. The improved feasibility procedure outperforms the algorithm of Goyal and Kusy not only by time and average deviation, but the maximum deviation is also much smaller. When precision is more important, we can apply the golden-section search heuristic to obtain a solution for the above problems with a deviation of almost zero percent on average. When optimality must be guaranteed or when running time is less important, Lipschitz optimization can be applied to obtain a solution with arbitrary precision.

Results for the block replacement model For the solution of (R) we applied one iteration of the multiple-interval golden-section search heuristic, that is, we do not double the number of subintervals until no improvement is found. Since in the previous subsection it turned out that four subintervals are mostly sufficient to find a solution for problem (Q), we chose the number four here as well.

In Table 11.4 the relevant results of the 4200 problem instances are given (for the renewal function in the block replacement model we used the approximation of Smeitink and Dekker 1990). The solutions of the IFP and of the algorithm of Goyal and Kusy are now compared with the values of $v(Q)$ obtained by the multiple-interval golden-section search heuristic.

The average running time of the relaxation is again very small. It is larger than the average running time of, for example, the inspection model, since the golden-section search is not applied once but four times, according to one iteration of the multiple-interval golden-section search heuristic.

From the results it is evident that the IFP outperforms the iterative algorithm of Goyal and Kusy. The average deviation is 0.658% for the algorithm of Goyal and Kusy and only 0.051% for the IFP. Besides, the maximum deviation for the IFP is quite moderate, 5.921%, whereas for the algorithm of Goyal and Kusy this can be as large as 39.680%. It can happen that the algorithm of Goyal and Kusy sometimes performs slightly better than the IFP, reflected in the minimum deviations of - 0.222% and -0.002%, respectively.

The golden-section search heuristic applied to solve problem (Q) needed again four intervals in most cases. The average running time of the heuristic is 10.26 seconds. Remember that the solutions of the algorithm of Goyal and Kusy and of the IFP are compared with the solutions according to the golden-section search heuristic. Notice that the negative deviations of -0.222% and -0.002% imply that both the algorithm of Goyal and Kusy and the IFP can in some cases be better than the golden-section search heuristic, though the differences are small. This implies that the golden-section search heuristic is not optimal, but that was already clear from the results in the previous subsection. However, in most cases the heuristic is better than the other two approaches, regarding the average deviations of 0.658% and 0.051% for the algorithm of Goyal and Kusy and the IFP compared to the heuristic.

The conclusion here is again that when a solution is required in little time, we can solve the relaxation and apply the improved feasibility procedure; this is better than the algorithm of Goyal and Kusy (in particular the maximum deviation is much smaller). When precision is more important, we can apply the golden-section search heuristic, at the cost of more time.

6.   CONCLUSIONS

In this paper we presented a general approach for the coordination of maintenance frequencies. We extended an approach by Goyal et al. that deals with components with a very specific deterioration structure and does not indicate how good the obtained solutions are. Extension of this approach enabled incorporation of well-known maintenance models like minimal repair, inspection and block replacement . We presented an alternative solution approach that can solve these models to optimality (except the block replacement model, for which our approach is used as a heuristic).

The solution of a relaxed problem followed by the application of a feasibility procedure yields a solution in little time and less than one percent above the minimal value. This approach outperforms the approach of Goyal et al. When precision is more important, a heuristic based on a golden-section search can be applied to obtain a solution with a deviation of almost zero percent. For the special cases of Goyal et al., the minimal-repair model and the inspection model, application of a procedure using a dynamic Lipschitz constant yields a solution with an arbitrarily small deviation from an optimal solution, with running times somewhat larger than those of the golden-section search heuristic.

In the solution approach of this paper many maintenance-optimization models can be incorporated. Not only the minimal-repair, inspection and block replacement models, but many others can be handled as well. It is also possible to combine easily *different* maintenance activities, for example to combine the inspection of a component with the replacement of another. Altogether, the approach presented here is a flexible and powerful tool for the coordination of maintenance frequencies for multiple components.

Appendix: Determination of Lipschitz Constant

We will prove here that the objective function of problem (Q) is Lipschitz on the interval $[T_l, T_u]$ for the special cases of Goyal et al., the minimal-repair model with an increasing rate of occurrence of failures, and the inspection model. Furthermore, we derive a Lipschitz constant L.

It is obvious that if L_i is a Lipschitz constant for the function $g_i(\cdot)$ (see (11.3)), then the Lipschitz constant L for the objective function of (Q) equals

$$L = S + \sum_{i=1}^{n} L_i, \tag{11.9}$$

with S the set-up cost. Consequently, we have to find an expression for L_i.

To do so, consider an arbitrary $i \in \{1, \ldots, n\}$ and determine which of the intervals $I_i^{(k)}$ (see Section 3.) overlap with the interval $[T_l, T_u]$. Clearly, this is for each k with $\lfloor T_l x_i^* \rfloor \le k \le \lfloor T_u x_i^* \rfloor$. Now define $L_i^{(k)}$ as the Lipschitz constant of $g_i(\cdot)$ on $I_i^{(k)}$ for each of these $k \ge 1$. If $\lfloor T_l x_i^* \rfloor = 0$, then let $L_i^{(0)}$ be the Lipschitz constant of $g_i(\cdot)$ on $[1/x_n^*, 1/x_i^*]$. We will show that

$$L_i = \max\{L_i^{(k)}\}, \quad \text{where } k \text{ ranges from } \max\{0, \lfloor T_l x_i^* \rfloor\} \text{ to } \lfloor T_u x_i^* \rfloor. \tag{11.10}$$

To prove this, observe first that if t_1, t_2 belong to the same interval $I_i^{(k)}$, then by definition

$$|g_i(t_1) - g_i(t_2)| \le L_i^{(k)} |t_1 - t_2| \le L_i |t_1 - t_2|.$$

If t_1, t_2 do not belong to the same interval then assume without loss of generality that $g_i(t_1) \ge g_i(t_2)$. For $t_1 < t_2$ with t_1 belonging to $I_i^{(k)}$ it then follows that

$$
\begin{aligned}
0 \le g_i(t_1) - g_i(t_2) \; & \le \; g_i(t_1) - \Phi_i(x_i^*) \\
& = \; g_i(t_1) - g_i((k+1)/x_i^*) \\
& \le \; L_i^{(k)}((k+1)/x_i^* - t_1) \\
& \le \; L_i^{(k)}(t_2 - t_1) \\
& \le \; L_i|t_1 - t_2|.
\end{aligned}
$$

The other case $t_2 < t_1$ can be derived in a similar way and so we have shown that

$$
|g_i(t_1) - g_i(t_2)| \le L_i|t_1 - t_2|,
$$

with L_i according to (11.10).

If we now find an expression for the Lipschitz constant $L_i^{(k)}$, then with (11.9) and (11.10) we have an expression for the Lipschitz constant L. In the proof of Lemma 2 we showed that if $M_i(t)$ is convex on $(0, \infty)$, then $\Phi_i(1/t)$ is also convex on $(0, \infty)$. We saw in the proof of Theorem 1 that $M_i(t)$ is convex on $(0, \infty)$ for the special cases of Goyal et al., the minimal-repair model with an increasing rate of occurrence of failures, and the inspection model. Consequently, for these models $\Phi_i(1/t)$ is convex on $(0, \infty)$. This implies that the derivative of the function $\Phi_i(1/t)$ is increasing, and consequently we obtain that for all $t_1 \le t_2 \in [1/x_n^*, 1/x_i^*]$:

$$
\begin{aligned}
|g_i(t_1) - g_i(t_2)| \; & = \; |\Phi_i(1/t_1) - \Phi_i(1/t_2)| \\
& \le \; -\frac{d}{dt}\Phi_i(1/t)\bigg|_{t=t_1} \cdot |t_1 - t_2| \\
& \le \; -\frac{d}{dt}\Phi_i(1/t)\bigg|_{t=1/x_n^*} \cdot |t_1 - t_2| \\
& = \; (x_n^*)^2 \Phi_i'(x_n^*)|t_1 - t_2|,
\end{aligned}
$$

so that

$$
L_i^{(0)} = (x_n^*)^2 \Phi_i'(x_n^*). \tag{11.11}
$$

By the same argument we find that for $k \ge 1$

$$
L_i^{(k)} = \max\left\{ -\frac{d}{dt}\Phi_i((k+1)/t)\bigg|_{t=k/x_i^*}, \; \frac{d}{dt}\Phi_i(k/t)\bigg|_{t=(k+1)/x_i^*} \right\},
$$

and so

$$
L_i^{(k)} = \max\left\{ \frac{k+1}{k^2}(x_i^*)^2\Phi_i'\left(\frac{k+1}{k}x_i^*\right), \; -\frac{k}{(k+1)^2}(x_i^*)^2\Phi_i'\left(\frac{k}{k+1}x_i^*\right) \right\}. \tag{11.12}
$$

Notice that both arguments in (11.12) are decreasing in k since $\Phi_i'(\cdot)$ is increasing. This implies that $L_i^{(k)}$ is maximal for $k = 1$ or $k = 0$. Consequently, (11.10) becomes

$$L_i = \begin{cases} L_i^{(\lfloor T_l x_i^* \rfloor)} & \text{if } \lfloor T_l x_i^* \rfloor \geq 1, \\ \max\{L_i^{(1)}, L_i^{(0)}\} & \text{if } \lfloor T_l x_i^* \rfloor = 0, \end{cases}$$

with $L_i^{(k)}$ given by (11.11) and (11.12).

This analysis also shows that the Lipschitz constant L is decreasing in T. That is, if L_1, L_2 are the Lipschitz constants for the objective function of (Q) on the intervals $[T_1, T_u]$ and $[T_2, T_u]$ respectively, with $T_1 \leq T_2 \leq T_u$, then $L_1 \geq L_2$.

References

[1] Backert, W., and Rippin, D.W.T. The Determination of Maintenance Strategies for Plants Subject to Breakdown. *Computers and Chemical Engineering*, **9** (1985) 113-126.

[2] Bazaraa, M.S., Sherali, H.D., and Shetty, C.M. Nonlinear Programming. *Theory and Algorithms*, 1993.

[3] Cho D.I., and Parlar, M. A Survey of Maintenance Models for Multi-unit Systems. *European Journal of Operational Research*, **51** (1991) 1-23.

[4] Dekker, R. Integration Optimisation, Priority Setting, Planning and Combining of Maintenance activities. *European Journal of Operational Research*, **82** (1995) 225-240.

[5] Dekker, R., Frenk, J.B.G., and Wildeman, R.E. How to Determine Maintenance Frequencies for Multi-Component Systems? A general Approach. *In reliability and Maintenence of Complex Systems*, Springer-Verlag, Berlin. 239-280, 1996.

[6] Dekker, R., Frenk, J.B.G., and Wildeman, R.E. How to determine maintenance frequencies for multi-component systems? A general approach. *Proceedings of the NATO-ASI*, Kemer-Antalya, Turkey, 1995.

[7] Dekker, R., and Roelvink, I.F.K. Marginal cost criteria for preventive replacement of a group of components. *European Journal of Operational Research*, **84** (1995) 467-480.

[8] Dekker, R., Van Der Duyn Schouten, F.A., and Wildeman, R.E. A Review of Multi-Component Maintenance Models with Economic Dependence. *Zeitschrift fur Operations Research*, **45** (1997) 411-435.

[9] Duffuaa, S.O., and Ben-Daya, M. An Extended Model for the Joint Overhaul Scheduling Problem. *International Journal of Operations & Production Management*, **14** (1994) 37-43.

[10] Goyal, S.K. Determination of Economic Packaging Frequency for Items Jointly Replenished. *Management Science*, **20** (1973) 293-298.

[11] Goyal, S.K., Determination of Optimum Packaging Frequency for items Jointly Replenished. *Management Science*, **21** (1974) 436-443.

[12] Goyal, S.K., and Gunashekharan, A. Determining Economic Maintenance Frequency of a Transport fleet. *International Journal of Systems Science*, **4** (1992) 655-659.

[13] Goyal, S.K., and Kusy, M.I. Determining Economic Maintenance Frequency for a Family of Machines. *Journal of the Operational Research Society*, **36** (1985) 1125-1128.

[14] Goyal, S.K., and Satir, A.T. Joint Replenishment Inventory Control: Deterministic and Stochastic Models. *European Journal of Operational Research*, **38** (1989) 2-13.

[15] Hiriart-Urruty, J.-B., and Lemarichal, C. Convex Analysis and Minimization Algorithms I: *Fundamentals*, **305** of A series of comprehensive studies in mathematics. Springler-Verlag, Berlin. 1993.

[16] Horst, R.A., and Pardalos. P.M. Handbook of Global Optimization. *Kluwer Academic Publishers*, 1995.

[17] Howard, R. Dynamic Programming and Markov Process. *Wiley*, New York, 1960.

[18] Martos, B. Nonlinear Programming Theory and Methods. *Akademiai Kiado*, Budapest, 1975.

[19] Smeitink, E., and Dekker, R. A Simple Approximation to the Renewal Function. *IEEE Transactions on Reliability*, **39** (1990) 71-75.

[20] Van Der Duyn Schouten, F.A., Van Vlijmen, B., and Vos De Wael, S. Replacement Policies for Traffic Control Signals. *IMA Journal of Mathematic Applied in Business and Industry* , **9** (1998) 325-346.

[21] Van Der Duyn Schouten, F.A., and Vanneste, S.G. Analysis and Computation of (n,N) Strategies for Maintenance of a Two-Component System. *European Journal of Operations Research*, **48** (1990) 260-274.

[22] Van Dijikhuzen, G., and Van Harten, A. Optimal Clustering of Repititive Frequency-Constrained Maintenance Jobs with Shared Setups. *European Journal of Operations Research* , **99**(3) (1997) 552-564.

[23] Van Egmond, R., Dekker, R., and Wildeman, R.E. Correspondence on: Determining Economic Maintenance Frequency of a Transport Fleet. *International Journal of Systems Science*, **26** (1995) 1755-1757.

[24] Van Eijs, M.J.G., Heuts, R.M.J., and Kleijnen, J.P.C. Analysis and Comparison of Two Strtegies for Multi-Item Inventory Systems with Joint Replenishment Costs. *European Journal of Operations Research*, **59** (1992) 405-412.

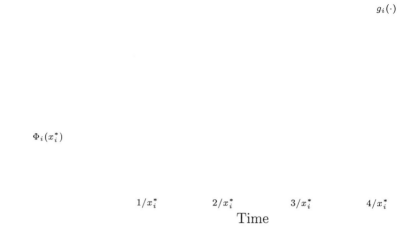

$g_i(\cdot)$

$\Phi_i(x_i^*)$

$1/x_i^*$ $2/x_i^*$ $3/x_i^*$ $4/x_i^*$

Time

Figure 11.1 An example of the function $g_i(\cdot)$. The thin lines are the graphs of the functions $\Phi_i(1/t), \Phi_i(2/t), \ldots, \Phi_i(5/t)$. The (bold) graph of $g_i(\cdot)$ is the lower envelope of these functions.

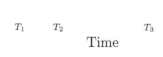

T_1 T_2 T_3

Time

Figure 11.2 An example of the objective function of problem (Q). There are three local minima: at T_1, T_2 and T_3.

$\Phi_i(x_i^*)$

$1/x_i^*$

Time

Figure 11.3 An example of the function $g_i^{(R)}(\cdot)$. Notice the similarity with the graph of $g_i(\cdot)$ in Figure 11.1.

$T(R)$

Time

Figure 11.4   An example of the objective function of problem (R).

Table 11.1   Values for the parameters in the four models

$n = 3, 5, 7, 10, 25, 50$
$S = 10, 50, 100, 200, 500, 750, 1000$
$c_i^p \in [1, 500]$ (random)

Special Case of Goyal and Kusy:	*Minimal-Repair Model:*
$f_i \in [15, 50]$ (random)	$\lambda_i \in [1, 20]$ (random)
$v_i \in [1, 20]$ (random)	$\beta_i \in [1.5, 4]$ (random)
$e \in [1, 4]$ (random)	$c_i^r \in [1, 250]$ (random)

Inspection Model:	*Block-Replacement Model:*
$\lambda_i \in [1, 20]$ (random)	$\lambda_i \in [1, 20]$ (random)
$\beta_i \in [1.5, 4]$ (random)	$\beta_i \in [1.5, 4]$ (random)
$c_i^u \in [c_i^p/\mu_i + 1, 1000]$ (random)	$c_i^f \in [2c_i^p/(1 - \sigma_i^2/\mu_i^2) + 1, 5000]$ (random)

The variables μ_i and σ_i in this table (for the block-replacement and inspection model) are the expectation and the standard deviation of the the lifetime distribution of component i. Notice that for the inspection model we take $c_i^u \geq c_i^p/\mu_i + 1$ and for the block-replacement model $c_i^f \geq 2c_i^p/(1 - \sigma_i^2/\mu_i^2) + 1$. This guarantees the existence of a finite minimum x_i^* for the individual average-cost function $\Phi_i(\cdot)$. In Dekker (1995) it is shown that for the inspection model a finite minimum for $\Phi_i(\cdot)$ exists if $c_i^u < c_i^u \mu_i$, and, a forteriori, if $c_i^u \geq c_i^p/\mu_i + 1$. For the block-replacement model it can be shown (see also Dekker 1995) that a finite minimum exists if $c_i^f > 2c_i^p/(1 - \sigma_i^2/\mu_i^2)$. Notice finally that since $\beta_i \geq 1$, the rate of occurrence of failure for the minimal repair model is increasing.

Table 11.2 Results of 4200 random examples for the special case of Goyal and Kusy, the minimal-repair model and the inspection model

	GoyKus	MinRep	Inspec
Average running time relaxation (sec.)	0.01	0.01	0.06
Improved Feasibility Procedure (IFP):			
Average running time IFP (sec.)	0.07	0.05	1.32
Average deviation IFP $(v(IFP)-v(Q))/v(Q)$	0.443%	0.065%	0.129%
Minimum deviation IFP	0.000%	0.000%	0.000%
Maximum deviation IFP	10.842%	4.250%	7.184%
Golden-Section Search (GSS):			
Average running time GSS (sec.)	0.72	0.41	11.81
Average deviation GSS $(v(GSS)-v(Q))/v(Q)$	0.001%	0.000%	0.000%
Minimum deviation GSS	0.000%	0.000%	-0.001%
Maximum deviation GSS	0.334%	0.152%	0.107%
Iterative Algorithm Goyal and Kusy (GK):			
Average running time GK (sec.)	0.07	0.12	4.64
Average deviation GK $(v(GK)-v(Q))/v(Q)$	0.829%	0.421%	1.253%
Minimum deviation GK	0.000%	-0.001%	0.000%
Maximum deviation GK	11.654%	18.289%	66.188%

Table 11.3 Percentage of the examples where the IFP and the algorithm of Goyal and Kusy generated deviations of more than 1% and 5%.

Algorithm	Deviation > 1%	Deviation > 5%
	Special Case of Goyal and Kusy	
IFP	12.86	1.79
Goyal and Kusy	27.50	2.10
	Minimal-Repair Model	
IFP	1.57	—
Goyal and Kusy	12.38	1.64
	Inspection Model	
IFP	3.12	0.05
Goyal and Kusy	26.50	6.69

Table 11.4   Results of 4200 random examples for the block-replacement model

Average running time relaxation (sec.)	0.23
Improved Feasibility Procedure (IFP):	
Average running time IFP (sec.)	1.30
Average deviation IFP $(v(\text{IFP})-v(Q))/v(Q)$	0.051%
Minimum deviation IFP	-0.002%
Maximum deviation IFP	5.921%
Golden-Section Search (GSS):	
Average running time GSS (sec.)	10.26
Iterative Algorithm Goyal and Kusy (GK):	
Average running time GK (sec.)	3.72
Average deviation GK $(v(\text{GK})-v(Q))/v(Q)$	0.658%
Minimum deviation GK	-0.222%
Maximum deviation GK	39.680%

Chapter 12

MAINTENANCE GROUPING IN MULTI-STEP MULTI-COMPONENT PRODUCTION SYSTEMS

Gerhard van Dijkhuizen
University of Twente
The Netherlands
G. C. vanDijkhuizen@sms.utwente.nl

Abstract   Since maintenance jobs often require one or more preparatory set-up activities to be carried out in advance, there is a perspective of significant savings in both set-up times and costs by conducting these jobs simultaneously (maintenance grouping). This chapter starts with an overview of commonly used maintenance grouping theories and practices at different planning levels. To this end, a clear distinction is made between long term (static), medium term (dynamic) and short term (opportunistic) grouping possibilities. Subsequently, and in much more detail, we investigate the possibilities for long term grouping (clustering) of preventive maintenance jobs in a multi-setup multi-component production system . More specifically, we consider the case where each component is maintained preventively at an integer multiple of a certain basis interval , which is the same for all components, and corrective maintenance is carried out in between whenever necessary. A general mathematical modeling framework is presented which allows for a large class of failure characteristics and preventive maintenance strategies for each component.

Keywords: Preventive Maintenance, Maintenance Grouping, Maintenance Frequency, Opportunistic Grouping, Direct Grouping, Indirect Grouping, Dynamic Grouping.

1.   INTRODUCTION

Unfortunately, almost all production systems in our society are subject to random failure of one or more of their components. In general, these failures may severely affect the performance and profitability of

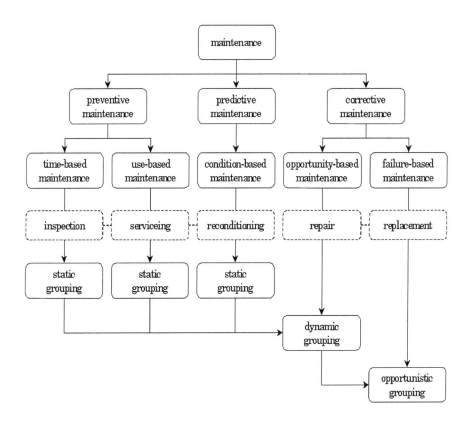

Figure 12.1 Maintenance initiators, activities, and packages: an overall perspective.

such systems, since they often occur at inconvenient times, and usually involve high recovery and consequential costs. Nowadays, it is widely recognized that the influence of failures on system performance is not just to be taken for granted. More than ever, manufacturing industries have realized that there is a huge potential of efficiency improvements, if the number of component failures could be reduced and/or components could be repaired or replaced before they fail. In the past few decades, this has given rise to an increased effort to introduce technological innovations (e.g. design for reliability , design for maintainability) , and better preventive maintenance concepts in practice as well. At the same time, a growing interest can be observed in theoretical literature concerning the modeling and optimization of maintenance and reliability in failure prone systems.

Traditionally, production and operations managers have rarely viewed the maintenance function as a competitive factor in their firm's business strategies. Nowadays, as many industries are moving towards just-in-time production, and at the same time rely on highly mechanized and automated production systems and processes, the strategic importance of maintenance is widely recognized. More than ever, maintenance is being considered as a basis factor to satisfy production needs, rather than a necessary evil. Therefore, it should be managed together with production on an equal basis, with an open eye for their interactions [37]. Although many steps can be undertaken to maintain or improve the overall performance of a production system , only a few of them are normally considered to be the responsibility of the maintenance department. The most fundamental decision problems that are faced by maintenance management are:

- which items are to be maintained?

- what kind of maintenance must be conducted?

- when should these activities take place?

In general, these decisions are expressed in terms of an overall maintenance concept, prescribing which maintenance activities must be carried out at which times, and under which conditions [22]. In view of such a maintenance concept, a clear distinction is usually made between maintenance initiators, maintenance activities, and maintenance packages (see Figure 12.1). Within safety and legislative restrictions, a maintenance concept should be based upon an overall attempt to minimize both direct maintenance costs (e.g. labour, materials), and indirect maintenance costs (e.g. production loss, deterioration costs). In practice, estimation of indirect maintenance costs is usually very difficult. Nevertheless, they are often much larger than direct maintenance costs [31].

1.1   Maintenance Initiators. As a starting point, there has to be some kind of control mechanism with which the need for maintenance is initiated. Simply stated, maintenance initiators used in practice can be categorized into preventive maintenance (before failure), predictive maintenance (just before failure), and corrective maintenance (upon or after failure). More specifically, the most commonly applied maintenance initiators are:

- time-based maintenance,

- use-based maintenance,

- condition-based maintenance,

- opportunity-based maintenance,

- failure-based maintenance.

Preventive maintenance is planned and performed before failure , and is either time-based (e.g. on a weekly basis), use-based (e.g. based on running hours), or a combination of both. Predictive maintenance aims at the initiation of preventive maintenance just before failure , and is mostly condition-based (e.g. if too much vibration is observed). Finally, corrective maintenance is performed upon or after failure , and is either opportunity-based (not urgent) or failure-based (urgent) .

1.2   Maintenance Activitities. Excluding the influence of technological improvements in equipment design and layout (e.g. modifications), the most basic maintenance activities can be classified as follows:

- inspection ,

- serviceing,

- reconditioning ,

- repair,

- replacement.

As a starting point, most items will be inspected regularly, in order to detect any signs of reduced effectiveness and/or impending failure . Additionally, items will normally be serviced at regular intervals (e.g. readjusted, lubricated, cleaned) in order to ensure continued effective operation in the future. Moreover, reconditioning activities will often be carried out in order to sustain satisfactory operation of items or equipment before they fail. Upon failure , repairs will normally be required to restore the equipment into satisfactory operation. Finally, replacement of items and equipment will occur when they are no longer capable of proper functioning, or are beyond economic repair.

1.3   Maintenance Packages. Since maintenance activities often require one or more preparatory set-up activities (e.g. crew travelling, equipment rental, dismantling), there is a perspective of significant gains if they can be carried out simultaneously (maintenance grouping). In general, three different types of maintenance grouping can be identified:

- static grouping ,

- dynamic grouping ,

- opportunistic grouping .

In the long term, planned preventive maintenance activities are combined into so-called preventive maintenance packages , each of which is more or less treated as a single maintenance activity in an operational planning phase (static grouping). In the medium term, planned preventive maintenance activities can be combined with each other, and with plannable corrective maintenance activities as well (dynamic grouping). In the short term, unplanned corrective maintenance activities can be combined with planned preventive and/or corrective maintenance activities (opportunistic grouping). Although each grouping strategy takes place at a different planning level, their mutual objective is to improve efficiency in terms of reducing set-up times and costs in an operational planning phase.

2.   LITERATURE REVIEW

Simply stated, the overall objective of maintenance optimization models is to determine the frequency and/or timing of preventive and/or corrective maintenance activities, in order to arrive at an optimal balance between the costs and benefits of both. In this section, we will briefly discuss some relevant literature on maintenance modeling and optimization. For clarity, this discussion is divided into mathematical models for single unit systems, and mathematical models for multiple unit systems, since the latter implies the existence of economies of scale in conducting maintenance activities simultaneously.

2.1   Maintenance Models for Single Components.

The general philosophy of most maintenance optimization models for single unit systems, is to decide at each feasible moment whether it is cost-effective to carry out preventive maintenance now, or to postpone it to the next feasible moment, e.g. see [6] and [19]. As a consequence, the main differences between these models originate from their interpretation of "feasible moments", or equivalently the mechanism with which preventive maintenance is, or can be, activated. In this respect, a clear distinction must be made between continuous review models, periodic review models, and opportunistic review models. The reader is referred to [27], [30], [33], and [34] for a more comprehensive review on existing literature. Here, we will only mention some important references.

2.1.1   Continuous Review Models. In continuous review models, it is assumed that the condition of the system can be monitored continuously. As a consequence, preventive maintenance is usually of a predictive, condition-based nature. According to [28], there are five basic techniques typically used in condition monitoring: vibration monitoring, process parameter monitoring, thermography, tribology and visual inspection . Continuous monitoring of those parameters that allow the accurate prediction of failure will permit precise scheduling of repairs without the costs of emergency downtime. Mathematical models in this area derive their value from finding the parameters, and corresponding threshold values, with which the occurrence of failures can be predicted accurately. The reader is referred to e.g. [5] for an introduction into the practice, methods and applications of condition monitoring techniques.

2.1.2   Periodic Review Models. In periodic review models, it is assumed that the condition of the system cannot be monitored continuously, as is the case in continuous review models, but only through periodic inspection at fixed costs. In these models, inspections are usually carried out at regular intervals , and are either time-based or use-based. In general, use-based maintenance policies outperform time-based maintenance policies in view of efficiency. On the other hand, time-based maintenance policies do have the advantage that one does not have to keep track of individual component ages, as a result of which they can easily be implemented and executed in a practical context. Mathematical models in this area are usually concerned with finding the optimal maintenance interval , either time-based or use-based, in order to arrive at an optimal balance between the costs and benefits of preventive and corrective maintenance . Well-known maintenance models of the use-based type are the age replacement and minimal repair model [2]. Classical examples of time-based maintenance models are the block replacement model [4], the modified block replacement model [8], the standard inspection model [3], and the delay time model [10].

2.1.3   Opportunistic Review Models. In opportunistic review models, it is assumed that inspections cannot be carried out at any time, as is the case in periodic review models, but only at so-called maintenance opportunities. The underlying observation behind these models is that in many practical situations, preventive maintenance on non-critical units is delayed to some moment in time where the unit is not required for production. In general, such opportunities may arise due to e.g. random breakdowns and/or withdrawn production orders. Because of the random occurrence of opportunities, and because of their sometimes restricted duration, traditional maintenance models fail to make effective use of them. Mathematical models in this area are primarily used to determine whether a maintenance activity must be conducted at a given opportunity, or whether it must be postponed to the next one, e.g. see [7], [16], [14], [40] and [41].

2.2 Maintenance Models for Multi-Component Systems.

The justification of most maintenance optimization models for multiple unit systems is a potential of reductions in set-up costs and/or times if maintenance activities are carried out simultaneously (maintenance grouping). As we explained in the previous section, mathematical models in this area can be categorized into static grouping, dynamic grouping, and opportunistic grouping strategies. Although each grouping strategy takes place at a different planning level, their mutual objective is to improve efficiency in terms of reducing set-up times and costs in an operational planning phase. The reader is referred to [9] and [18] for an extensive and up-to-date literature review on maintenance models for multi-unit systems with economic dependence. Here, we will restrict ourselves to some important references.

2.2.1 Static Grouping Models.

Static grouping refers to the combination of planned preventive maintenance activities in a strategical planning phase. In this respect, a clear distinction must be made between direct and indirect grouping models. In direct grouping models, the collection of preventive maintenance activities is partitioned into several maintenance packages, each of which is executed at an interval that is optimal for that package. In indirect grouping models, maintenance packages are not determined in advance, but are formed indirectly whenever the maintenance of different units coincides. To achieve this, each maintenance activity is carried out at an integer multiple of a certain basis interval. Basically, static grouping models attempt to find the optimal balance between the costs of deviating from the optimal preventive maintenance intervals for individual units, and the benefits of combining preventive maintenance activities on different units. Typical examples of static grouping models can be found in e.g. [21], [24], [23], [15], [39] and [38].

2.2.2 Dynamic Grouping Models.

Dynamic grouping refers to the combination of planned preventive maintenance activities with each other, and/or with plannable corrective maintenance activities, in a tactical planning phase. Of course, the latter is only possible if the repair of failed units can be postponed to a more suitable moment in time, e.g. because standby units are available, or the unit does not affect the system as a whole. The main difficulty of dynamic grouping models is that the failure of a unit cannot be predicted in advance. Therefore, dynamic grouping models often make use of a so-called rolling horizon approach. More specifically, they use a finite horizon in order to arrive at a sequence of decisions, but only implement the first one. Basically, mathematical models for dynamic grouping derive their value from finding an optimal balance between the costs of postponing corrective maintenance activities, and the benefits of combining them with other preventive and/or corrective maintenance activities. Typical examples of dynamic grouping models are presented in e.g. [1], [32], [26], and [44].

2.2.3   Opportunistic Grouping Models. Opportunistic grouping refers to the combination of planned maintenance activities with unplanned maintenance activities in an operational planning phase. In these models, the failure of a particular unit is used as an opportunity for planned maintenance on other units. In practice, this means that opportunistic maintenance grouping is difficult to manage, since it affects the plannable nature of preventive maintenance . Nevertheless, if all the preparations needed for preventive maintenance have been made in advance, it certainly is an effective method to reduce set-up costs and times in an operational planning phase. Basically, mathematical models for opportunistic grouping attempt to find an optimal balance between the costs of advancing planned maintenance activities, and the benefits of combining them with other unplanned maintenance activities. Typical examples of opportunistic grouping models can be found in e.g. [25], [29], [35], [36], [17], and [43].

3.   A MULTI-SETUP MULTI-COMPONENT PRODUCTION SYSTEM

Typically, mathematical models for the modeling and optimization of maintenance in multi-component production systems restrict themselves to preventive maintenance jobs with identical, so-called common set-up activities, e.g. see [24] and [15]. The latter implies that creating an occasion for preventive maintenance on one or more components requires a fixed set-up cost , irrespective of how many and which components are maintained. Although this might be an interesting approach from a theoretical point of view, it is obvious that nowadays production systems are usually much more complicated. Ideally, however, preventive maintenance models should at least account for multiple set-up activities and components, allowing different set-up costs for different components, or groups of components. On the other hand, it seems virtually impossible to support a separate, independent data structure for each possible group of components, that could arise in the most general situation.

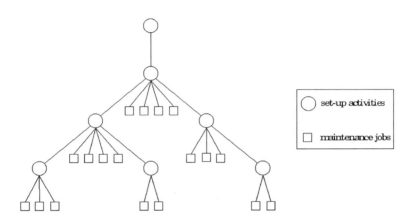

Figure 12.2 Tree-like structure of set-up activities and maintenance jobs (example).

In order to arrive at a compromise, a powerful modeling framework has been developed, in which a variety of complex set-up structures can be modeled to a proper level of detail. Within this modeling framework , it is assumed that the collection of set-up activities can be ordered hierarchically into a tree -like structure , in which each maintenance job can be associated with exactly one set-up activity (see Figure 12.2). Considering this tree -like structure , it is now immediately clear that some maintenance jobs may not share all set-up activities, but only a subset of them. Obviously, these possibilities for shared set-up activities provide a richer and more realistic modeling framework in comparison with the requirement of completely coinciding paths, as is the case with common set-ups .

To a certain extent, the above-mentioned concept of shared set-up activities originates from [20], who developed a somewhat similar but less powerful modeling framework , in which maintenance jobs can only

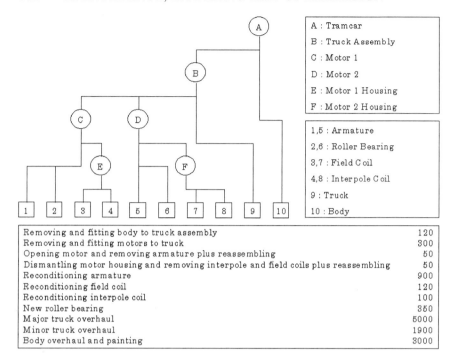

| A : Tramcar |
| B : Truck Assembly |
| C : Motor 1 |
| D : Motor 2 |
| E : Motor 1 Housing |
| F : Motor 2 Housing |

| 1,5 : Armature |
| 2,6 : Roller Bearing |
| 3,7 : Field Coil |
| 4,8 : Interpole Coil |
| 9 : Truck |
| 10 : Body |

Removing and fitting body to truck assembly	120
Removing and fitting motors to truck	300
Opening motor and removing armature plus reassembling	50
Dismantling motor housing and removing interpole and field coils plus reassembling	50
Reconditioning armature	900
Reconditioning field coil	120
Reconditioning interpole coil	100
New roller bearing	350
Major truck overhaul	5000
Minor truck overhaul	1900
Body overhaul and painting	3000

Figure 12.3 Multiple interrelated set-up and maintenance activities with associated costs for a tramcar (Sculli and Suraweera, 1979).

be defined at the lowest-level set-up activities. Practical examples of shared, but not common, set-ups can be found in various areas, e.g. in aircraft maintenance, maintenance of nuclear power plants, off-shore maintenance, and even tramcar maintenance (see Figure 12.3). Summarizing, the notion of shared but not common set-up activities seems to be common sense in practice, but at the same time an underexposed point of view in existing literature.

In the remainder of this chapter, we will present a powerful modeling framework which allows for the precise scheduling of preventive maintenance activities in terms of a repetitive maintenance cycle, thereby taking into account several opportunities for efficiency improvement by reducing set-up times and/or costs. The framework is capable of dealing with extremely complex set-up structures, and a large class of preventive maintenance strategies for each component as well. Therefore, it is well-suited for implementation in existing or future decision support systems for maintenance modeling and optimization.

4.   MODELING FRAMEWORK

Consider a production system consisting of m set-up activities, denoted $\mathcal{I} = \{1, ..., m\}$, and n components, denoted $\mathcal{J} = \{1, ..., n\}$, which are organized in a tree -like structure . Creating an opportunity for preventive maintenance on component $j \in \mathcal{J}$ requires a collection $I_j \subseteq \mathcal{I}$ of preparatory set-up activities to be carried out in advance, with corresponding set-up costs $\sum_{i \in I_j} s_i$. Here, $s_i > 0$ denotes the individual cost of set-up activity $i \in \mathcal{I}$. If components are maintained simultaneously, the corresponding set-up activities can be combined. More specifically, preventive maintenance on a subset of components $U \subseteq \mathcal{J}$ involves a set-up cost $s(U)$, which depends completely on the set of required set-up activities:

$$s(U) = \sum_{\substack{i \in \bigcup_{j \in U} I_j}} s_i. \tag{12.1}$$

Within this setting, we consider preventive maintenance activities of the **block type.** To be specific, each component is maintained preventively at fixed intervals (e.g. daily, weekly, monthly, yearly), whereas intermediate corrective maintenance activities are carried out upon failure . With $\Phi_j(x)$, we denote the expected maintenance costs (exclusive of preventive set-up costs) of component $j \in \mathcal{J}$ per unit of time, if maintained preventively every $x > 0$ time units (or any other measurable quantity e.g. running hours). For notational convenience, and without loss of generality, we restrict ourselves to cost functions of the following type:

$$\Phi_j(x) = \frac{c_j + M_j(x)}{x}. \tag{12.2}$$

Here, $c_j > 0$ reflects the expected cost of preventive maintenance on component $j \in \mathcal{J}$. Moreover, $M_j(x)$ denotes the expected cumulative deterioration costs (due to failures , repairs , etc.), x time units after its last preventive maintenance . By doing this, a variety of maintenance models can be incorporated, allowing different models for each component . The reader is referred to [13] for an extensive list of block-type models. Here, we only mention some important ones:

- in the **standard block replacement** model [4], a component is replaced correctively upon failure , and preventively at fixed intervals of length $x > 0$;

- in the **modified block replacement** model [8], a component is replaced correctively upon failure , and preventively at fixed intervals of length $x > 0$, but only if its age exceeds a certain threshold value $y < x$;

- in the **minimal repair** model [2], a component is replaced preventively at fixed intervals of length $x > 0$, with intermediate failure repairs occurring whenever necessary, restoring the component into a state as good as before failure ;

- in the **standard inspection** model [3], a component is inspected preventively at fixed intervals of length $x > 0$, followed by a corrective replacement if it turns out to have failed upon inspection ;

- in the **delay time** model [11], a component passes through a visible defective state before it actually fails somewhat later; in line with this, the component is replaced correctively upon failure , and inspected preventively at fixed intervals of length $x > 0$, followed by a preventive replacement if it turns out to be defective at the time of inspection .

In our modeling framework , we assume that the initiation of preventive maintenance on different components is based on the same system parameter x, e.g. calendar or operating time, and that the deterioration cost functions $M_j(x)$ are all available. Moreover, we assume that the individual cumulative deterioration costs $M_j(.)$ for each component are totally independent of all other components. Of course, other assumptions would lead to other interesting versions of the problem under consideration, but are left for future research.

4.1   Further Notation and Assumptions.

In the sequel, we will use the following notation and assumptions to characterize the mutual relationships between set-up activities and components. First of all, we denote with $J_i = \{j \in \mathcal{J} \mid i \in I_j\}$ and $J_i^* \subseteq J_i$ the set of components that require resp. are attached to set-up activity $i \in \mathcal{I}$. In a similar way, we let $S_i \subset \mathcal{I}$ and $S_i^* \subset \mathcal{I}$ denote the set-up activities $i' \in \mathcal{I}$ that require resp. are attached to set-up activity $i \in \mathcal{I}$. As an illustrative example, consider a production system consisting of $m = 3$ set-up activities and $n = 3$ components, and assume that component j requires all set-up activities $i \leq j$ to be carried out in advance. Then it is easily verified that:

$$I_1 = \{1\}, \; I_2 = \{1, 2\}, \; I_3 = \{1, 2, 3\},$$
$$J_1 = \{1, 2, 3\}, \; J_2 = \{2, 3\}, \; J_3 = \{3\},$$
$$J_1^* = \{1\}, \; J_2^* = \{2\}, \; J_3^* = \{3\}, \qquad (12.3)$$
$$S_1 = \{2, 3\}, \; S_2 = \{3\}, \; S_3 = \varnothing,$$
$$S_1^* = \{2\}, \; S_2^* = \{3\}, \; S_3^* = \varnothing.$$

For notational convenience, and without loss of generality, we assume that $i' > i$ for all $i \in \mathcal{I}$ and $i' \in S_i$ in the sequel. Moreover, and in line with the above, we assume that $I_1 \cap ... \cap I_n = \{1\}$, i.e. there exists exactly one common set-up activity $i = 1$. If $I_1 \cap ... \cap I_n = \emptyset$, the problem can be decomposed into two or more (smaller) subproblems, which can be treated and solved separately. If $|I_1 \cap ... \cap I_n| = k > 1$, the common set-up activities $i \in \{1, ..., k\}$ can as well be replaced by a single common set-up activity, with corresponding set-up costs $s_1 + ... + s_k > 0$. For similar reasons, we can assume that $J_i \neq \emptyset$ for all $i \in \mathcal{I}$, since obviously set-up activities with no components can as well be neglected without affecting the problem.

4.2 Problem Definition.
The coordination of preventive maintenance activities is now modeled as follows: preventive maintenance on component $j \in \mathcal{J}$ is carried out at integer multiples of $k_j \cdot t$ time units, i.e. at times $\{0, k_j \cdot t, 2 \cdot k_j \cdot t, ...\}$, where $k_j \in \mathbb{N}^*$ denotes the **maintenance frequency** of component $j \in \mathcal{J}$ relative to a **basis interval** of $t > 0$ time units. Obviously, this corresponds to a repetitive, periodic preventive **maintenance cycle** of $\operatorname{lcm}(k_1, ..., k_n) \cdot t$ time units, where $\operatorname{lcm}(k_1, ..., k_n)$ denotes the least common multiplier of the integers k_j ($j \in \mathcal{J}$), e.g. $\operatorname{lcm}(2, 3, 4) = 12$. In general, this leads to an optimization in $n + 1$ variables.

As a starting point, let us restrict ourselves to the case where the maintenance interval $t > 0$ is fixed, and the integer frequency k_j must be chosen from a finite set of possibilities, say $k_j \in \mathcal{K}$ for all $j \in \mathcal{J}$. This situation arises if t is a natural time limit (e.g. 1 day, week, or month) and one is interested in a maintenance cycle of prescribed length (e.g. 1 year). In general, this yields an optimization problem in n variables $(k_1, ..., k_n)$, with $|\mathcal{K}|^n$ different solutions, where our objective is to minimize average maintenance costs per unit of time:

$$(P) \quad \min_{k_j \in \mathcal{K}} \left\{ \sum_{i \in \mathcal{I}} \frac{s_i \cdot \Delta_i(\mathbf{k})}{t} + \sum_{j \in \mathcal{J}} \Phi_j(k_j \cdot t) \right\}. \qquad (12.4)$$

Here, $\Delta_i(\mathbf{k})$ represents the fraction of times that set-up activity $i \in \mathcal{I}$ is carried out at an **opportunity** for preventive maintenance (e.g.

Table 12.1   Example of a maintenance cycle for a production system consisting of $m = 3$ set-up activities and $n = 3$ components , where component j requires all set-up activities $i \leq j$ to be carried out in advance.

							components							3
	2		3	2				2	3		2			2
1	1	1	1	1	1	1	1	1	1	1	1	1	1	1
			3			*set-ups*		3						3
	2		2	2				2	2		2			2
1	1	1	1	1	1	1	1	1	1	1	1	1	1	1
						frequencies								5
	3		5	3				3	5		3			3
1	1	1	1	1	1	1	1	1	1	1	1	1	1	1
						opportunities								
1	2	3	4	5	6	7	8	9	10	11	12	13	14	15

$\Delta_1(\mathbf{k}) = 1$, $\Delta_2(\mathbf{k}) = 7/15$, and $\Delta_3(\mathbf{k}) = 1/5$ in Table 12.1). Since each occurrence of component $j \in \mathcal{J}$ implies the occurrence of all set-up activities $i \in I_j$, it is easily observed that $\max\{k_j^{-1} \mid j \in J_i\} \leq \Delta_i(\mathbf{k}) \leq 1$ for all $i \in \mathcal{I}$, implying that $\Delta_i(\mathbf{k}) = 1$ if $k_j = 1$ for some $j \in J_i$. Based on the principle of inclusion and exclusion, [12] derives the following general expression for $\Delta_i(\mathbf{k})$:

$$\Delta_i(\mathbf{k}) = \sum_{l=1}^{|J_i|} (-1)^{l+1} \sum_{\{j_1,\dots,j_l\} \subseteq J_i} \operatorname{lcm}(k_{j_1}, \dots, k_{j_l})^{-1}. \qquad (12.5)$$

Typically, the basis maintenance interval t is restricted to several days or weeks, whereas the corresponding maintenance cycle $\operatorname{lcm}(k_1, \dots, k_n) \cdot t$ varies from several weeks to several months, or even years. In case of a prescribed cycle length of $T = N \cdot t > 0$ time units, appropriate values for k_j must satisfy $k_j \mid N$, or stated otherwise $N \bmod k_j = 0$. Amongst others, this approach is particularly useful in calendar-based maintenance planning systems, by which workload and capacity profiles have to be matched on a regular basis. On the other hand, if there are no explicit constraints on the length of the basis maintenance interval and the corresponding maintenance cycle , our optimization problem becomes even more complex:

$$(Q) \quad \min_{t>0} \min_{k_j \in \mathbb{N}^*} \left\{ \sum_{i \in \mathcal{I}} \frac{s_i \cdot \Delta_i(\mathbf{k})}{t} + \sum_{j \in \mathcal{J}} \Phi_j(k_j \cdot t) \right\}. \qquad (12.6)$$

Examples of this type can be found in various industries, e.g. if the initiation of maintenance activities is based on cumulative operating time rather than calendar moments. In such cases, the need for well-defined maintenance intervals (e.g. multiples of 100 or 1000 running hours) is less restricted, and often based on intuitive or administrative reasoning only. The reader is referred to [38] for a detail discussion of problem (Q). Here, we will focus on problem (P), i.e. where the basis maintenance interval t and the set of possible maintenance frequencies $\mathcal{K}$ are fixed.

5. A MIXED INTEGER LINEAR PROGRAMMING FORMULATION

Let us now present a mixed integer linear programming formulation, with which problem (P) can be solved to optimality. The underlying observation behind this formulation is that, given the basis maintenance interval of $t > 0$ time units , the assignment of components $j \in \mathcal{J}$ to maintenance frequencies $k \in \mathcal{K}$ will always result in a preventive maintenance cycle of at most $\text{lcm}(\mathcal{K}) \cdot t$ time units . In line with this, we denote with $\mathcal{L} = \{1, ..., \text{lcm}(\mathcal{K})\}$ the set of so-called maintenance opportunities $\{0, t, 2 \cdot t, ...\}$ within this preventive maintenance cycle (e.g. $|\mathcal{L}| = 15$ in Table 12.1).

To continue our analysis, we denote with $K_l = \{k \in \mathcal{K} \mid l \bmod k = 0\}$ the set of maintenance frequencies that correspond with maintenance opportunity $l \in \mathcal{L}$ (e.g. $K_5 = K_{10} = \{1, 5\}$ in Table 12.1). The problem now consists of assigning components $j \in \mathcal{J}$ to maintenance frequencies $k \in \mathcal{K}$, in such a way that the costs of the corresponding maintenance cycle are minimized. In our model, the assignment of set-up activities $i \in \mathcal{I}$ to maintenance opportunities $l \in \mathcal{L}$ is comprised into variables $x_{il} \in \{0, 1\}$, whereas the assignment of components $j \in \mathcal{J}$ to maintenance frequencies $k \in \mathcal{K}$ is represented by variables $y_{jk} \in \{0, 1\}$:

$$x_{il} = \begin{cases} 1 & \text{if set-up } i \in \mathcal{I} \text{ is assigned to opportunity } l \in \mathcal{L}, \\ 0 & \text{otherwise,} \end{cases} \qquad (12.7)$$

$$y_{jk} = \begin{cases} 1 & \text{if component } j \in \mathcal{J} \text{ is assigned to frequency } k \in \mathcal{K}, \\ 0 & \text{otherwise.} \end{cases}$$
$$(12.8)$$

With $a_{il} = s_i/(\text{lcm}(\mathcal{K}) \cdot t)$, we denote the average cost per unit of time associated with the assignment of set-up activity $i \in \mathcal{I}$ to maintenance opportunity $l \in \mathcal{L}$. Similarly, $b_{jk} = \Phi_j(k \cdot t)$ denotes the average cost per unit of time associated with the assignment of component $j \in \mathcal{J}$ to maintenance frequency $k \in \mathcal{K}$. With this in mind, problem (P) can now be formulated in terms of a mixed integer linear program , where our objective is to minimize average maintenance costs per unit of time:

$$\min \sum_{i \in \mathcal{I}} \sum_{l \in \mathcal{L}} a_{il} \cdot x_{il} + \sum_{j \in \mathcal{J}} \sum_{k \in \mathcal{K}} b_{jk} \cdot y_{jk} \qquad (12.9)$$

First of all, we need a restriction to ensure that the assignment of component $j \in \mathcal{J}$ to maintenance frequency $k \in \mathcal{K}$ causes the corresponding set-up activity to be carried out at the corresponding maintenance opportunities too.

$$x_{il} \geq y_{jk} \qquad \forall i \in \mathcal{I}, l \in \mathcal{L}, j \in J_i^*, k \in K_l \qquad (12.10)$$

For similar reasons, we need a restriction to make sure that the assignment of set-up activity $i \in \mathcal{I}$ to maintenance opportunity $l \in \mathcal{L}$ causes the parental set-up activities to be carried out at the same maintenance opportunity too.

$$x_{il} \geq x_{i'l} \qquad \forall i \in \mathcal{I}, l \in \mathcal{L}, i' \in S_i^* \qquad (12.11)$$

Finally, we need a set of restrictions to make sure that exactly one maintenance frequency is assigned to each component :

$$\sum_{k \in \mathcal{K}} y_{jk} = 1 \qquad \forall j \in \mathcal{J} \qquad (12.12)$$

Note that the structure of these equations is such that $x_{il} \in \{0,1\}$ for all $i \in \mathcal{I}$ and $l \in \mathcal{L}$ as soon as $y_{jk} \in \{0,1\}$. In other words, we do not need to formulate explicit integrality constraints for x_{il}.

5.1   Reduction of Maintenance Opportunities.
In general, the number of maintenance opportunities lcm($\mathcal{K}$) grows exponentially with the set of possible maintenance frequencies $\mathcal{K}$. Therefore, a more efficient mixed integer linear programming formulation has been developed, which is based on the following observations:

- maintenance opportunities $l \in \mathcal{L}$ with $K_l = \varnothing$ can as well be neglected, since evidently $x_{il} = 0$ for all $i \in \mathcal{I}$ in an optimal solution;

- maintenance opportunities $l_1, l_2 \in \mathcal{L}$, with $K_{l_1} = K_{l_2}$ and $l_1 \neq l_2$, can as well be replaced by a single maintenance opportunity l with $K_l = K_{l_1} = K_{l_2}$ and corresponding cost $a_{il} = a_{l_1} + a_{l_2}$ for all $i \in \mathcal{I}$, since evidently $x_{il_1} = x_{il_2}$ for all $i \in \mathcal{I}$ in an optimal solution.

Table 12.2 Reductions in problem size for several period sets $\mathcal{K} = \{1, ..., n\}$.

n	1	2	3	5	10	15	25
$|\mathcal{L}|$	1	2	6	60	2520	360360	26771144400
$|\mathcal{L}^*|$	1	2	4	12	48	192	2880

Obviously, these observations may lead to significant reductions in the problem size. To be specific, let us denote with $\mathcal{L}^*$ the reduced set of maintenance opportunities, and with $\eta(U)$ the number of times that precisely cluster $U \subseteq \mathcal{K}$ shows up in the maintenance cycle (e.g. $\eta(\{1,3\}) = 4$ and $\eta(\{1,5\}) = 2$ in Table 12.1). Then the number of opportunities $|\mathcal{L}^*|$ to be considered in the reduced version of the problem equals the number of clusters $U \subseteq \mathcal{K}$ with $\eta(U) > 0$. Based on the principle of inclusion and exclusion, $\eta(U)$ yields an expression which is similar to $\Delta_i(\mathbf{k})$:

$$\eta(U) = \mathrm{lcm}(\mathcal{K}) \cdot \sum_{l=|U|}^{|\mathcal{K}|} (-1)^{l-|U|} \sum_{U \subseteq \{k_1,...,k_l\} \subseteq \mathcal{K}} \mathrm{lcm}(k_1, ..., k_l)^{-1}. \quad (12.13)$$

In general, the determination of $\eta(U)$ with the use of this equation is a complex problem, since the number of clusters $U \subseteq \mathcal{K}$ grows exponentially with the number of possible maintenance frequencies $\mathcal{K}$. Therefore, a more efficient method has been constructed, which is described in more detail in [38]. Here, we restrict ourselves to maintenance cycles with a prescribed cycle length of $T = N \cdot t$, and feasible maintenance frequencies $k \mid N$. In view of keeping things simple, we assume that enumeration of all maintenance opportunities up to $\mathrm{lcm}(\mathcal{K}) = N$ can be done in reasonable time. Clearly, this approach leaves us with an alternative problem formulation with significantly less maintenance opportunities, and thus variables and constraints (see Table 12.2).

5.2 Reduction of Maintenance Frequencies.

When taking a closer look at the problem under consideration, it is immediately clear that another reduction in problem size can be obtained by observing that a component will never be executed with a certain frequency, as soon as integer multiples of that particular frequency involve lower individual maintenance costs. More specifically:

■ if $b_{jk_1} > b_{jk_2}$ and $k_1 \mid k_2$ for some $j \in \mathcal{J}$ and $k_1, k_2 \in \mathcal{K}$, then maintenance frequency k_1 can as well be neglected for component j, since obviously $y_{jk_1} = 0$ in an optimal solution.

Of course, this nice structural property should also be further exploited in the formulation of problem (P) as a mixed integer linear program , in particular if the number of components and/or maintenance frequencies grows (too) large. However, it depends on the individual maintenance cost functions whether these observations will lead to significant reductions in the problem size.

5.3   Computational Results.
The mixed integer linear programming formulation presented above was tested on a series of randomly created test problems of various types. For a detailed description of these test problems, we refer to [38]. Here, we only mention that the LP-relaxation generated an integer, and thus a feasible solution to problem (P), in approximately 68% of all test problems. Moreover, an optimal integer solution was usually found within a few iterations of the consecutive branch and bound algorithm. Probably, this phenomenon originates from the fact that problem (P) is closely related to a standard assignment problem, which is known to possess the above-mentioned integrality property. Apparently, the underlying hierarchical structure of set-up activities and components does not conflict too much with this nice and useful property of the standard assignment problem.

6.   AN ILLUSTRATIVE EXAMPLE

Let us now present an illustrative numerical example, in order to clarify the mixed integer linear programming formulation presented in the previous section. Our main objective here is to illustrate how the set of maintenance frequencies is translated into a set of maintenance opportunities, each of which corresponds to a subset of maintenance frequencies. Let us consider an arbitrary production system , and assume that preventive maintenance is carried out during weekends, i.e. the basic maintenance interval equals $t = 1$ week. A yearly maintenance cycle must be determined, which in practice means that a maintenance cycle of length $T = 48$ weeks must be determined. The remaining 4 weeks of the year are reserved for a major overhaul of the production system . Hence, the feasible maintenance frequencies for this production system are 1, 2, 3, 4, 6, 8 ,12, 24 and 48.

Assuming that each component is maintained preventively at the end of each maintenance cycle , let us now determine which weeks of the year correspond to which maintenance frequencies, and vice versa.

Obviously, maintenance frequency $k = 1$ corresponds to each week in the maintenance cycle , whereas e.g. maintenance frequency $k = 12$ corresponds to weeks 12, 24, 36 and 48. By doing this, a regular pattern of maintenance moments can be constructed, as is shown more explicitly in Figure 12.4.

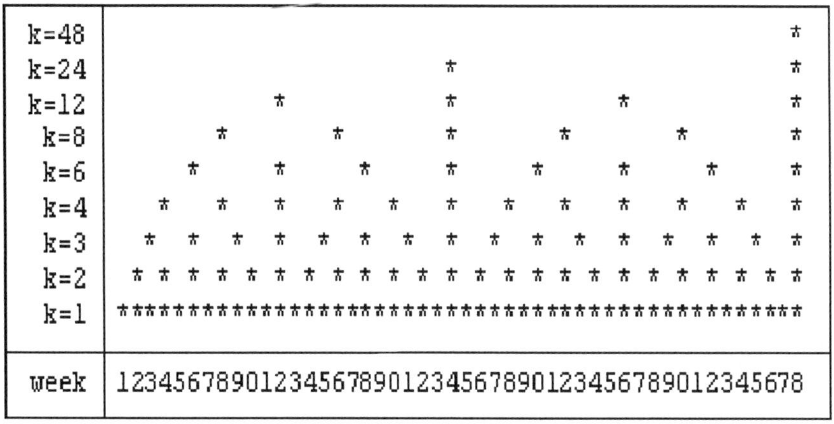

```
k=48                                                      *
k=24                           *                          *
k=12              *            *            *             *
 k=8         *         *            *            *        *
 k=6       *      *       *     *       *     *      *    *
 k=4     *    *   *    *   *    *   *   *    *   *    *    *
 k=3    *   *  *   *  *   *  *   *  *   *   *   *  *   *  *
 k=2   * * * * * * * * * * * * * * * * * * * * * * * * * *
 k=1  *******************************************************

week  1234567890123456789012345678901234567890123456 78
```

Figure 12.4 How to create maintenance opportunities from maintenance frequencies(example).

Let us now take a closer look at this figure, and determine how many maintenance opportunities remain after implementation of the problem reduction techniques that we discussed in the previous section. For instance, there are 16 weeks in the maintenance cycle which correspond to maintenance frequency $k = 1$ only. The other way around, there is only one week in the maintenance cycle that corresponds to the entire set of maintenance frequencies. It is left as an exercise to the reader that eventually 9 maintenance opportunities can be distinguished, each of which corresponds to a different set of maintenance frequencies (see Table 12.3).

7.   CONCLUDING REMARKS

In this chapter, we presented a powerful modeling framework for the coordinated planning of set-up activities and preventive maintenance jobs in a multi-setup multi-component production system . This modeling framework could easily be adopted in decision support systems

Table 12.3 Maintenance opportunities, corresponding clusters and frequencies of occurence for the example of Figure 12.4.

opportunity	corresponding seperation	corresponding weeks
1	{1}	{1,5,7,11,13,17,19,23,25,29,31,35,37,41,43,47}
2	{1,2}	{2,10,14,22,26,34,38,46}
3	{1,3}	{3,9,15,21,27,33,39,45}
4	{1,2,4}	{4,20,28,44}
5	{1,2,3,6}	{6,18,30,42}
6	{1,2,4,8}	{8,16,32,40}
7	{1,2,3,4,6,12}	{12,36}
8	{1,2,3,4,6,8,12,24}	{24}
9	{1,2,3,4,6,8,12,24,48}	{48}

for maintenance optimization, since it explicitly takes into account the hierarchically organized structure of modern/current production systems , and allows for a large class of preventive maintenance strategies for each component . Nevertheless, one could still think of a variety of interesting and promising model extensions. To conclude this chapter, we will briefly discuss a number of these.

So far, we have assumed a completely additive cost structure in deriving the costs associated with the simultaneous execution of set-up activities and preventive maintenance jobs . Simply stated, this assumption arises from the underlying reasoning that preventive maintenance jobs are carried out sequentially by one and the same maintenance crew. There are situations, however, where preventive maintenance jobs can be carried out in parallel, i.e. by different maintenance crews, and other measures may seem more useful. Especially if it is total time rather than total costs that matters, it could be wise to define the costs of a cluster as the maximal time required to complete a maintenance job in that cluster.

Another interesting extension to the modeling framework presented here is to incorporate some possibilities for workload balancing . Although it can be proven that - from a cost efficiency point of view - it is always optimal to combine all activities at least once during the maintenance cycle [42], this certainly is not a necessary property for optimality. One could easily think of situations where the optimality in terms of cost efficiency is not affected, but simultaneous execution of all maintenance jobs never does occur in the maintenance cycle . By doing this, it could be possible to construct a more balanced maintenance cycle against no additional costs.

References

[1] Assaf, D., and Shanthikumar, J.G. Optimal Group Maintenance Policies with Continuous and Periodic inspections. *Management Science*, **33** (1987) 1440-1452.

[2] Barlow, R. and Hunter, L.C. Optimum Preventive Maintenance Policies. *Operations Research*, **8** (1960) 90-100.

[3] Barlow, R.E.,Hunter, L.C., and Proschan, F. Optimum Checking Procedures. *Journal of the Society for Industrial and Applied Mathematics*, **4** (1963) 1078-1095.

[4] Barlow, R.E., and Proschan, F. *Mathematical Theory of Reliability*. Wiley, New York, 1965.

[5] Barron, R. *Engineering Condition Monitoring : Practice, Methods and Applications*. Longman, Harlow, 1996.

[6] Berg, M. Marginal Cost Analysis for Preventive Replacement Policies. *European Journal of Operational Research*, **4** (1980) 136-142.

[7] Berg, M. A Preventive Replacement Policy for Units Subject to Intermittent Demand. *Operations Research*, **32** (1984) 584-595.

[8] Berg M. and Epstein, B. A Modified Block Replacement Policy. *Naval Research Logistics Quarterly*, **23** (1976) 15-24.

[9] Cho, D.I., and Parlar M. A Survey of Maintenance Models for Multi-Unit Systems. *European Journal of Operational Research*, **51** (1991) 1-23.

[10] Christer, A.H. Modelling Inspection Policies for Building Maintenance. *Journal of the Operational Research Society*, **33** (1982) 723-732.

[11] A.H. Christer and Waller, W.M. Delay Time Models for Industrial Maintenance Problems. *European Journal of Operational Research*, **35** (1984) 401-406.

[12] Dagpunar, J.S. Formulation of a Multi Item Single Supplier Inventory Problem. *Journal of the Operational Research Society*, **33** (1982) 285-286.

[13] Dekker, R. Integrating Optimisation, Priority Setting, Planning and Combining of Maintenance Activities. *European Journal of Operational Research*, **82** (1995) 225-240.

[14] Dekker, R., and Dijkstra, M.C. Opportunity-Based Age Replacement: Exponentially Distributed Times Between Opportunities. *Naval Research Logistics*, **39** (1992) 175-190.

[15] Dekker R., Frenk, H., and Wildeman, R.E. How to Determine Maintenance Frequencies for a Multi-Component System: a General Approach. In S. zekici, editor, *Reliability and Maintenance of Complex Systems*, Springer Verlag, Berlin, (1996) 251-268.

[16] Dekker, R. and Smeitink, E. Opportunity-Based Block Replacement. *European Journal of Operational Research*, **53** (1991) 46-63.

[17] Dekker, R., and Smeitink, E. Preventive Maintenance at Opportunities of Restricted Duration. *Naval Research Logistics*, **41** (1994) 335-353.

[18] Dekker, R., Wildeman, R.E., and Van der Duyn Schouten, F.A. A Review of Multi-Component Maintenance Models with Economic Dependence. *Mathematical Methods of Operations Research*, **45** (1997) 411-435.

[19] Frenk, H., Dekker, R., and Kleijn, M. A Unified Treatment of Single Component Replacement Models. *Mathematical Methods of Operations Research*, **45** (1997) 437-454.

[20] Gertsbakh, I.B. Optimum Choice of Preventive Maintenance Times for a Hierarchical System. *Automated Control Computer Sciences*, **6** (1972) 24-30.

[21] Gertsbakh, I.B. *Models of Preventive Maintenance*. North-Holland, Oxford, 1977.

[22] Gits, C.W. *On the Maintenance Concept for a Technical System: a Framework for Design*. PhD thesis, University of Eindhoven. The Netherlands, 1984.

[23] Goyal, S.K., and Gunasekaran, A. Determining Economic Maintenance Frequency of a Transport Fleet. *International Journal of Systems Science*, **4** (1992) 655-659.

[24] Goyal, S.K., and Kusy, M.I. Determining Economic Maintenance Frequencies for a Family of Machines. *Journal of the Operational Research Society*, **36** (1985) 1125-1128.

[25] Haurie, A., and L'Ecuyer, P. A Stochastical Control Approach to Group Preventive Replacement in a Multi-Component System. *IEEE Transactions on Automatic Control*, **27** (1982) 387-393.

[26] Jansen, J. and Vander Duyn Schouten, F.A. Maintenance Optimization on Parallel Production Units. *IMA Journal of Mathematics Applied in Business and Industry*, **6** (1995) 113-134.

[27] McCall, J.J. Maintenance Policies for Stochastically Failing Equipment: a Survey. *Management Science*, **11** (1965) 493-524.

[28] Niebel, B.W. *Engineering Maintenance Management.* Dekker, New York, 2nd edition, 1994.

[29] Ozekici, S. Optimal Periodic Replacement of Multi-Component Reliability Systems. *Operations Research*, **36**,1988 542-552.

[30] Pierskalla, W.P., and Voelker, J.A. A Survey of Maintenance Models: the Control and Surveillance of Deteriorating Systems. *Naval Research Logistics Quarterly*, **23** (1976) 353-388.

[31] Pintelon, L.M., and Gelders, L.F. Maintenance Management Decision Making. *European Journal of Operational Research*, **58** (1992) 301-317.

[32] Ritchken, P. and J.G. Wilson. (m,T) Group Maintenance Policies. *Management Science*, **36** 632-639.

[33] Sherif, Y.S., and Smith, M.L. Optimal Maintenance Models for Systems Subject to Failure - a Review. *Naval Research Logistics Quarterly*, **28** (1990) 47-74.

[34] Valdez-Flores, C., and Feldman, R.M. A Survey of Preventive Maintenance Models for Stochastically Deteriorating Single-Unit Systems. *Naval Research Logistics*, **36**, (1989) 419-446.

[35] Van der Duyn Schouten, F.A., and Vanneste, S.G. Analysis and Computation of (n,N)-Strategies for Maintenance of a Two-Component System. *European Journal of Operational Research*, **48** (1990) 260-274.

[36] F.A. Van der Duyn Schouten and S.G. Vanneste. Two Simple Control Policies for a Multi-Component Maintenance System. *Operations Research*, **41** (1993) 1125-1136.

[37] Van Dijkhuizen, G.C. *Maintenance Meets Production: On the Ups and Downs of a Repairable System.* PhD thesis, University of Twente. Enschede, The Netherlands, 1998.

[38] Van Dijkhuizen, G.C. and Van Harten, A. Coordinated Planning of Preventive Maintenance in Multi-Setup Multi-Component Systems. Technical report, University of Twente, The Netherlands. Submitted to *Management Science*, 1997.

[39] Van Dijkhuizen, G.C., and Van Harten, A. Optimal Clustering of Frequency-Constrained Maintenance Jobs with Shared Set-Ups. *European Journal of Operational Research*, **99** (1997) 552-564.

[40] Van Dijkhuizen, G.C., and Van Harten, A. Two-Stage Generalized Age Maintenance of a Queue-Like Production System. *European Journal of Operational Research*, **108** (1998) 363-378.

[41] Van Dijkhuizen, G.C., and Van Harten, A. Two-Stage Maintenance of a Production System with Exponentially Distributed on-and-off Periods. *International Transactions in Operational Research*, **5** (1998) 79-85.

[42] Van Harten, A., and Van Dijkhuizen, G.C. , On a Number Theoretic Property Optimizing Maintenance Grouping. Technical report, University of Twente, The Netherlands. Submitted to *European Journal of Operational Research*, 1999.

[43] Wijnmalen, D.J.D., and Hontelez, J.A.M. Coordinated Condition-Based Repair Strategies for Components of a Multi-Component Maintenance System with Discounts. *European Journal of Operational Research*, **98** (1997) 52-63.

[44] Wildeman, R.E., Dekker, R. and Smit, A.C.J.M. A Dynamic Policy for Grouping Maintenance Activities. *European Journal of Operational Research*, **99** (1997) 530-551.

IV

CONDITION BASED MAINTENANCE

Chapter 13

ON-LINE SURVEILLANCE AND MONITORING

Hai S. Jeong

Seowon University, Department of Applied Statistics Cheongju, Chonbuk 361-742, Korea

hsjeong@dragon.seowon.ac.kr

Elsayed A. Elsayed

Rutgers University, Department of Industrial Engineering Piscataway, NJ 08854-8018, USA

elsayed@rci.rutgers.edu

Abstract   This chapter presents approaches for condition monitoring and fault diagnostic systems. It describes in detail methodologies for the limit value determination that correspond to a predetermined reliability level. Applicability of various condition monitoring parameters are also discussed.

Keywords: Predictive maintenance, On-condition maintenance, On-line surveillance and monitoring peak level, Proportional hazards model (PHM) Vibration monitoring, Corrosion monitoring, Fluid monitoring

1.   INTRODUCTION

Maintenance actions or policies can be classified as corrective maintenance , preventive maintenance and on-condition maintenance which is also called predictive maintenance (Chan *et al.*, 1997). Maintenance actions are dependent on many factors such as the failure rate of the machine, the cost associated with downtime, the cost of repair and the expected life of the machine. For example, a maintenance policy which requires no repairs, replacements or preventive maintenance until failure allows for maximum run-time between repairs. Although it allows for maximum run-time between repairs it is neither economical nor efficient as it may result in a catastrophic failure that requires extensive repair time and cost. Another widely used maintenance policy is to maintain the machine according to a predetermined schedule, whether a problem is apparent or not. On a scheduled basis, machines are removed from operation, disassembled, inspected for defective parts and, repaired accordingly. Actual repair costs can be reduced in this manner, but production loss may increase if the machine is complex and requires days or even weeks to maintain. This preventive maintenance also may create machine problems where none existed before. Becker *et al.* (1998) cite a 1990 report from Electric Power Research Institute (EPRI) which states that one-third of the money spent on preventive maintenance in the electric power industry (which that year amounted to $ 60 billion) was wasted. Obviously, if a machine failure can be predicted and the machine can be taken off-line to make only the necessary repairs, a tremendous cost saving can be made. Predictive maintenance can also be done when failure modes for the machine can be identified and monitored for increased intensity and when the machine can be shut down at a fixed control limit before critical fault levels are reached.

Predictive maintenance results in two benefits. The first benefit is the result of taking a machine off-line at a predetermined time, which allows production loss to be minimized by scheduling production around the downtime. Since defective components can be predetermined, repair parts can be ordered and manpower scheduled for the maintenance accordingly. Moreover, sensors for monitoring the machines eliminate the time for diagnostics thus reducing the time to perform the actual repair. The second benefit is that only defective parts need to be repaired or replaced and the components in good working order are left as is, thus, minimizing repair costs and downtime.

Global competition to increase production output and to improve quality is spurring manufacturing companies to use condition monitoring and fault diagnostic systems for predictive maintenance. In many practical situations, it is critical to continuously monitor the condition of the system in order to detect and distinguish faults occurring in machinery. The recent developments in sensors, chemical and physical nondestructive testing (NDT), and sophisticated measurement techniques have facilitated the continuous monitoring of the system performance. When linked with microcomputer based control equipment, performance monitoring and predictive control of complete systems have become more practical and affordable.

There are three main tasks to be fulfilled for predictive maintenance . The first task is to find the condition parameter which can describe the condition of the machine. A condition parameter could be any characteristic such as vibration , sound, temperature, corrosion, crack growth, wear and lubricant condition. The second task is to monitor the condition parameter and to assess the current machine condition from the measured data. The final task is to determine the limit value, S_L, of the condition parameter and its two components: the alarm value S_a and the breakdown value S_b. If a running machine reaches the alarm value it is an indication that it is experiencing intensive wearing. Hence the type and advancement of the fault must be identified in order to prepare the maintenance procedure. If a machine reaches the breakdown value, S_b, the shutdown of a machine for maintenance becomes necessary.

In this chapter, leaving aside the choice of the condition parameter, we will consider the condition parameter analysis techniques for fault detection and diagnostics in the machinery and briefly discuss reliability prediction and the limit value determination in condition monitoring .

2.   RELIABILITY PREDICTION AND THE LIMIT VALUE DETERMINATION IN CONDITION MONITORING

As discussed above, condition monitoring requires observing a condition parameter with time. When the parameter's value degrades and reaches a predetermined limit, S_L, then a corrective action is taken accordingly. In this section, we present three approaches for determining the limit values of typical condition parameter.

2.1   Reliability Estimation Based on Degradation Data.   The breakdown value, S_b, can be determined by observing the changes in the condition parameter with time. These observations can then be used in a degradation model to estimate the value of S_b corresponding to an acceptable reliability value of the system as described below.

It is assumed that the effect of the degradation phenomenon on the product performance or the condition parameter can be expressed by a random variable called degradation criterion. Typical criteria include the amount of wear of mechanical parts such as shafts and bearings, the drift of a resistor, output power drop of light emitting diodes, fatigue-crack-growth, the gradual corrosion of a reinforcing bar, and the propagation delay of an electronic chip. These criteria contain useful information about product reliability. It is clear that units with the same age would have different degradation criterion levels. Figure 13.1 is a plot of the crack-length measurements versus time (in million cycles) from Bogdanoff and Kozin (1985). There are 21 sample paths, one for

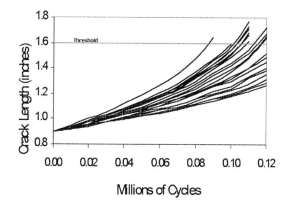

Figure 13.1 Fatigue-crack-growth data from Bogdanoff and Kozin (1985).

each of 21 test units. From this plot, we can find that the degradation criterion is a time-dependent random variable that can follow different distributions at different distinct times.

In general, the degradation criterion, X, may follow a distribution which changes with time in the type of the distribution family and its parameters as shown in Figure 13.2. The solid curve represents the mean of the degradation criterion versus time and the areas under the density functions and above the threshold level line represent the failure probability at the corresponding times.

Eghbali and Elsayed (1997) develop a statistical approach based on degradation data. They assume the degradation criterion follows the same distribution family but its parameters may change with time. Furthermore, it is assumed that the degradation paths are monotonic functions of time; they are either Monotonically Increasing Degradation Paths (MIDP) or Monotonically Decreasing Degradation Paths (MDDP) . They define:

X degradation criterion (random variable), $x \geq 0$

t time

$f(x; t)$ probability density function (pdf) of the degradation criterion, X, at a given time t

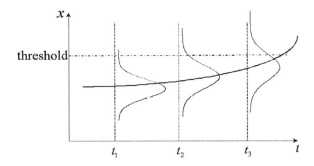

Figure 13.2 $f(x;t)$ versus t for MIDP.

$\bar{F}(x;t)$ complement of the cumulative distribution function (CDF)

$\lambda(x;t)$ failure rate function of $f(x;t)$, referred to as the *degradation failure rate function*

$x), q(t)$ positive functions

a, b, γ constants

$R_x(t) = P(T_x > t)$, reliability at time t and threshold degradation level, x

$\Lambda_x(t) = -\ln R_x(t)$

They assume that the degradation hazard function $\lambda(x;t)$ can be represented by two functions as

$$\lambda(x;t) = g(x)q(t), \quad g(x) > 0, q(t) > 0,$$

where $g(x)$ and $q(t)$ are positive functions of the degradation criterion and time, respectively. This assumption ensures that the distribution family of the degradation criterion does not change with time. They consider the following functions:

$$q(t) = \frac{1}{b \exp(-at)}, \quad g(x) = \gamma x^{\gamma-1} \text{ and } b \gg 0.$$

The parametric form of $q(t)$ is determined based on graphing the degradation data versus time. Moreover, the distribution of the degradation data is studied in order to determine $g(x)$. It is realized that

$g(x) = \gamma x^{\gamma-1}$ is a good fit for most of the degradation data. The corresponding degradation criterion distribution for this degradation hazard function is a Weibull distribution with a time-dependent scale parameter,

$$f(x;t) = \frac{\gamma}{\theta(t)} x^{\gamma-1} \exp[\frac{-x^\gamma}{\theta(t)}], \quad t > 0$$

where $\theta(t) = b\exp(-at)$ is the scale parameter, x is the degradation criterion at which a failure occurs and the corresponding reliability function can then be determined as

$$R_x(t) = P(X > x;t) = \exp[\frac{-x^\gamma}{b\exp(-at)}].$$

Conversely, if the desired reliability at time t, $R_x(t)$ is given, the level of the degradation criterion, x, which is equivalent to the breakdown value, S_b, can be determined. Hence we use the above equation to determine S_b corresponding to $R_{S_b}(t)$.

Important studies that have used degradation data to assess reliability can be found in Gertsbackh and Kordonskiy (1969), Nelson (1981), Bogdanoff and Kozin (1985), Carey and Koenig (1991), Lu and Meeker (1993), Chick and Mendel (1996), Feinberg and Widom (1996), Lu *et al.* (1997), Meeker *et al.* (1998), and Ettouney and Elsayed (1999).

2.2   Determination of the Limit Values in Condition Monitoring.

The last stage of condition monitoring is the inferences on the machine condition which is based on values of the condition parameter, S. One of them is to assess the symptom limit value, S_L and its two components: the alarm value S_a and the breakdown value S_b. If a running machine reaches the alarm value it is an indication that it experiences an intensive wearing. Hence the type and advancement of the fault must be identified in order to prepare the maintenance procedure. If a machine reaches the second limit value, S_b, the shutdown of a machine for maintenance becomes necessary. The knowledge of these two limit values is of great importance for critical machines which run continuously with automatic monitoring and shutdown systems. However in most cases of diagnostic implementation, for large and expensive machinery in particular, it is difficult to perform active diagnostic experiments, which means establishing the S_L on the basis of the known machine condition. Hence the determination of S_L is possible only as the result of passive diagnostic experiments, where the values of S are observed on the group of running machines without knowledge of their condition.

Detailed description of condition inference techniques and the use of statistical methods to estimate the limit symptom value, S_L, can be found in Cempel (1984, 1985, 1987, 1990). A simple solution for determining S_L is given by Dabrowski (1981). It is determined in the way its tail probability does not exceed a given small level, α ; $\Pr(S > S_L) \leq \alpha$. Another possible way of determining S_L from passive experimental data is based on the Neyman-Pearson technique of the statistical decision theory. It minimizes the number of breakdowns at an assumed and allowed percent of needless repairs, A, by means of a proper choice of the breakdown symptom value, S_b. According to Cempel(1985), this condition of minimizing the breakdown number can be written as follows

$$A = P_g \int_{S_b}^{\infty} p(s)ds, \tag{13.1}$$

where $p(s)$ is the pdf of the condition parameter, S, and P_g is the probability of good machine condition.

Cempel (1987) treated observed symptoms as an outcome of a Weibull type stochastic process and estimated S_b using Equation (4.1). Additionally he defined the alarm symptom value, S_a, and estimated it using the following equation.

$$A = P_g \int_{S_a}^{S_b} p(s)ds.$$

2.3 Reliability Prediction based on Condition Monitoring.

The classical approach to the calculation of reliability characteristics is based on the probability distribution of time to failure. Recently a new approach based on relevant condition parameter was introduced by Knezevic (1987), which describes the condition of the system and calculates the reliability characteristics at every instant of operating time. A condition parameter could be any characteristic such as crack, corrosion, vibration *etc.*, that is directly or indirectly connected with an item of the system and its performance, and describes the condition of the item during the operating life. In particular, he defined a *relevant condition parameter* (RCP) as a condition parameter, whose numerical value fully describes and quantifies the condition of the item at every instant of operating time. The change of RCP during operation is a random process denoted as $X(t)$. Applying the RCP reliability approach, at any instant of operating time the reliability of the system is equal to the probability of the RCP being within the tolerance range expressed as:

$$R_X(t) = P(X^{in} < X(t) < X^{lim}),$$

where X^{in} is the initial value of RCP, and X^{lim} is the limit value of RCP. The prescribed limits, X^{in} and X^{lim} can either be fixed or stochastically follow a specific probability distribution depending upon failure mechanism. RCP's can be classified into the following four categories.

1. Single X^{in}, single X^{lim}

2. Single X^{in}, stochastic X^{lim}

3. Stochastic X^{in}, single X^{lim}

4. Stochastic X^{in}, stochastic X^{lim}

The first category has been dealt with by Knezevic (1987, 1989). The second type was reported by El-Haram (1995). The last two cases were presented in Saranga and Knezevic (1999).

This approach is used to monitor the system throughout its life. The approach can be applied in Health Usage Monitoring Systems (HUMS) for fitment to aero engines (Massam and Cruickshank, 1997) to quantify the condition of the engines. Aircraft such as the Tornado and the Hawk are already fitted with HUMS and recent research involves the development of HUMS into more sophisticated Fatigue and Usage Monitoring Systems (FUMS) (Saranga and Knezevic, 1999).

Recently, in an attempt to develop a realistic approach to determine a maintenance policy, a proportional hazards model (PHM) has been used to schedule for maintenance. Performance of a system is influenced not only by the operating time, but by other factors. These influencing factors include operating condition (*e.g.*, vibration levels, temperature, pressure, levels of metal particles in engine oil, humidity, dust) and operating history (*e.g.*, number of previous overhauls, time since last failure and maintenance). They are generally referred to as covariates or explanatory variables. Given all the explanatory variables, PHM can be used to identify the explanatory factors of interest and to schedule for maintenance

$$h(t, \mathbf{z}) = h_0(t)e^{\beta \mathbf{z}},$$

where $h(t, \mathbf{z})$ is the hazard function at time t for observations with covariate vector, $\mathbf{z}$, $h_0(t)$ is an unspecified baseline hazard function (*i.e.* the hazard function when all covariates are zero), and β is a vector of unknown regression coefficients. This model assumes that the covariates act multiplicatively on the hazard function, so that for different values of the explanatory variables the hazard functions at each time are proportional to each other. The use of the PHM in maintenance decision making is described in Jardine *et al.* (1989), Love and Geo (1991a, b) and Kobbacy *et al.* (1997).

3. VIBRATION MONITORING

Most industrial machinery operates by means of motors and other moving parts which will eventually cause faults. These faults may cause the machine to break down or degrade its performance. Generally, when a machine develops a fault, it gives a signal in various forms, *e.g.* changes in vibration , pressure, oil characteristics, etc. In order to keep the machine performing at its best, techniques such as fault diagnostics need to be utilized. In rotating machinery such as gear boxes and bearings, a vibration signal is commonly used for fault diagnostics. This is due to the fact that when a machine or a structural component is in good condition, its vibration profile has a normal characteristic shape, which it changes as a fault begins to develop (Paya *et al.*, 1997).

3.1   Vibration. Vibration is a force, alternating in direction, exerted upon the machine at regular intervals related to the speed of rotation and causing the physical displacement of some portion of the machine in response to this force (Callaway, 1982). The vibration is defined and is measured by its frequency, amplitude and phase angle. The smallest interval of time to complete a vibration cycle is a period. Vibration frequency is the number of cycles completed in a predetermined amount of time, expressed in cycles per minute (cpm), cycles per second (Hz) or revolutions per minute (rpm). Vibration amplitude is how much vibration is present from a reference value of zero and measured as displacement, velocity or acceleration. Phase angle is the fractional part of a period between a reference (zero vibration amplitude) and a particular time of interest (some vibration amplitude). It is measured in degrees using a circle as a complete period of vibration (360°).

Examples of the problems causing vibrations include imbalance of parts, torque vibrations, bent shafts, misalignment of couplings and bearings, loose fittings, worn bearings, worn or eccentric gears, worn drive belts or chains, improper tension on drive belts or chains and resonance. The vibration levels generated by these problems induce greater stress, which, in turn, causes more vibration.

Each machine has a characteristic vibration or vibration signature composed of a large number of harmonic vibrations of different amplitudes. The effects of component wear and failure on these harmonic vibrations differ widely depending on the contribution made by a particular component to the overall signature of the machine. For example, in a multi-cylinder engine, the major harmonic vibrations are calculated from the number of working strokes per revolution. Thus, misfiring of one or more cylinders would produce significantly different vibrations than the original signature of the engine which can be easily detected by an accelerometer. The accelerometer is an electro-mechanical transducer which produces an electric output proportional to the vibratory acceleration to which it is exposed (Niebel, 1994).

In general, there is no single transducer which is capable of measuring the extremly wide range of signatures. Therefore, transducers are developed for different frequency, amplitude, velocity, and acceleration ranges. For example, the bearing probe, being a displacement measuring device, is sensitive only to low frequency large amplitude vibrations. This makes it useful only as a vibration indicator for gear box vibrations and turbine blades (Elsayed, 1996).

3.2   Vibration Analysis.   Condition monitoring techniques based on vibration data analysis are classified into two types; the first, *analytic* methods, are suitable for diagnosis of specific faults while the second, *discriminant* methods, simply relate the general condition to a single number or series of numbers. Although *analytic* methods can yield considerable information, they require skilled interpretation of data and a large part of the success of these methods is due to the analyst. *Discriminant* methods require less skilled operators although this may lead to a reduction in the reliability of the techniques. We now describe these methods in detail.

3.2.1   Analytic Methods.   Each type of fault generates a particular vibration. The most widely known and applied technique to detect and distinguish the faults occurring in machinery is the spectral analysis. The classical technique for spectral analysis is Fourier transform . The essence of the Fourier transform of a waveform is to decompose or separate the waveform into a sum of sinusoids of different frequencies. The pictorial representation of the Fourier transform is a diagram which displays the amplitude and frequency of each of the determined sinusoids.

The Fourier transform identifies or distinguishes the different frequency sinusoids (and their respective amplitudes) which combine to form an arbitrary waveform. Mathematically, this relationship is stated as

$$S(f) \;=\; \int_{-\infty}^{\infty} s(t)\exp(-j2\pi ft)dt, \quad \text{direct Fourier transform}$$

$$s(t) \;=\; \int_{-\infty}^{\infty} S(f)\exp(j2\pi ft)df, \quad \text{inverse Fourier transform,}$$

where $s(t)$ is the waveform signal function to be analyzed, $S(f)$ is the Fourier transform of $s(t)$, and $j = \sqrt{-1}$. This defines the (complex) frequency component $S(f)$ of $s(t)$ for each frequency f. The above equations show that a time function $s(t)$ can be analyzed to give an associated frequency function $S(f)$ that contains all of the information in $s(t)$. $S(f)$ is usually called the spectrum of $s(t)$.

The discrete equivalent of the above transform is the discrete Fourier transform (DFT) of the discrete signal function $s(k), k = 0, 1, \cdots, N-1$, sampled from $s(t)$:

discrete Fourier transform

$$S(n) = \sum_{k=0}^{N-1} s(k) \exp(-\frac{j2\pi nk}{N}), \quad n = 0, 1, \cdots, N - 1.$$

inverse discrete Fourier transform

$$s(k) = \frac{1}{N} \sum_{n=0}^{N-1} S(n) \exp(\frac{j2\pi nk}{N}), \quad k = 0, 1, \cdots, N - 1.$$

The above transforms are based on the fact that if the wave form $s(t)$ is sampled at a frequency of at least twice the largest frequency component of $s(t)$, there is no loss of information as a result of sampling (Brigham, 1974). The DFTs are obtainable using commercially available spectrum analyzers that use the so-called fast Fourier transform (FFT), which is a highly efficient implementation of the DFT, described by Cooley and Tukey (1965).

One of the important advantages of examining spectra as opposed to overall vibration level in the time domain, is that changes can often be detected at an earlier stage which means we can follow the trend more reliably over a longer period. For example, in applying spectral analysis to detection of bearing faults, several frequencies can be predicted as being associated with particular faults. The relationships determining these frequencies can be obtained by the following expressions (Taylor, 1980) :

$$f_i = \frac{f_r \cdot N}{2}(1.0 + \frac{d}{p}cos\phi)$$

$$f_o = \frac{f_r \cdot N}{2}(1.0 - \frac{d}{p}cos\phi)$$

$$f_b = \frac{f_r \cdot p}{2d}[1.0 - (\frac{d}{p})^2 cos^2\phi],$$

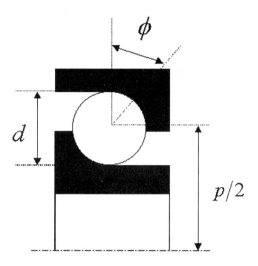

Figure 13.3 Ball bearing rotation.

where f_r is the rotating frequency of the shaft, f_i is the ball pass frequency of the inner race, f_o is the ball pass frequency of the outer race, f_b is the ball spin frequency, d is the ball diameter, p is the pitch diameter, N is the number of balls and ϕ is the contact angle as shown in Figure 13.3. In the expressions above, all frequencies are expressed in Hz, assuming that the inner race rotates and outer is fixed.

Ball pass frequencies are generated as the balls or rollers pass over a defect on the race-ways. Ball spin frequency is generated when a defect on the ball or roller strikes the race-ways. Hence, defective race-ways can be identified by a narrow band spike at the ball pass frequency of the race on which the defect exists. The spectrum of an outer race defect is shown in Figure 13.4 (Taylor, 1980).

Trend analysis may be performed on these characteristic frequencies and limits may be set to indicate the need for component replacement. However, in practice, numerous effects such as resonance, speed vibration, inter-modulation and side-banding, can obscure the vibration signature and in such instances more sophisticated analyses may be required.

Once ensuing vibration signals have been detected by a transducer mounted on the casing of the machine, they are next amplified suffi-

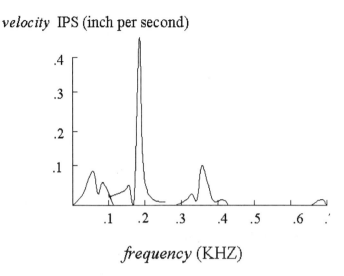

velocity IPS (inch per second)

frequency (KHZ)

Figure 13.4 Outer race defect. The ball pass frequency on the outer race is 172 Hz. The spectral line at 172 Hz is the representation of defect on the outer race of the bearing (Taylor, 1980).

ciently for analysis. The transducer produces an electronic signal which is characteristic of a machine's mechanical performance. Since the signals from the transducer are noisy, they are passed through appropriate filters for elimination of a large part of the background noise. The filtered signal can then be enveloped to remove the high frequency content leaving a train of impulses caused by the impacts. Finally, the amplitude spectrum of the enveloped signal is calculated. A fault in the bearing can then be detected by the appearance in the spectrum of a component at the impact frequency corresponding to a defect on the inner race, outer race or one of the rolling elements. There are many filters such as low-pass, high-pass, band-pass, band elimination multi-band pass filters, *etc.* For example, the band-pass filtering has the advantage of eliminating much of the unwanted vibration from the signal leaving just the vibration excited by the bearing. Thus this technique is useful for complex machines such as gear-boxes where the vibration produced by a damaged bearing may be an order of magnitude smaller than the vibration produced by the meshing of normal gears (McFadden and Smith,

1984). The basic steps for the spectral analysis for the signals are shown in Figure 13.5.

There are many other data representation techniques for processing the raw vibration signal acquired from the sensors. Using Windowed Fourier Transforms (WFT) for detection of gear failure by vibration analysis is studied by McFadden (1987), Wang and McFadden (1993), Staszewski and Tomlinson (1993). Cepstrum analysis is studied by Randall (1982). This technique is appropriate for a variety of diagnostic purposes, because of source separation properties (Pau, 1981). The study of the Wigner-Ville distribution in condition monitoring is presented in McFadden and Wang (1990), Forrester (1992), Staszewski *et al.* (1997). Most of these methods are suitable for non-transient signals. However, recently a new method of signal processing technique has been developed, called wavelet transforms. Wavelet transforms are used to represent all possible types of transients in vibration signals generated by the machinery faults. It is shown that wavelet transforms provide a powerful tool for condition monitoring and fault diagnosis. Research on the wavelet transforms began in the early 1990s (Daubechie (1991), Chui (1992)), and the literature is still growing rapidly. The wavelet transform of a time signal is an expansion of the signal in terms of a family of functions generated from a single function called the 'mother wavelet', by 'dilation', and by 'translation'. Compared with the Fourier transform, whose coefficients provide global frequency information about a signal, wavelet transforms allow the time resolution to be adapted for the analysis of the local frequency of the signal. Procedures based on wavelet theory, such as the fast wavelet transform (FWT) described in Kamarthi and Pittner (1997), lead to superior signal analysis in a wide range of applications. It can be used widely in signal processing, speech identification and image compressing fields, *etc.* The applications of wavelet transforms to machinery fault diagnostics have also been demonstrated in Newland (1993), Mcfadden (1994), Staszewski and Tomlinson (1994), Petrilli *et al.* (1995), Wang and McFadden (1996), Paya *et al.* (1996, 1997). Ma *et al.* (1997) presented a method to design a filter with combination wavelets.

3.2.2   Discriminant Methods. Discriminant Methods relate the machine condition to a single value or a series of values. These methods should be appreciated in the view of the objective to provide a simple and inexpensive fault detection technique. Each of the following values has its own physical meaning and contribution to the interpretation of defect phenomena.

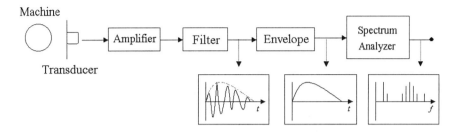

Figure 13.5 Signal analysis by spectrum.

(a) Peak Level

Attention can be focused on the peak level characteristic of a signal. With the established baseline peak levels, changes in peak level can be a good indication of incipient damage (Weichbordt and Bowden, 1970).

An obvious measure of the machine condition can be obtained by observing changes of the statistical moments. To obtain the statistical moments, we should define the probability density function (pdf) of amplitude, X. The amplitude characteristics of a vibration signal $s(t)$ (assumed to be a stationary random process) can be expressed in terms of pdf (Bendat *et al.*, 1971). This is estimated by determining the duration a signal remains in a set of amplitude windows and, for a typical window at amplitude x and of width Δx, the instantaneous probability is:

$$P(x \leq s(t) \leq x + \Delta x) = \sum_{i=1}^{j} \frac{\Delta t_i}{T}, \qquad (13.2)$$

where T is the sampled duration, Δt_i is the i-th time interval the signal remains in the window at amplitude x and of width Δx and j is the number of the intervals in the window at amplitude x and of width Δx. Solution of equation (4.2) for all x as Δx approaches zero, results in an estimate of pdf for the amplitude, $p(x)$.

A series of statistical moments can be used to indicate the shape of the pdf and the nth moment is defined by the general integral:

$$\mu_n = \int_{-\infty}^{\infty} x^n p(x) dx, \quad n = 1, 2, \cdots m.$$

The first and second moments are those most well known; $n = 1$ gives the mean, μ, and $n = 2$ gives the mean square. The second moment about μ, which is also called variance or mean square, is expressed by

$$\sigma^2 = \int_{-\infty}^{\infty} (x - \mu)^2 p(x) dx.$$

(b) RMS

Perhaps the simplest and most common approach for vibration monitoring is to measure the overall intensity of a wide-band vibration signal. This is normally achieved with an estimate of the root mean square (RMS) displacement, velocity or acceleration of the time record. It has been found, from experience covering many types of rotating machinery, that RMS velocity is a very good indicator of vibration severity over a wide frequency range (Wheeler, 1968 and Monk, 1972). Operational standards by RMS have been developed, recommending vibration boundary levels for satisfactory or unsatisfactory running conditions. Such standards include VDI 2056 and ISO-2732. Since the measured values of RMS and peak level are dependent upon the factors on which a machine operates, such as rotating speed and loading, it is difficult to define the condition of a machine from these measurements only.

(c) Impact Index

The impact index is the ratio of RMS and peak level of the vibration signal. It attempts to give a qualitative assessment of a machine condition, and has been found to be partially insensitive to the factors on which a machine operates. In monitoring the vibrational waveform from a localized fault, a periodic peak is seen to occur in the signal. As the discrete fault grows, the amplitude of this peak increases, but the RMS value is affected little since the peak is of short duration. Weichbordt and Bowden (1970) found that the impact index indeed relates directly to bearing condition.

(d) Kurtosis

Another development in the state of the art technology of detecting faults occuring in machinery is a statistically based parameter termed Kurtosis. The Kurtosis coefficient is defined as :

$$K = \frac{\mu_4}{\sigma^4}$$

where μ_4 is the fourth moment of X about μ, and σ is the standard deviation of X, namely,

$$\mu_4 = \int_{-\infty}^{\infty} (x - \mu)^4 p(x) dx.$$

The Kurtosis technique has the major advantage that the estimator takes a value which is independent of load or speed conditions. For an undamaged bearing Kurtosis remains close to 3 (±8 percent) whereas the RMS and peak level vary by ±50 percent and ±65 percent respectively for the same range of load and speed (Dyer and Stewart, 1978). It also has been found that the Kurtosis remains constant for an undamaged bearing irrespective of load and speed, yet changes with damage.

4. SOUND RECOGNITION AND ACOUSTIC EMISSION

Sound recognition is used to detect a wide range of abnormal occurrences in manufacturing processes. The sound recognition system recognizes various operational sounds, including stationary and shock sounds, using a speech recognition technique, and then compares them with the expected normal operational sounds (Takata and Ahn, 1987).

The operational sound is collected by an unidirectional condenser microphone which is set near the component or machine to be monitored. When the sound of an abnormal operation is generated, such as the sound of tool breakage or the sound of worn out motor bearings, the features of the sound signal are extracted and a sound pattern is formed. The sound pattern is then compared with the standard patterns through pattern matching techniques and the most similar standard pattern is selected. The failure or fault corresponding to this category is then recognized and diagnosed.

Acoustic emission (AE) is defined as the transient elastic energy spontaneously released from materials undergoing deformation, fracture or both. The released energy produces high-frequency acoustic signals. The strength of the signals depends on parameters such as the rate of deformation, the volume of the participating material and the magnitude of the applied stress. AE is used in many applications such as non-destructive evaluation and materials research.

AE is a passive monitoring technique which, rather than requiring 100% scanning (such as radiography, ultrasonic, or eddy currents), can inspect the entire structure volumetrically using remote sensors electronically tied to a computer for analysis. It can, therefore, be more efficient with respect to preparation and inspection time, considerably reducing costs. In many applications it is also more sensitive than other non-destructive testing (NDT) techniques (Matthews and Hay, 1983).

The first known comprehensive investigation of AE was performed by Kaiser (1953). In fact, Kaiser characterized a basic irreversibility phenomenon which bears his name. In the Kaiser Effect, when a material is stressed to a given level and the stress is removed, upon reapplication of stress there is no detectable emission at a fixed sensitivity level until previously applied stress levels have been exceeded. Advances in both materials science and electronics technology have contributed to bring AE to the forefront of new NDT methods. The signals can be detected by sensors often placed several feet away from the source of signal generation. There are presently many AE transducers and sensors that can be utilized for specific applications.

4.1   Types of Transducers and Sensors.   Piezoelectric transducers have been the primary means of gathering AE information. They can be selected for specific applications by choosing a specific resonant frequency or by selecting one with a broad frequency band.

Capacitor microphones, which measure the capacitance changes in a DC biased electrostatic element located near the surface due to surface displacement, are used almost exclusively in the laboratory. They are noted for their flat frequency response and can be utilized in the finite spectral analysis.

Electromagnetic sensors are used to detect the eddy currents produced by the displacements accompanying the passage of an elastic wave.

Optical interferometers are used to measure changes in optical path length caused by surface displacement. They have a small focused beam $\cong$10mm and make measurement possible in hostile environments. They can be calibrated accurately and make repeatable finite measurements. But the band width depends on the electronics employed and they are cost prohibitive.

4.2   Analysis of Acoustic Emission Data.   Due to the complexity of many parameters that are recorded and presented to the monitor, it would be impossible to list all of the information available for analysis in the confines of this section. Therefore, key parameters will be discussed to give insight into the data available during an AE monitoring.

Once an AE signal has been detected from a structure or specimen by a transducer, it is sufficiently amplified for analysis. The analysis generally involves establishing relationships between signal parameters and the physical variables to which the structure is subjected, such as load, stress, strain, time, etc. The signals are also passed through appropriate filters for elimination of a large part of the background noise. Most AE signals can be classified either as burst emission (signal amplitudes appreciably larger than background, of short duration, and well separated in time of occurrence) or as continuous emission (signal amplitudes slightly higher than background and events closely spaced in time to form a single waveform). Crack nucleation and propagation are often considered as bursts. Other defects show up as continuous and can be difficult to distinguish from pseudo sources when looking at the waveform alone. These pseudo sources are liquid and gas leakage, crack closure, loose particles and loose parts, slag cracking, frictional rubbing, cavitation, and any sudden volume change such as boiling, freezing, melting and chemical reaction (Duesing, 1989). Depending on the application, signal analysis may be either in the time domain or frequency domain as discussed below.

4.2.1 Time Domain Analysis.

A number of techniques can be employed to analyze AE signals in the time domain.

(a) Ring Down Count and Count Rate

Perhaps the most commonly used technique for time domain analysis is the determination of the ring down count and count rate. These are related to the cumulative damage and the rate of occurrence of the damage respectively (Bassim and Houssny-Eman, 1983). After amplification, the AE event is observed on the oscilloscope screen and is distinguishable above the ever present background or electrical noise. A threshold is set above this noise by cutting off the system's counting detector when the signal is below a certain amplitude. Figure 13.6 illustrates this procedure and shows the information contained in each wave packet or event. The following terms are used:

Event The envelope of each signal from initial threshold crossing to the last threshold crossing.

Duration The amount of time involved with the above event.

Ring down Count The number of peaks in a given event.

Count Rate The number of counts / unit time.

Rise Time The time from first threshold crossing to the peak amplitude.

(b) Energy Analysis

Another time domain approach is the measurement of the energy of the AE signal. Since AE is attributed to the release of energy in the material due to changes in the state of stresses, measurement of the energy content of the signal is believed to be an indication of this energy release. Direct energy analysis can be achieved by measuring the area under the envelope of the square of the signal using a digital integrator in accordance with the following equation:

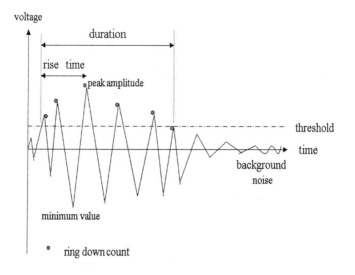

Figure 13.6   Information for the AE event.

$$E = \frac{1}{R} \int_0^T V^2(t)dt,$$

where E is the electrical energy, $V(t)$ is the output voltage of the sensor, and R is the resistance of the measuring circuit. Another approach is the determination of the energy of the signal is by measuring the RMS of the amplitude using an RMS voltmeter. The RMS voltage V_{rms} can be expressed by the following equation:

$$V_{rms}^2 = \frac{1}{T} \int_0^T V^2(t)dt,$$

where T is the time duration of the whole digitized waveform over which the voltage signal is averaged. If the exact value of the energy is of no interest, but rather its relative value, then comparing the V_{rms} is sufficient. That is to say, the system only needs to distinguish between no damage, damage, and intense damage (Derakhshan *et al.*, 1991). The advantage of this method is that energy measurements can be directly related to an important physical parameter, which is merely approximate as in the case of the ring down counts.

(c) Amplitude and its Distribution

The amplitude of the acoustic signal is indicative of the intensity of damage in the material which causes a signal to occur. It is however greatly affected by the position of the transducer with respect to the source as well as the response of the transducer. Even with this drawback, the observed amplitude can be considered as a measure of the released energy. This is because, in general, the greater the energy released, the greater the energy observed within a given narrow band of frequencies. Furthermore, the observed amplitude is proportional to the square root of the energy incident on the sensor within the frequency band of the instrument response; thus the square of the observed amplitude is a rough measure of the energy released at the emission source (Nakamura *et al.*, 1971).

Particular emphasis has been placed on obtaining amplitude distributions of emission signals. The amplitude distributions observed in the study of Nakamura *et al.* (1971) have a great significance in terms of the fracture properties of the material. They use the amplitude distribution to seek emission characteristics which might be useful both for determining the criticality of the crack which emits the signals and for gaining an understanding of the emission mechanisms. The amplitude distribution is useful because the growth of a crack is a nearly stochastic process in which the amount of each growth increment is determined by the balance between the amount of energy required to produce additional crack surface areas and the amount of energy available, both of which have their own statistical distributions along a crack tip in a heterogeneous medium. Since the balance should be influenced by the stress level at the crack tip, some variation of emission amplitude distribution with stress level is expected. The amplitude distribution is also related to the way a crack propagates and for this reason can be expected to supply important data for understanding fracture mechanisms. The study on amplitude distribution analysis can be found in Sun *et al.* (1991) *etc.*

The other parameters in the time domain include location of peak, location of minimum amplitudes, location of maximum slope, mean value, variance, and counts above specific threshold, which can be used in pattern recognition.

4.2.2   Frequency Domain Analysis. Frequency domain analysis can also be applied to identify the sources of emissions in materials and structures. Apart from amplitude related techniques the only other widely used means of examining acoustic emission signals is spectral analysis. Recent advances in electronic signal recording and processing have extended the range over which spectral analysis can be accurately performed to encompass the primary frequencies of interest in stress wave analysis. In particular, the development of the videotape recorder and high speed transient digitizer have revolutionized the AE spectral analysis. Two methods of spectral analysis of the recorded data are widely used. Commercial spectrum analyzers are relatively inexpensive, readily available, and require only a repetitive analog signal whose repetition rate is faster than the scan rate of the analyzer, a condition easily met by the stop action mode of the videotape recorder or the analog output of the transient digitizer. Digital methods of spectral analysis, usually via the FFT, are more versatile but somewhat more expensive as they require a relatively complicated interface between the digitizer and the computer for operation (Kline, 1983).

The basic features used in pattern recognition in the frequency domain include peak frequency, maximum amplitude, minimum amplitude, area, maximum slope, frequency at maximum slope, frequency at minimum amplitudes, 6dB down bandwidth, mean value, secondary peak frequency, RMS, and variance.

5.   TEMPERATURE MONITORING

Elevation in component or equipment temperature is frequently an indication of potential problems. For example, most of the failures of electric motors are attributed to excessive heat which is generated by antifriction bearings. The bearing life is dependent on its maintenance schedule and their operating conditions. Similarly, hot spots(in electric boards), which are usually caused by excessive currents, indicate that failure is imminent. Therefore, a measure of temperature variation can be effectively used in monitoring components and equipment for predictive maintenance purposes. In most electrical equipment, the limiting components are made from polymeric materials and they age because of thermal degradation . The rate at which they age can be calculated using the activation energy for the degradation process, which is obtained from accelerated ageing tests.

The effect of temperature on the device is generally modeled using the Arrhenius reaction rate equation given by

$$r = Ae^{-(E_a/kT)}, \tag{13.3}$$

where

$$
\begin{aligned}
r &= \text{the speed of reaction,} \\
A &= \text{an unknown nonthermal constant,} \\
E_a &= \text{the activation enerage (eV),}
\end{aligned}
$$

k = the Boltzman Constant $(8.62310^{-5}$ eV/$^{\circ}$K),

T = the temperature in Kelvin.

Assuming that device life is proportional to the inverse reaction rate of the process, then Equation (4.3) can be rewritten as

$$L = Ae^{+(E_a/kT)},$$

where L is the nominal life of the device. Using this Arrhenius model , the life distribution related to the monitored temperature can be estimated. Also, we can obtain the failure rate and temperature behaviour curve over time. From here we can determine the maintenance schedule as discussed in Section 2.

There is a wide range of instruments for measuring variations in temperature such as mercury thermometers which are capable of measuring temperatures in the range of -35 to 900° F and thermocouples which can provide accurate measurements up to about 1,400° F. Optical pyrometers, where the intensity of the radiation is compared optically with a heated filament, are useful for the measurements of very high temperatures (1,000 to 5,000° F).

Recent advances in computers made the use of the infrared temperature measure possible for many applications that are difficult or impractical to contact with other instruments (Niebel, 1994). The infrared emissions are the shortest wavelengths of all radiant energy and are visible with special instrumentation. Clearly the intensity of the infrared emission from an object is a function of its surface temperature. Therefore when a sensing head is aimed at the object whose surface temperature is being measured, we can calculate the surface temperature and provides a color graphic display of temperature distribution.

6.   FLUID MONITORING

Analysis of equipment fluids such as oil can reveal important information about the equipment wear and performance. It can also be used to predict the reliability and expected remaining life of parts of the equipment. Measuring oil quality is usually done with a complex chemical laboratory benchmark procedure that measures several parameters indicating oil degradation . These factors include the particle count, the types of particles, and total acid number. As the equipment operates, minute particles of metal are produced from the oil covered parts. The particles remain in suspension in the oil and are not removed by the oil filters due to their small size. The particle count will increase as equipment parts wear out. There are several methods that can identify the particle count and the types of particles in the oil. The two most commonly used methods are atomic absorption and spectrographic emission.

With the atomic absorption method , a small sample of oil is burnt and the flame is analyzed through a light source which is particular for each element. The method is very accurate and can obtain particle count as low as 0.1 parts per million (ppm). However, the analysis is tedious and time consuming except when the type of particle is known (Cumming, 1990).

Spectrographic emission is similar to the atomic absorption method in burning a small sample of oil. It has the advantage that all quantities of all the materials can be read at one burn. However it is only capable of detecting particle counts of 1 ppm or higher. Moreover, spectrometry is unable to give adequate warning in situations when the failure mode is characterized by the generation of large particles from rapidly deteriorating surfaces (Eisentraut *et al.*, 1978).

Sensors can also be used to measure the dielectric constant of oil, which correlates the acidity, an indicator of oil degradation . The dielectric constant correlates well with a total acid number and is easier to measure than other properties, making it suitable as the basis of an on-board sensing system (DeGaspari, 1999).

A different approach to monitoring engine oil quality has been developed by a branch of the automotive industry where a mathematical model uses the engine's computers to infer the rate of oil degradation from data already being collected by various systems within the vehicle. Schwartz *et al.* (1987) found that oil temperature, vehicle mileage, engine revolutions, and changes in the physical and chemical properties of oil during use all provided an indication of oil degradation . Based on these measurements, they developed a mathematical model which relates oil life to oil temperature and either vehicle mileage or engine revolutions.

A combined approach of monitoring driving conditions and using sensors has also been taken by another section of the automotive industry. They developed a passenger car maintenance system which calculates oil change intervals based on driver-specific data, and supplements that information with a sensor that continuously monitors oil level, oil temperature, and the dielectric number of the engine oil (DeGaspari, 1999).

Jardine *et al.* (1989) studied an interesting examination of the method of proportional hazards modelling (PHM) to determine whether or not PHM could improve on the accuracy of the oil-analyst/expert system in determining the risk of failure of a diesel engine.

7.   CORROSION MONITORING

Corrosion is a degradation mechanism of many metallic components . Clearly, monitoring the rate of degradation , *i.e.*, the amount of corrosion, has a major impact on the preventive maintenance schedule and the availability of the system. There are many techniques for monitoring corrosion such as visual, ultrasonic thickness monitoring, electrochemical noise, impedance measurements, and thin layer activation.

We briefly describe one of the most effective on-line corrosion monitoring techniques, namely, Thin Layer Activation (TLA). The principle of TLA is that trace quantities (1 in 10^{10}) of a radioisotope are generated in a thin surface layer of the component under study by an incident high energy ion beam. Loss of the material (due to corrosion) from the surface of the component can be readily detected by a simple γ-ray monitor (Asher et al. 1983). The reduction in activity is converted to give a depth of corrosion directly and, provided that the corrosion is not highly localized, this gives a reliable measurement of the average loss of material over the surface.

Corrosion causes degradation in the system's performance and it becomes necessary to determine the time for the performance to degrade to a threshold value. Reliability prediction using the degradation data can be obtained accordingly as described in Ettouney and Elsayed (1999).

8.   NEW APPROACHES IN CONDITION MONITORING

8.1   Prognostics and Health-Management System.

New advances in sensor technology and failure analysis are instigating a revolution in the way large electromechanical systems such as aircraft, helicopters, ships, power plant, and many industrial operations will be maintained in the future. For industry and the military, the 21st century will bring about the age of prognostics and health-management systems (Becker *et al.*, 1998).

So far, we mentioned fault diagnostics, the detection of an existing problem or failure in a system in order to correct it. Today's advances are raising the bar toward machine prognostics, where failure modes and the remaining life of a system can be predicted. This will result not only more efficient operations and reduced maintenance cost, but also the saving of lives. Both industry and the military have shown an interest in this emerging technology (Becker *et al.*, 1998).

Machine prognostics essentially involves taking data from sensors and probes that are placed on various system components to record specific condition parameters, and feeding these data into a computer program so that potential system faults and failures can be identified, tracked, and predicted (see Figure 13.7). The aim of prognostics is to stop disabling or fatal failures before they happen. The concept of prognostics goes beyond diagnostics, in which the sensed data are simply monitored for the occurrence of anomalies or failure that are then corrected. The prognostics process is analogous to the way physicians deal with medical problems. First the problem is detected; then a diagnosis is made about the failure mode and its severity. It is also important to predict the evolution of the failure in order to estimate the remaining useful life of the machine (Becker *et al.*, 1998).

Industry is already incorporating some machine prognostic technology into their programs which are based on the technological advances in such several areas as sensing, modeling to predict the behavior of systems, fusing data gathered by many sensors, and developing of the automatic feature extraction algorithms necessary to make meaningful predictions.

8.2   Automatic Feature Extraction.

Within the framework of condition monitoring it is generally insufficient to characterize a breakdown by an excessive value of just a single parameter. The breakdowns may also result from the joint occurrence of deviations of such parameters, in spite of the parameters remaining within specified tolerance values. Knowing automatic classification of these deviation vectors one can naturally classify the degraded performance with this. This enables us to connect the monitoring device to a condition indicator which activates an alarm whenever the level of degraded performance reaches a predetermined level. The essential precondition to diagnose the machine condition by this method is that sufficient prior knowledge or historical fault information is available for the interpretation of the data. In most applications this precondition is not satisfied; especially for new machines or new processes. Jin and Shi (1998) proposed an automatic feature extraction methodology for the development of a feature-based diagnostic system using waveform signals with limited or with no prior fault information.

8.3   Other Diagnostic Methods.

Components and systems can be monitored in order to perform maintenance and replacements by observing some of the critical characteristics using a variety of sensors or microsensors. For example, pneumatic and hydraulic systems can be monitored by observing pressure, density of the flow, rate of flow, and temperature change. Similarly, electrical components or systems can

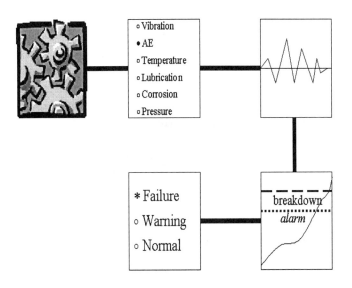

Figure 13.7 Prognostic and health-management system.

be monitored by observing the change in resistance, capacitance, volt, current, temperature, and magnetic field intensity. Mechanical components and systems can be monitored by measuring velocities, stress, angular movements, shock impulse, temperature, and force (Elsayed, 1996).

Recent technological advances in measurements and sensors resulted in observing characteristics that were difficult or impossible to observe, such as odor sensing. At this point of time silicon microsensors have been developed that are capable of mimicking the human sense of sight (e.g., a CCD), touch (e.g., a tactile sensor array), and hearing (e.g., a silicon microphone). Sensors to mimic the human sense of smell to discriminate between different odor types or notes are at the early stage of development. Nevertheless, some commercial odor discriminating sensors are now available such as the Fox 2000 or Intelligent Nose (Alpha MOS, France). Such instruments are based upon an array of six sintered metal oxide gas sensors which respond to a wide range of odorants. The array signals are processed using an artificial neural network (ANN) technique. The electronic nose is first trained on known odors, then the ANN can predict the nature of the unknown odors with a high success rate (Gardner, 1994).

The improvements in sensors' accuracy and the significant reduction in their cost have resulted in their use in a wide variety of applications. For example, most of the automobiles are now equipped with electronic diagnostic systems which provide signals indicating the times to service the engine, replace the oil filter, and check engine fluids.

Most importantly, the advances in microcomputers, microprocessors, and sensors can now offer significant benefits to the area of preventive maintenance and replacements. Many components, systems, and entire plants can now be continuously monitored for sources of disturbances and potential failures. Moreover, on-line measurements, analysis, and control of properties and characteristics, which have been traditionally performed off-line, result in monitoring of a wider range of components and systems than ever before.

9.   CONCLUSIONS

This chapter presents different approaches for condition monitoring and fault diagnostics systems for predictive maintenance . It describes methodologies for determining the limit value of the condition parameter (or criterion) which corresponds to a predetermined reliability level. It then provides details of the most commonly used condition parameters, starting with vibration, acoustic emission and concluding with feature extraction. Applicability of such condition parameters and advantages and disadvantages are also discussed.

References

[1] Asher, J., Conlon, T. W., Tofield, B. C. and Wilkins, N. J. M. Thin-Layer Activation - A New Plant Corrosion-Monitoring Technique. *On-Line Monitoring of Continuous Process Plants* edited by Butcher, D.W., Ellis Horwood Limited, West Sussex, England, 95-105, 1983.

[2] Bassim, M. N. and Houssny-Emam, M. Time and Frequency Analysis of Acoustic Emission Signals. *Acoustic Emission* edited by Matthews, J. R., 139-163, 1983.

[3] Becker, K. C., Byington, C. S., Forbes, N. A. and Nickerson G. W. Predicting and Preventing Machine Failures. *The Industrial Physi-*

cist December (1998) 20-23.

[4] Bendat, J. S. and Piersol, A. G. *Random Data: Analysis and Measurement Procedures.*, Wiley-Interscience, 1971.

[5] Bogdanoff, J. L. and Kozin, F. *Probabilistic Models of Cumulative Damage.*, New York, John Wiley, 1985.

[6] Brigham, E. O. *The Fast Fourier Transform.*, Prentice-Hall, Inc., 1974.

[7] Callaway, B. (1982) Narrow Analysis - An Enhancement to Productivity. *1982 Proceedings Annual Reliability and Maintainability Symposium*, 62-67.

[8] Carey, M. B. and Koenig, R. H. Reliability Assessment Based on Accelerated Degradation: A Case Study. *IEEE Transactions on Reliability* , **40** (1991) 499-506.

[9] Cempel, C. The Vibration Symptom Limit Value in Condition Monitoring. *Condition Monitoring '84* - Proceedings of International Conference on Condition Monitoring held at Univ. College of Swansea, 328-339, 1984.

[10] Cempel, C. Determination of Vibration Symptom Limit Value in Diagnostics of Machinery. *Maintenance Management International* 5, 297-304, 1985.

[11] Cempel, C. Passive Diagnostic and Reliability Experiment and its Application in Machines Condition Monitoring. *Condition Monitoring '87* - Proceedings of International Conference on Condition Monitoring held at Univ. College of Swansea, 202-219, 1987.

[12] Cempel, C. Condition Assessment and Forecasting form Plant Diagnostic Data with Pareto Model. *Condition Monitoring and Diagnostic Engineering Management* - Proceedings of COMADEM 90, 403-406, 1990.

[13] Chan, L., Mui, L. and Woo, C. Reliability Analysis and Maintenance Policy of Radiators for a Large Fleet of Buses. *Quality & Reliability Engineering International* **13**(3) (1997) 117-126.

[14] Chick, S. E. and Mendel, M. B. An Engineering Basis for Statistical Lifetime Models with an Application to Tribology. *IEEE Transactions on Reliability* **45** (1996) 208-214.

[15] Chui, C. K. *An Introduction to Wavelets.* Academic Press, San Diego, 1992.

[16] Cooley, J. W. and Tukey, J. W. An Algorithm for the Machine Calculation of Complex Fourier Series. *Mathematics of Computation* **19**(90) (1965) 297-301.

[17] Cumming, A. C. D. Condition Monitoring Today and Tomorrow - An Airline Perspective. In *Condition Monitoring and Diagnostic Engineering Management* edited by Rao, R., Au, J., and Griffiths, B., Chapman and Hall, London: United Kingdom, 1-7, 1990.

[18] Dabrowski, H. Condition Assessment of Airborne Equipment by Means of Vibroacoustic Technique. Ph.D. thesis, Warsaw Technical Univ, 1981.

[19] Daubechie, I. The Wavelet Transform, Time-Frequency Localization and Signal Processing *IEEE Transactions on Information Theory.* **36** (1991) 961-1005.

[20] Derakhshan O., Houghton, J. R., Johnes, R. K. and March, P. A. Cavitation Monitoring of Hydroturbines with RMS Acoustic Emission Measurement. *Acoustic Emission: Current Practice and Future Directions*, ASTM STP 1077, 305-315, 1991.

[21] Duesing, L. A. Acoustic Emission Testing of Composite Materials. *1989 Proceedings Annual Reliability and Maintainability Symposium*, 128-134, 1989.

[22] Dyer, D. and Stewart R. M. Detection of Rolling Element Bearing Damage by Statistical Vibration Analysis. *Journal of Mechanical Design* **100** (1978) 229-235.

[23] Eghbali, G. and Elsayed, E. A. Reliability Estimation Based on Degradation Data. *Rutgers University, Department of Industrial Engineering, Working Paper*, 97-117, 1997.

[24] Eisentraut, K. J., Thornton, T. J., Rhine, W. E., Cnstandy, S. B., Brown, J. R. and Fair, P. S. Comparison of the Analysis Capability of Plasma Source Spectrometers vs. Rotating Disc Atomic Emission and Atomic Absorption Spectrometry of Wear Particles in Oil: Effect of Wear Metal Particle Size. Presented at the *1st International Symposium on Oil Analysis*, Erding, Germany, July 4-6, 1978.

[25] El-Haram, M. Integrated Approach to Condition Based Reliability Assessment and Maintenance Planning. *University of Exeter, Ph.D. Thesis*, 1995.

[26] Elsayed, E. A. *Reliability Engineering.* New York, Addison Wesley Longman, New York, 1996.

[27] Ettouney, M. and Elsayed, E. A. Reliability Estimation of Degraded Structural Components Subject to Corrosion. *Rutgers University, Department of Industrial Engineering, Working Paper*, 99-113, 1999.

[28] Feinberg, A. A. and Widom, A. Connecting Parametric Aging to Catastrophic Failure through Thermodynamics. *IEEE Transactions on Reliability* **45** (1996) 28-33.

[29] Forrester, B. D. Time-Frequency Analysis in Machine Fault Detection. In *Time-Frequency Signal Analysis* edited by Boashash, B., Melbourne, Longman Cheshire, 1992.

[30] Gardner, J. W. *Microsensors Principles and Applications* , John Wiley & Sons, New York, 1994.

[31] Gertsbackh, I. B. and Kordonskiy, K. B. *Models of Failure.* English Translation from the Russian Version. Springer-Verlag, New York, 1969.

[32] Jardine, A. K. S., Ralston, P., Reid, N. and Stafford, J. Proportional Hazards Analysis of Diesel Engine Failure Data. *Quality & Reliability Engineering International* **3**(5) (1989) 207-216.

[33] Jin, J. and Shi, J. Automatic Feature Extraction of Waveform Signals for In-Process Diagnostic Performance Improvement. *1998 IEEE International Conference on Systems, Man, and Cybernetics,* 4716-4721, 1998.

[34] Kaiser, J. Untersuchungen Uger Das Auftreten Gerauschen Beim Zugversuch Archiv Fir Das Eisenhuttenwesen, 24, 1953.

[35] Kamarthi, S. V. and Pittner, S. Fourier and Wavelet Transform for Flank Wear Estimation - A Comparison. *Mechanical Systems and Signal Processing,* **11**(6) (1997) 791-809.

[36] Kline, R. A. Acoustic Emission Signal Characterization. *Acoustic Emission* edited by Matthews, J. R., 105-138, 1983.

[37] Knezevic, J. Condition Parameter Based Approach to Calculation of Reliability Characteristics. *Reliability Engineering* 19(1) (1987) 29-39.

[38] Knezevic, J. Required Reliability Level as the Optimization Criterion. *Maintenance Management International,* 6(4) (1987) 249-256.

[39] Knezevic, J. Real-Time Operational Reliability. *Reliability Data Collection and Use in Risk and Availability Assessment* edited by Colombari, V., Proceedings of the 6th EUREDATA Conference, Siena, Italy, 710-715, 1989.

[40] Kobbacy, K. A. H., Fawzi, B. B., Percy, D. F. and Ascher, H. E. A Full History Proportional Hazard Model for Preventive Maintenance Scheduling. *Quality & Reliability Engineering International,* 13(4) (1997) 187-198.

[41] Love, C. E. and Guo, R. Using Proportional Hazard Modelling in Plant Maintenance. *Quality & Reliability Engineering International,* 7(1) (1991) 7-17.

[42] Love, C. E. and Guo, R. Hazard Modelling to Bad-as-Old Failure Data. *Quality & Reliability Engineering International* 7(3) (1991) 149-157.

[43] Lu, C. J. and Meeker, W. Q. Using Degradation Measures to Estimate a Time-to-Failure Distribution. *Technometrics,* 35(2) (1993) 161-173.

[44] Lu, J., Park, J. and Yang, Q. Statistical Inference of a Time-to-Failure Distribution Derived from Linear Degradation Data. *Technometrics* 39(4) (1997) 391-400.

[45] Ma, J. C., Luo, L. and Wu, Q. B. A Filter Design Method Based on Combination Wavelets. *Mechanical Systems and Signal Processing* 11(5) (1997) 767-772.

[46] Massam, M. and Cruickshank, J. L. Impact of an Integrated Health Management and Support System (IHMaSS) on Operational Effectiveness of Future Military Aircraft. *Proceedings of 7th International M.I.R.C.E. Symposium on System Operational Effectiveness,* Exeter, 1997.

[47] Matthews, J. R. and Hay D. R. Acoustic Emission Evaluation. *Acoustic Emission* edited by Matthews, J. R., 1-14, 1983.

[48] McFadden, P. D. Examination of a Technique for the Analysis Early Detection of Failure in Gears by Signal Processing of the Time Domain Average of the Meshing Vibration. *Mechanical System and Signal Processing* 1 (1987) 173-183.

[49] McFadden, P. D. Application of the Wavelet Transform to Early Detection of Gear Failure by Vibration Analysis. *Proceedings of International Conference on Condition Monitoring* held at Univ. College of Swansea, Wales,1994.

[50] McFadden, P. D. and Smith, J. D. Information from the Vibration of Rolling Bearing. *Condition Monitoring '84* - Proceedings of Inter-

national Conference on Condition Monitoring held at Univ. College of Swansea, 178-190, 1984.

[51] McFadden, P. D. and Wang, W. J. Time-Frequency Domain Analysis of Vibration signals for Machinery Diagnostics: (1) Introduction to Wigner-Ville Distribution. *Technical Report, Report No OUEL 1859/90*, Dept. of Engineering Science, Univ. of Oxford, 1990.

[52] Meeker, W. Q., Escobar, L. A. and LU, C. J. Accelerated Degradation Tests: Modeling and Analysis. *Technometrics*, **40**(2) (1998) 89-99.

[53] Monk, R. Vibration Measurement Gives Early Warning of Mechanical Faults. *Processing Engineering* Nov. (1972) 135-137.

[54] Nakamura, Y., Veach, C. L. and Mccauley B. O. Amplitude Distribution of Acoustic Emission Signals. *Acoustic Emission*, ASTM Special Technical Publication 505, 164-186, 1971.

[55] Nelson, W. Analysis of Performance-Degradation Data from Accelerated Test. *IEEE Transactions on Reliability*, **30** (1981) 149-155.

[56] Newland, D. E. *An Introduction to Random Vibration, Spectral and Wavelet Analysis* 3rd Edition, Longman, Harlow and John Wiley, New York, 1993.

[57] Niebel, B. W. *Engineering Maintenance Management.* Marcel Dekker, New York, 1994.

[58] Pau, L. F. *Failure Diagnosis and Performance Monitoring.* Marcel Dekker, New York, 1981.

[59] Paya, B. A., Esat, I. I. and Badi, M. N. Fault Classification in Gear-boxes Using Neural Networks. *Energy Technology Conference & Exhibition (ETCE) 96*, Symposium of Structural Dynamics/Vibration/Buckling, 167-174, 1996.

[60] Paya, B. A., Esat, I. I. and Badi, M. N. Artificial Neural Network Based Fault Diagnostics of Rotating Machinery Using Wavelet Transforms as a Preprocessor. *Mechanical Systems and Signal Processing* **11** (5) (1997) 751-765.

[61] Petrilli, O., Paya, B. A., Esat, I. I. and Badi, M. N. Neural Network Based Fault Detection Using Different Signal Precessing Techniques as Pre-Preprocessor. *JN-American Society of Mechanical Engineers, PD Publication* PY 70, 97-101, 1995.

[62] Randall, R. B. Cepstrum Analysis and Gear-box Fault Diagnosis. 1982 *Bruel & Kjaer Application Note*, 233-280, 1982.

[63] Saranga, H. and Knezevic, J. Reliability Prediction Using the Concept of Relevant-Condition Parameters. *1999 Proceedings Annual Reliability and Maintainability Symposium*, 32-37, 1999.

[64] Schwartz, S. E. and Smolenski, D. JR. Development of an Automatic Engine Oil-Change Indicator System. *SAE International Congress and Exposition*, Detroit, MI, USA, 1987.

[65] Staszewski, W. J. and Tomlinson, G. R. Application of the Moving Window Procedure in a Spur Gear. *COMEDEM-93*, Bristol, England, 21-23, 1993.

[66] Staszewski, W. J. and Tomlinson, G. R. Application of the wavelet transform to Fault Detection in a Spur Gear. *Mechanical Systems and Signal Processing*, 8 (1994) 289-307.

[67] Staszewski, W. J., Worden, K. and Tomlinson, G. R. Time-Frequency Analysis in Ger-box Fault Detection Using The Wigner-Ville Distribution and Pattern Recognition. *Mechanical Systems and Signal Processing*, 11(5) (1997) 673-692.

[68] Sun, X., Hardy, Jr. and Rao, M. V. M. S. Acoustic Emission Monitoring and Analysis Procedures Utilized During Deformation Studies on Geologic Materials. *Acoustic Emission: Current Practice and Future Directions*, ASTM STP 1077, 365-380, 1991.

[69] Takata, S. and Ahn, J. H. Overall Monitoring System by Means of Sound Recognition. In *Diagnostic and Preventive Maintenance Strategies in Manufacturing Systems* edited by Milacic, V.R., and McWaters, J.F., North-Holland, Amsterdam, 99-111, 1987.

[70] Taylor, J. I. Identification of Bearing Defects by Spectral Analysis. *Journal of Mechanical Design*, 102 (1980) 199-204.

[71] Wang, W. J. and McFadden, P. D. Early Detection of Gear Failure by Vibration Analysis, I: Calculation of the Time-Frequency Distribution. *Mechanical System and Signal Processing*, 7(3) (1993) 193-203.

[72] Wang, W. J. and McFadden, P. D. Application of Wavelets to Gearbox Vibration Signals for Fault Detection. *Journal of Sound and Vibration*, 192(5) (1996) 927-939.

[73] Weichbordt, B. and Bowden, J. Instrumentation for Predicting Bearing Damage. *General Electric Company Report,* March, 1970, S-70-1021, AD 869 633, 1970.

[74] Wheeler, P. G. Bearing Analysis Equipment Keeps Downtime Down. *Plant Engineering,* **25** (1968) 87-89.

Chapter 14

MAINTENANCE SCHEDULING USING MONITORED PARAMETER VALUES

Dhananjay Kumar

Reliability Engineer

Nokia Svenska AB

PO BOx 1070

164 64 Kista Sweden

Abstract   Applications of the reliability methods in maintenance scheduling have been widely investigated in the literature considering failure times. However, information obtained using condition monitoring devices, whenever possible is being used more and more in industries for maintenance scheduling . This trend is accelerated by the availability of reliable sensors and a rapid development in information technologies. The monitored parameter values (MPV) may explain the failure characteristics and influence the maintenance scheduling of a system. There are several reliability models that can be used to model MPV for maintenance scheduling. These models include regression models, proportional hazards family and accelerated failure time family. The latter two appear to be suitable for practical applications.

The paper describes the models that can be used to model the MPV. Some guidelines for selection of suitable models for a given dataset are also discussed. These models are used to estimate the relative importance of the MPV in explaining the failure characteristics. Once the relative importance of the MPV is estimated, either graphical, numerical, or analytical methods can be used for maintenance scheduling based on the MPV. Maintenance cost models that include planned and unplanned maintenance costs are further extended to include the MPV. Graphical methods such as the total time on tests-plot or the cumulative intensity plot can be used to determine the optimum maintenance interval.

The applications of reliability models and graphical methods for determination of the optimum maintenance time interval are illustrated with field failure data. The proposed approach can be used for repairable as well as non-repairable systems.

Keywords: : Proportional Hazards Model, Proportional Intensity Model , Accelerated Failure Time Models, Total Time on Test-Plotting, Maintenance Policy, Maintenance scheduling , Monitored Parameters.

1.   INTRODUCTION

Applications of the reliability methods in maintenance scheduling have been widely investigated in the literature considering failure times only (see a survey by Valdez-Flores & Feldman, 1989; Cho & Parkar, 1991). However, information obtained using condition monitoring devices, whenever possible, is being used more and more in industries for maintenance scheduling . This trend is accelerated by the availability of reliable sensors and a rapid development in information technologies. Presently, maintenance scheduling of a system is decided mainly based on experience, manufacturers' recommendations, characteristics of , or reliability analysis. The maintenance scheduling of a system may be influenced by the surrounding environment (e.g. temperature, humidity, and dust), conditions indicating parameters (e.g. vibration, and pressure), design modifications, skill of operators, operating time since the last repair, failure history, types and number of repairs carried out on the system. In reliability analysis, all these influencing factors are referred to as covariates, risk factors, explanatory variables or concomitant variables. The term monitored parameter values (MPV) is used here to denote all such factors which may explain the failure characteristics and influence the maintenance scheduling of a system.

There are some reliability models that can be used to model MPV for maintenance scheduling. One possibility for analysing the MPV is to use the regression models which are generally used in biostatistics. Applications of some of these models in the reliability contexts are presented in Crowder et al. (1991). In these models for reliability application, estimation of hazard rate or intensity function based on the MPV is generally the primary aim of modelling. The hazard rate, $h(x; \mathbf{z})$, of a component or a system is a measure of the proneness to fail shortly after the time x provided it has survived up to time x in the presence of the MPV represented by vector $\mathbf{z}$. The hazard rate considering $\mathbf{z}$ is defined as

$$h(x; \mathbf{z}) = \lim_{\Delta x \to 0} \frac{\Pr\{x \le X < x + \Delta x \mid X \ge x; \mathbf{z}\}}{\Delta x} \tag{14.1}$$

where X denotes a random variable and x is the local time (time between successive failures or time to failure).

The intensity function based on the MPV is a measure of the probability of failure in a short time interval and is defined as (Prentice et al. 1981)

$$\nu\left\{t|\,N\left(t\right),\mathbf{z}\right\} = \lim_{\Delta t \to 0} \frac{\Pr\left\{t \leq T_{n(t)+1} < t + \Delta t|\,N\left(t\right),\mathbf{z}\right\}}{\Delta t} \qquad (14.2)$$

where $N(t) = \{n(u) : u \leq t\}$ is the counting process for which $n(u)$ is the number of failures of an individual in the interval $(0, t]$, $T_{n(t)}$, is the cumulative time of occurrence of the *n:th* failure and t is the global time (cumulative of chronologically ordered time between successive failures).

2.   RELIABILITY MODELS

Several reliability models for estimating $h(x; \mathbf{z})$ and $v\{t|N(t), \mathbf{z}\}$ have been proposed. These models largely owe their origin to methods used in biostatistics. There is an extensive literature on the analysis of influence of treatment methods, age, sex, or geographical location on the individuals under study (see Kalbfleisch & Prentice, 1980, Lawless, 1982; Cox & Oakes, 1984; McCullagh & Nelder, 1989). The methodology applied in biostatistics forms the basic concept in reliability models incorporating MPV. The basic approach in most of these models is to assume a baseline hazard rate or baseline intensity function which equals the hazard rate or intensity function of a system when the MPV is equal to zero.

Most of these models can be broadly classified into two groups on the basis of approaches used (Lawless, 1983) namely, parameteric and non (or semi)-parametric regression models. In the former group of models, the lifetime of a system is assumed to have a specific distribution that depends on the MPV e.g. Weibull regression models, (Smith, 1991) and the log-logistic model (Bennet, 1983a). The general approach in the later group of models is to decompose the hazard rate into two parts, one as a function of time, and the other as a function of the MPV. Examples of non (or semi)-parametric models are proportional hazards models (Cox, 1972a) and proportional odds models (McCullagh, 1980; Bennet, 1983b).

If the classification criterion is how the MPV explains the lifetime, most of these models can be broadly classified as multiplicative and accelerated failure time models. In multiplicative models, an MPV is assumed to accelerate or decelerate the hazard rate or intensity function (or their transformations), e.g. the proportional hazards model. In the accelerated failure time models, an MPV is assumed to accelerate or decelerate the failure time (or some transformation of time), e.g. Weibull and log-logistics regression models. A generalised mathematical form that includes both types of models is discussed in Lawless (1986). There are some theoretical and computational difficulties associated while estimating parameters of some of these models. Models from the proportional hazards family and accelerated failure time family appear to be suitable for practical applications due to the particular method used for estimating the parameters of these models.

3.   PROPORTIONAL HAZARDS MODEL

The proportional hazards model (PHM) was first introduced by Cox (1972a). Since then various applications of the PHM in reliability analysis have been presented. Application of the PHM to various type of reliability data and some of the graphical methods for goodness-of-fit tests are summarised in Bendell et al. (1991). A review of the PHM along with the available computer programs can be found in Kumar & Klefsj (1994). The basic approach in the proportional hazards modelling is to assume that the hazard rate of a system consists of two multiplicative factors, the baseline hazard rate, $h_0(x)$, and generally an exponential function representing the importance of the MPV in explaining the failure characteristics. These monitored parameter values can be lubricant pressure, temperature, particle contents in hydraulics or vibration levels. Hence, the hazard rate of a system can be written as (Cox, 1972a)

$$h(x; \mathbf{z}) = h_0(x) \exp(\mathbf{z}\beta) \qquad (14.3)$$

where $\mathbf{z}$ is a row vector consisting of the MPV and β is a column vector consisting of the corresponding regression coefficients. The unknown parameter β defines the importance of the MPV on the failure process. In the case of continuously or periodically monitored parameters during a life cycle, an approximation such as slope of the curve followed by the variable or changes in the slope or an appropriate function should be used for $\mathbf{z}$. The baseline hazard rate, $h_0(x)$ represents the hazard rate that a system would experience if the effects of all the MPV are equal to zero. In some applications, a parametric form of the baseline hazard rate in the PHM is also assumed (e.g. Jardine et al., 1987). The basic assumption of this model is that any two hazard rates corresponding to any two MPV sets are proportional and independent of time.

The regression coefficient β can be estimated without estimating the baseline hazard rate. No specific distribution is needed to be assumed about failure data. The partial likelihood function is generally used for estimating the regression vector β (Cox, 1975; Kalbfleisch & Prentice, 1980; Cox & Oakes, 1984):

$$L(\beta) = \prod_{i=1}^{k} \frac{\exp(\mathbf{z}_i\beta)}{\left[\sum_{n\in\lambda(x_i)} \exp(\mathbf{z}_n\beta)\right]^{d_i}} \qquad (14.4)$$

where d_i is a small number of tied failures compared to the number of failure times n in the risk set $\lambda(x_i)$ at time x_i and k is the number of uncensored observations. These observations may arise from failures of less than n repairable systems with multiple failures or from n non-repairable systems. It should be noted that (14.4) is based on the assumptions of independent and identical distribution of failure times, conditional on the MPV. This assumption should be tested for repairable systems. Here failure times of a repairable system correspond to the time between successive failures.

Different methods have been suggested for estimating the baseline hazard rate (see Breslow, 1974; Kalbfleisch & Prentice, 1980). The following estimate of the cumulative baseline hazard rate suggested by Breslow (1974) is easier to calculate:

$$H_0(x) = \sum_{i:x_i\leq x} \frac{d_i}{\sum_{n\in\lambda(x_i)} \exp(\mathbf{z}_n\beta)} \qquad (14.5)$$

The reliability function, $R(x; \mathbf{z})$ is related to the baseline reliability function, $R_0(x)$ and the cumulative baseline hazard rate as follows

$$R(x;\mathbf{z}) = \exp[-H_0(t)\exp(\mathbf{z}\beta)] = [R_0(x)]^{\exp(\mathbf{z}\beta)} \qquad (14.6)$$

Hence we can get an estimate of the total reliability function, $R(x; \mathbf{z})$, at any desired time points. This estimate of $R(x; \mathbf{z})$ is then used in the maintenance scheduling model.

4.   PROPORTIONAL INTENSITY MODEL

The discrete values of a single monitored parameter or combinations of discrete values of a set of the monitored parameter can be used for grouping a data set. The number of groups that can be formed is termed as number of strata of a data set. If a stratum specific PHM is assumed, the corresponding model is called stratified PHM. In this model, it is assumed that hazard rates are proportional within the same stratum but not necessarily across strata. Two general classes of models of the stratified PHM type were presented by Prentice et al. (1981) for analysing the importance of the MPV in explaining the multiple failures of repairable systems. These generalisations are based on the way time scales are considered for estimating β . Prentice et al. (1981) considered two time scales, namely global time t and the time from immediately preceding failure, $t - t_{n(t)}$. In these models, it is assumed that a system enters stratum j immediately following the $(j - 1) : th$ failure, $j = 1, 2, ..., r$ and it enters stratum 1 at $t = 0$. The following two models were suggested by Prentice et al. (1981):

$$v\left(t\,|N\left(t\right),\mathbf{z}\right) = v_{0j}\left(t\right)\exp\left(\mathbf{z}\beta_j\right) \tag{14.7}$$

$$v\left(t\,|N\left(t\right),\mathbf{z}\right) = v_{0j}\left(t - t_{n(t)}\right)\exp\left(\mathbf{z}\beta_j\right) \tag{14.8}$$

where, $v\left(t\,|N\left(t\right),\mathbf{z}\right)$ and $v_{oj}(t)$ are the intensity function and the baseline intensity function, and β_j is the regression coefficient for the $j : th$ stratum. One of the advantages of model (14.7) and (14.8) is that depending on situations, one may assume the same baseline intensity function $v_{oj}(t) = v_o(t)$ or $\beta_j = \beta$ for all $j \geq 1$. Cox (1972b) considered a special case of (14.7) in which $v_{oj}(t) = v_o(t)$ i.e.,

$$v\left(t, \mathbf{z}\right) = v_0\left(t\right)\exp\left(\mathbf{z}\beta\right) \tag{14.9}$$

The regression vector β can be estimated, using a non-parametric approach, without making any specific assumption about the baseline intensity function. The approximation of the likelihood function corresponding to model (7) can be written as (Prentice et al., 1981)

$$L\left(\beta\right) = \prod_{i=1}^{r}\prod_{j=1}^{k_i} \frac{\exp\left(\mathbf{z}_{ij}\beta_i\right)}{\left[\displaystyle\sum_{(k,l)\in\lambda(t_{ij})} \exp\left(z_{kl}\beta_k\right)\right]^{d_{ij}}}, \tag{14.10}$$

where r is the number of strata , k_i is the number of observations in the $i : th$ stratum, $\lambda(t_{ij})$ is the risk set corresponding to those items which were functioning and uncensored just prior to the observed failure at time t_{ij}, and d_{ij} is the number of tied observed failures at time t_{ij}. Lawless (1987) suggested that each one of similar repairable systems should form a stratum, if failures of the systems occur according to the non-homogeneous Poisson process. Love & Guo (1991) extended this approach for modelling multiple failures of a single repairable system that is in bad-as-old condition after unplanned maintenance and good-as-new condition after planned maintenance. Further extensions are proposed in Guo & Love (1992) for modelling the situation when a repairable system is between good-as-new and bad-as-old conditions.

5.   ACCELERATED FAILURE TIME MODELS

The accelerated failure time models (AFTM) includes all the parametric regression models. The AFTM in the form of hazard rate can be written as (Kalbfleisch & Prentice, 1980)

$$h(x; \mathbf{z}) = h_0(x\psi(\mathbf{z}\beta))\,\psi(\mathbf{z}\beta) \qquad (14.11)$$

The MPV is related to the hazard rate in such a manner that the hazard rate increases or decreases according to whether $\psi > 1$ or $\psi < 1$. The estimated value of β acts multiplicatively on time so that its effect is to accelerate or decelerate failure time relative to baseline hazard rate. In the PHM, the estimated value of β acts multiplicatively on the baseline hazard rate and its effect is to accelerate or decelerate the hazard rate relative to the baseline hazard rate.

The AFTM can also be represented in terms of location-scale regression models or regression of random variables (see Kalbfleisch & Prentice, 1980; Lawless, 1982; and Cox & Oakes; 1984). Engineering applications of the AFTM are discussed in detail by Nelson (1990). The exponential and Weibull regression models are two particular models which belong to the PHM and the AFTM (see Kalbfleisch & Prentice, 1980 and Cox & Oakes, 1984).

Likelihood methods can be used to estimate parameters of the AFTM (see Kalbfleisch & Prentice, 1980; Lawless, 1982; Cox & Oakes, 1984). A detailed treatment of estimation and interpretation for the AFTM can also be found in Pike (1966), Louis (1981), and Newby (1988).

6.   SOME OTHER MODELS

Some other types of models which can be considerd for applications are Aalen's regression model (Aalen, 1980; 1989; Kumar & Westberg, 1996), the proportional odds model (McCullagh, 1980; Bennet, 1983b), The log-logistic model (Bennet, 1983a), the extended hazard regression model (Ciampi & Etezadi-Amoli, 1985; Etezadi-Amoli & Ciampi, 1987), and the additive hazards model (Pijnenburg, 1991; Newby, 1994). All these models have theoretical or practical limitations (Kumar & Westberg, 1997a).

Some other types of models are lesser known and some of these have unresolved theoretical problems (see Newby, 1993). Such models proposed in the literature have been classified into six groups by Crowder et al. (1991) such as time shift models, polynomial hazards models, and multiphase hazards models. There are some more reliability models incorporating MPV, for example the frailty model (Vaupel et al.

1979; Lancaster, 1990; Ridder, 1990), and the virtual age model (Newby, 1993).

7.   SELECTION OF MODELS

Before selecting any reliability model incorporating MPV for a given data set, the data should be explored for the group of data corresponding to discrete values of a single MPV or combination of discrete values of a set of MPV. Preliminary investigations of the data to explore the presence of trends or other patterns or outliers can help in revealing the structure of the data. The structure of data can provide intuition about the type of model that can fit the data. This will provide a guideline on how to formulate the MPV. Simple graphical plots such as comparisons of average failure times corresponding to the MPV may help in the selection and formulation of the MPV (see Kumar & Westberg, 1997a). Further discussions on exploratory data analysis can be found in the papers by Bendell & Walls (1985) and Ansell & Philips (1989; 1990).

Some graphical plots corresponding to discrete values of a single monitored parameter or combination of discrete values of a set of monitored parameters can be used to get an intuition of the type of model that can be fitted to the data. Plots of the logarithm of the cumulative hazard rate versus time for each group of data should be compared. the product-limit method (Kaplan-Meier, 1958) may be used to obtain an empirical estimate of the cumulative hazard rate. These plots should roughly resemble vertically shifted copies of the logarithm of the cumulative hazard rate for each group, i.e. parallel in the vertical direction (see Figure 14.1) for the PHM to be an appropriate model. If these plots appear horizontally shifted copies, i.e. parallel in the horizontal direction (see Figure 14.1), the AFTM may be an appropriate model (Lawless, 87, Crowder et al., 1991). If the horizontal distance between the plotted curves is roughly equal, then the AFTM fits the data. If the horizontal axis corresponds to $log x$ instead of x, and the plotted curves are roughly linear and roughly parallel, the Weibull distribution can be assumed for the baseline hazard rate. If the plotted curves are not parallel, the stratified PHM may be suitable. If the plotted curves are parallel but neither in a vertical nor in a horizontal direction, some mixed or Weibull regression model may be appropriate.

Assuming the PHM or the AFTM for a particular data set may lead to different results. But, for uncensored data, the assumption of the PHM in place of the AFTM or vice versa does not have significant effect on the relative importance of the MPV. However, in the case of censored data it is important to consider the appropriate model (Solomon, 1984).

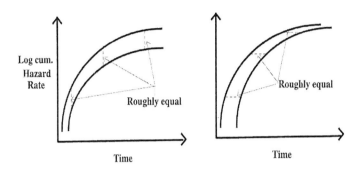

Figure 14.1   An illustration of the pattern of the log cumulative hazard rate.

The final choice of type of models to be used for a given data set depends on the objective of the reliability analysis and the results of the goodness-of-fit tests of the selected model. An initial intuition about the type of the model that can be fitted to a given data set can be obtained by the cumulative hazard rate plots. In case of two or more MPV some iteration between informal methods, model fitting and goodness-of-fit tests may be necessary.

8.   MAINTENANCE SCHEDULING UNDER THE AGE REPLACEMENT POLICY

In the   maintenance policy is called the Age Replacement Policy (ARP), it is assumed that a system is renewed by planned and unplanned maintenance either through replacements or by each repair being carried out after failure. Further it is assumed that unplanned maintenance is carried out after a failure and planned maintenance is carried out after the system has been working for x_0 units of time since the last planned or unplanned maintenance. Figure 14.2 illustrates the age replacement policy and the pattern of the hazard rate function when assuming as good as new condition. It is assumed that a system has identical hazard rate after repairs. The hazard rate at time x is a measure of the proneness of a component or a system which has survived up to time x to fail shortly after x. Here x denotes the local time.

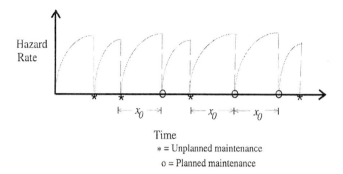

Time
* = Unplanned maintenance
o = Planned maintenance

Figure 14.2 The pattern of the hazard rate function assuming a good-as-new condition.

Let c be the average planned maintenance cost and $c + a$ be the average unplanned maintenance cost, i.e. a is the additional cost due to a sudden failure of the system. The long-term average cost per unit time is then given by (see Hyland & Rausand, 1992)

$$C_A(x_0) = \frac{c + a \cdot F(x_0)}{\int_0^{x_0} R(x)dx} \qquad (14.12)$$

where $F(x_0)$ is the cumulative distribution function and $R(x) = 1 - F(x)$ is the reliability function. If information about MPV is available and the reliability analysis indicates that some of the monitored parameters can help in explaining the failure behaviour, the above maintenance cost equation can be extended to include this situation (see Love & Guo, 1991):

$$C_A(x_0; \mathbf{z}) = \frac{c + a \cdot F(x_0; \mathbf{z})}{\int_0^{x_0} R(x; \mathbf{z})dx}$$

$$= \frac{a}{\mu} \frac{c/a + F(x_0; \mathbf{z})}{\frac{1}{\mu} \int_0^{F^{-1}(F(x_0;\mathbf{z}))} (R(x; \mathbf{z}))\, dx}$$

$$= \text{constant} \frac{\dfrac{c/a + u}{F^{-1}(u)}}{\dfrac{1}{\mu} \int\limits_0^{} (1 - u)\, dt}$$

$$= \text{constant} \frac{c/a + u}{\varphi(u)} \tag{14.13}$$

where $u = F(x; \mathbf{z})$, $0 \leq u \leq 1$; $\mu = \int\limits_0^\infty R(x; \mathbf{z})\, dx$;

and $\varphi(u) = \frac{1}{\mu} \int\limits_0^{F^{-1}(u)} (1 - u)\, dx$.

The optimum maintenance time interval that will give minimum long-term average maintenance cost per unit time can be found by minimising the cost function $C_A(x_0; \mathbf{z})$ in (14.13) for a fixed value of a monitored parameter. One may use a numerical method (see Love & Guo, 1991) or a graphical method based on the Total Time on Test (TTT) plot (see Kumar & Westberg, 1997b) to obtain an estimate of the optimal maintenance time intervals. One may use the proportional hazards model (Cox, 1972a) or any other model to obtain an estimate of the reliability and distribution functions based on the MPV.

9.   TOTAL TIME ON TEST (TTT)-PLOTTING

The TTT-plot and its theoretical counterpart, the scaled TTT transform were introduced by Barlow & Campo (1975). Since then various applications of TTT-plotting have been presented (Bergman & Klefsj, 1982; Klefsj, 1986, Kumar et al. 1989; Dohi et al., 1995). The TTT-plot gives a picture of the failure data which is independent of the scale and is situated completely within a unit square with corners in $(0, 0)$, $(0, 1)$, $(1, 0)$ and $(1, 1)$. Let $0 = x_{(1)} \leq x_{(2)} \leq \Lambda \leq x_{(n)}$, denote an ordered and complete sample from a life distribution $F(x; \mathbf{z})$. Let $S(x_i; \mathbf{z_i})$ be the total time generated in ages by a failure at a time less than or equal to x_i. Then the TTT value at any time x_i is defined as (see Barlow & Campo, 1975)

$$S(x_i; \mathbf{z}_i) = n \int\limits_0^{x_i} R(x; \mathbf{z})\, dx = n \int\limits_0^{x_i} (1 - u)\, dx \tag{14.14}$$

where $S(x_0; z_0) = 0$. The scaled TTT-transform is defined as

$$\varphi(u_i) = \frac{1}{\mu} \int_0^{F^{-1}(u_i)} (1-u) \, dx = \frac{n \int_0^{x_i} (1-u) \, dx}{n\mu}$$

$$= \frac{n \int_0^{x_i} (1-u) \, dx}{n \int_0^{\infty} (1-u) \, dx}$$

$$\varphi(u_i) = \frac{S(x_i; z_i)}{S(x_n; z_n)} \qquad (14.15)$$

The TTT-plot is obtained by plotting $(u, \varphi(u))$ and joining the points by line segments. If we draw a line from the point $(-c/a, 0)$ and tangent to the TTT-plot, its slope will be given by

$$\text{Slope} = \frac{\varphi(u)}{c/a + u} \qquad (14.16)$$

A comparison of (14.13) and (14.16) shows that the slope of a line drawn from the point $(-c/a, 0)$ on the TTT-plot is inversely proportional to the average unit maintenance cost given by (14.13). Therefore, the largest possible slope will correspond to the minimum average unit maintenance (or replacement) cost. If the line with the largest possible slope is tangent to the TTT-plot at the point $(u_0, \varphi_0(u))$, the value of the failure time, X_0 which has been used to calculate $(u_0, \varphi_0(u))$ will be an estimate of the optimum maintenance (or replacement) time interval. Generally planned maintenance will not be economical to perform if the line is tangent to the point $(1, 1)$.

To consider the MPV, the TTT value at time t_i can be estimated based on the estimate of the reliability function given by (14.6) using the PHM, that is,

$$S(x_i; z_i) = n \sum_{j=1}^{i} (x_j - x_{j-1}) R(x_{j-1}; z_{j-1}), \qquad (14.17)$$

for $i = 1, 2, ..., n$ and $R(x_0; z_0) = 1$,

where z_i is the MPV at time x_i.

Consideration of values of monitored parameters in the TTT-plot is very useful in maintenance scheduling under different situations. The

TTT-plot should be obtained by considering the situations for which an optimum maintenance (or replacement) time interval is to be estimated. Then a line with the largest possible slope should be drawn tangent to the TTT-plot and passing through the point $(-c/a, 0)$ on the horizontal axis. The value of the failure time corresponding to the point at which the line touches the TTT-plot will be an estimate of the optimum maintenance (or replacement) time interval.

Suppose that more than one parameter has been found significant in the PHM analysis. These parameters can either be condition monitored parameters or any explanatory factors characterising the system failure. The TTT-plot should be obtained considering all the significant parameters to obtain an estimate of the optimum maintenance (or replacement) time interval. In some situations, we may be interested in estimating the optimum maintenance time interval, X_0 corresponding to one particular monitored parameter, say z_1. The reliability function, $R(x; z_1)$, corresponding to z_1 is estimated using the relation (14.6), i.e.,

$$R(x; z_1) = [R_0(x)]^{\exp(\beta_1 z_1)} \tag{14.18}$$

The above relation can be used in estimating $R(x; z_1)$ at all failure times and a TTT-plot can be obtained to estimate optimum maintenance time interval X_0 corresponding to the parameter z_1. If we are interested in estimating X_0 corresponding to two or more particular parameters, we should estimate the reliability function in the same way as the above equation based on each individual parameters. Then, the corresponding TTT-plots should be used for estimating X_0.

The value of a monitored parameter may remain the same during a particular interval between maintenance actions, but it may have a different value during another interval. One may be interested in estimating the optimum maintenance (or replacement) time intervals corresponding to different values of the monitored parameter. For example, let the monitored parameter, z_1, have three levels, say z_{10}, z_{11}, z_{12} and the MPV equal to z_{10} during some operational time until a failure occurs and equal to z_{11} during another operational time until a failure occurs. One may be interested in estimating the optimum maintenance time interval lengths, X_{10}, X_{11}, and X_{12} corresponding to the levels, z_{10}, z_{11}, and z_{12}, respectively. The reliability function is calculated at each failure time using the relation in (6), i.e.,

$$R(x; z_{1i}) = [R_0(x)]^{\exp(\beta_1 z_{1i})} \tag{14.19}$$

The TTT-plots are constructed as before using the corresponding estimates of the reliability function at all failure times. Lines with the largest possible slopes are drawn tangent to three TTT-plots from the point $(-c/a, 0)$. Let these lines be tangent to the TTT-plots at the points $u_{10}, \varphi_{10}(u)$, $u_{11}, \varphi_{11}(u)$ and $u_{12}, \varphi_{12}(u)$. The values of X_{10}, X_{11}, X_{12} will be equal to those failure times which are used in calculating these points, respectively. In the case of more than one parameter with different levels, TTT-plots should be obtained for all possible combinations of different levels of parameters.

There can be situations where planned and unplanned maintenance costs are different when failure occurs at different levels of a parameter. In such a case the cost ratios corresponding to different levels of the parameter will be different. The optimum maintenance time intervals corresponding to each parameters can be estimated as discussed before.

10.　　EXAMPLE

To illustrate the above concepts the data listed in the paper by Love & Guo (1991) is considered. The data consists of information about the times between replacements of pressure gauges which are replaced on failure or as part of planned maintenance . The data is listed in Table 14.1. The data has been arranged as per increasing magnitudes of the time intervals. The time points at which replacements occurred as part of planned maintenance are marked by asterisks.

Table 14.1 The interarrival times at which pressure gauges were replaced on failure or as part of planned maintenance (denoted by asterisks).

Observation Number	Time to failure	Monitored parameter, pressure	Observation Number	Time to failure	Monitored parameter, pressure
1	32	5	9	66	3
2	42	4	10	70	4
3	44*	5	11	70	5
4	47	5	12	77	4
5	51	5	13	95	3
6	53	4	14	101	3
7	60*	3	15	198	3
8	61*	4			

Source: Love & Guo (1991).

In the case of no information about monitored parameters, the approach proposed by Bergman (1977) can be used to get a TTT-plot (see Figure 14.3). Figure 14.3 shows the plot of $(u, \varphi(u))$. The reliability function is estimated using the product limit method

$$R(x) = \prod_j \frac{n-j}{n-j+1} \qquad (14.20)$$

where the product is made for those j where $x_j \leq x$ and x_j is an uncensored observation. The largest observed failure time x_n is always taken as an uncensored observation in order to get the plot to stop at the point $(1, 1)$. The distribution function is estimated using the relation $F(x) = 1 - R(x)$, and the TTT values and scaled TTT-transforms are calculated as follows

$$S(x_i) = n \sum_{j=1}^{i} (x_j - x_{j-1}) R(x_{j-1}), \qquad (14.21)$$

for $i = 1, 2, \ldots, n$, and $R(x_0) = 1$,

$$\varphi(u_i) = \frac{S(x_i)}{S(x_n)} \qquad (14.22)$$

To estimate the optimum maintenance or replacement time interval, a line with the largest possible slope is drawn tangent to the TTT-plot and passing through the point $(-c/a, 0)$ on the horizontal axis. For $c/a = 0.5$, the corresponding line touches the TTT-plot at the point $u_0, \varphi(u_0)$ which corresponds to the failure time that is equal to 42 hours. Hence the optimum replacement time intervals of pressure gauges will be 42 hours if $c/a = 0.5$. The estimated optimum maintenance or replacement time interval is the same even though the cost ratio varies in the range 0.16 to 0.56. Because, a line drawn from any point in the interval [-0.16, -0.56] on the horizontal axis will touch the TTT-plot at the same point which corresponds to 42 hours. This illustrates that it is easier to carry out a sensitivity analysis using a TTT-plot compared to using numerical methods.

In the presence of information about a monitored parameter, it is better to estimate the optimum maintenance or replacement time interval based on the values of the monitored parameter. It will enhance the effectiveness of maintenance scheduling . First of all, the magnitude of the effect of the monitored parameter on the hazard rate is estimated using the PHM. The effect of a parameter estimated in the PHM is a relative risk. Therefore, it is better to formulate a parameter value that includes zero. The pressure values were therefore coded as -1, 0, and 1

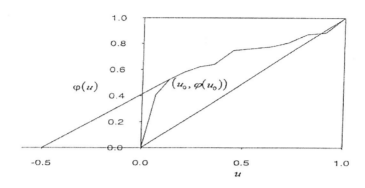

Figure 14.3 A TTT-plot to find the optimal replacement time intervals.

Table 14.2 Result of the PHM analysis.

Parameter,z	Co-efficient	Standard deviation	t-statistic	p-value
Pressure	1.188	0.493	2.41	0.016

for the pressure values equal to 3, 4, and 5, respectively. The result of the PHM analysis is listed in Table 14.2. The effects of the parameter pressure was estimated using equation (14.4). Hence the hazard rate of the pressure gauge can be written as

$$h(t; z) = h_0(t) \exp(1.188z) \tag{14.23}$$

where z is equal to -1, 0, or 1.

In order to estimate the optimum replacement time intervals when the coded value of the parameter is equal to only one of the three values, -1, 0, or 1, TTT-plots corresponding to each of the three values were obtained. The equation (14.19) was used to estimate the corresponding reliability functions. The corresponding TTT-plots are given in Figures 14.4, 14.5 and 14.6 for $c/a = 0.5$, lines with the largest slope and tangent to the TTT-plots are drawn from the point (-0.5, 0). These lines touch the TTT-plots at the points which correspond to failure times 66, 47 and 42 hours. Hence the optimum replacement time interval will be 66, 47 and 42 hours when the parameter z is equal to -1, 0, and 1, respectively, i.e. when the pressure is equal to 3, 4, and 5 units, respectively.

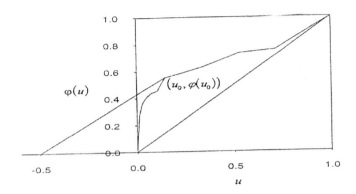

Figure 14.4 TTT-plot when parameter z=-1.

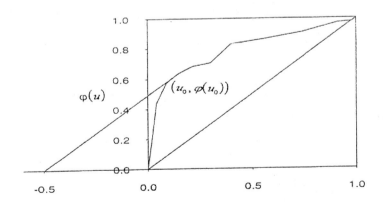

Figure 14.5 TTT-plot when parameter z=0.

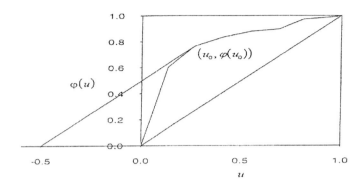

Figure 14.6   TTT-plot when parameter z=1.

The TTT-plot considering only failure times should be used when information about the monitored variable is not available or when the monitored variables are not important in explaining the failure characteristic of a system. Otherwise, the TTT-plot considering monitored variable values should be used while estimating the optimum maintenance time intervals.

11.   MAINTENANCE SCHEDULING UNDER THE BLOCK REPLACEMENT POLICY

Under the Block Replacement Policy with Minimal Repair (BRP-wMR), it is assumed that planned maintenance activities are carried out at regular time intervals, t_0, regardless of when a failure occurred previously or unplanned maintenance took place (see Figure 14.7). It is assumed that the system is renewed after planned maintenance and has the same intensity function after unplanned maintenance. The term intensity function at time t is a measure of the probability that a repairable system will fail in a small time interval $(t, t + \Delta t)$ after a certain time t. Here t denotes the global time.

Let c be the average cost for planned maintenance and d be the average cost for unplanned maintenance. Then the average cost for maintenance per unit time in the long run is defined as (see Barlow &

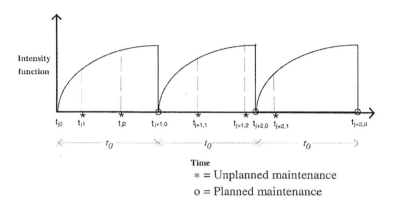

Time

* = Unplanned maintenance

o = Planned maintenance

Figure 14.7   An illustration of the block replacement policy with minimal repair.

Proschan, 1981)

$$C_B(t) = \frac{c + d \cdot E\left[N(t)\right]}{t} = d \cdot \frac{c/d + E\left[N(t)\right]}{t} \tag{14.24}$$

where $E[N(t)]$ is the expected number of failures in the time interval $(0, t]$. Since the counting process $N(t), t \geq 0$ is a non-homogeneous Poisson process with intensity function $v(t)$, then $E\left[N(t)\right] = V(t) = \int_0^t \nu(s)\, ds$ (Brown & proschan, 1983).

In order to minimise $C_B(t)$, analytical, numerical or graphical methods can be used. If the functional form of the cumulative intensity function $V(t)$ is known analytical methods can be used. Generally, only a non-parametric estimate of $V(t)$ is available such as (Nelson, 1969)

$$V(t) = \sum_{t_i \leq t} \frac{1}{n - i + 1} \tag{14.25}$$

where i is the reverse rank of all chronologically ordered failure times and the summation is done over those $i : s$ for which t_i is an observed failure time. By using a search procedure, we can numerically find the time point which minimizes $C_B(t)$. Graphically we can plot an estimate of $V(t)$ on the vertical axis against the chronologically ordered times to failure on the horizontal axis. A line drawn from the point $(0, -c/d)$ which is tangent to the plotted curve at time t_0 has the minimal slope

$$m = \frac{V(t) + c/d}{t} \tag{14.26}$$

Since $C_B(t) = d.m$, the average cost in the long run is minimised at time t_0.

In presence of MPV, the maintenance cost equation corresponding to (14.24) can be written as

$$C_B(t; \mathbf{z}) = \frac{c + d \cdot E[N(t; \mathbf{z})]}{t} \qquad (14.27)$$

Earlier, analytical and numerical methods to solve equation (14.27) corresponding to a certain MPV vector $\mathbf{z}$ have been presented by Love & Guo (1991). Alternatively, a graphical method can also be used to estimate the optimal maintenance time t_0 . The expected number of failures during the time interval $(0, t]$ is estimated using the relation (Cox & Lewis, 1966)

$$E[N(t; \mathbf{z})] = \int_0^t v_0(u) \exp(\mathbf{z}\beta) \, du$$

$$= \sum_{(i,j):t_{ij}<t} \exp(\mathbf{z}\beta) \int_{t_{i,j-1}}^{t_{ij}} v_0(u) du + \exp(\mathbf{z}_{IJ}\beta) \int_{t_{IJ}}^t v_0(u)\, du$$

$$= \sum_{(i,j):t_{ij}<t} \exp(\mathbf{z}\beta) [V_0(t_{ij}) - V_0(t_{i,j-1})] + \exp(\mathbf{z}_{IJ}\beta) [V_0(t) - V_0(t_{IJ})]$$

$$(14.28)$$

where (I, J) is the index of the largest time $t_{ij} < t$.

Based on the estimator of the cumulative baseline hazard rate in the PHM proposed by Breslow (1974) the cumulative baseline intensity function can be estimated as

$$V_0(t) = \sum_{(i,j):t_{ij}\leq t} \frac{d_{ij}}{\sum_{(k,\lambda)\in R(t_{ij})} \exp(z_{k\lambda}\beta)} \qquad (14.29)$$

12.   EXAMPLE

In Love & Guo (1991) operating data from a cement plant are given. Except from the time to failure in each planned maintenance cycle, three potential parameters were studied, the instantaneous change in pressure on the main bearing (D), the recirculating load on the bucket elevator (B), and the power demanded by the main motor (W). The data is presented in Table 14.3. The average cost for PM is \$352.08 (standard deviation 287.60) and the average cost for unplanned maintenance is

Table 14.3 Data from the cement plant, taken from Love & Guo (1991).

Time from last PM	Maintenance type	Cost ($)	D	B	W
	PM				
54	PM	93	12	10	800
133	failure	142	13	16	1200
147	PM	300	15	12	1000
72	failure	237	12	15	1100
105	failure	0	13	16	1200
115	PM	525	11	13	900
141	PM	493	16	13	1000
59	failure	427	8	16	1100
107	PM	48	9	11	800
59	PM	1115	8	10	900
36	failure	356	11	13	1000
210	PM	382	8	10	800
45	failure	37	10	19	1300
69	PM	128	12	14	1100
55	failure	37	13	18	1200
74	PM	93	15	12	800
124	failure	735	12	17	1100
147	failure	1983	13	16	1100
171	PM	350	11	13	900
40	failure	9	13	16	1100
77	failure	1262	14	17	1100
98	failure	142	12	15	1100
108	failure	167	12	15	1100
110	PM	457	16	14	1100
85	failure	166	8	19	1300
100	failure	144	12	15	1000
115	failure	24	13	16	1200
217	PM	474	9	11	900
25	failure	0	15	18	1200
50	failure	738	11	13	1100
55	PM	119	8	10	800

$367.00 (standard deviation 524.15). Note that there is a rather large uncertainty in the estimated costs.

The data set was analysed using the proportional intensity model , assuming that the failures data belong to only one stratum. The initial analysis in Table 14.4 indicated that MPV B was the only significant parameter. Further analysis showed that after eliminating insignificant coefficients, both parameters B and W could be found significant in a single parameter study.

Under final analysis in Table 14.4, it is shown that the parameter B is highly significant. This will give us the final model $v(t; \mathbf{z}) = v_0(t) \exp(0.479B)$. In Figure 14.8, the optimal PM cycle is estimated from the graphical analysis, using the failure times only. The optimal PM cycles for ten different values of the parameter B are estimated using graphical analyses. In Figure 14.9 some of the plots are shown. The results are summarised in Table 14.5.

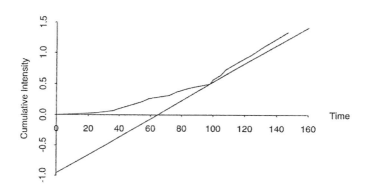

Figure 14.8   The optimal planned maintenance cycle length in hours.

We can see from Table 14.5 that we can form five different groups of the MPV B giving five different optimal PM cycles. As the MPV B is the measure of the influence that the recirculating load on the bucket elevator has on the failure process, it is possible to control that variable. If we can keep the recirculating load on the bucket elevator in the interval 15-18, the optimal strategy will be to conduct planned maintenance each 98 hours. The cost will vary with the value of the

Table 14.4 Result of the proportional intensity analysis.

INITIAL ANALYSIS				
MPV	Coefficient	Standard deviation	t-statistic	p-value
D	-0.013	0.127	-0.101	0.460
B	0.386	0.252	1.533	0.062
W	0.182	0.425	0.429	0.334
FINAL ANALYSIS				
MPV	Coefficient	Standard deviation	t-statistic	p-value
B	0.479	0.130	3.690	0.000

Table 14.5 The optimal PM cycle length in hours for different values of the MPV B.

Value of MPV B	Optimal PM cycle length in hours	Optimal hourly total maintenance cost ($)
No MPV	98	5.45
10 - 12	No Optimal	Not able to calculate
13	133	3.83
14	124	4.48
15 - 18	98	4.64 - 9.8
19	36	13

MPV, but the mean value in the interval will be $7.07 per hour. If the MPV B takes the values 10-12, no optimal PM cycle was found. This means that very few failures will occur at this low level, and that the most economical strategy is to repair the system at failures only and not conduct any planned maintenance action.

Note that the above conclusions are based on historical facts. After a new policy has been decided upon, it must constantly be studied and new analyses of the process must be conducted on a regular basis, eventually giving a new optimal PM cycle.

The graphical method has several advantages over the numerical method and the analytical method. One of the advantages is that sensitivity analysis can easily be conducted. Since the true PM and emergency costs are seldom known, some uncertainty in the quoting of them may not affect the optimal replacement time. This can directly be seen in Figure 14.10, where the case with no MPV is plotted. In the example the quote is $352.08/367 = 0.96$. In Figure 14.8 it is illustrated that the quote can vary between 0.19 and 1.20 and still give the same PM cycle length. If the quote is larger than 1.20, no planned maintenance is considered economical, since the line touches the plot at the last node. This means that the cost for unplanned maintenance is so much lower that the cost for planned maintenance, that the equipment only shall be repaired at failures, and not preventively renewed.

Figure 14.9 shows the optimal planned maintenance estimated from the plot of the estimated cumulative intensity function against the ordered cumulative time between failures, with the monitored parameter taking low (=10), average (=14.29) and high (=19) values from top to bottom. With low of value of the MPV, no optimal planned cycle was found. At average value, the line touches the plot at 98 hours and at high value, the line touches the plot at 36 hours. The plot in Figure 14.10 is based on failure times only. It indicates that some uncertainty in the costs for planned and planned maintenance will not affect the optimal choice of planned maintenance cycle. The solid line marks the optimal planned maintenance cycle, and the dotted lines mark the limit in the quote between planned and unplanned maintenance for which the optimal choice of planned maintenance cycle is unaffected. The quote can vary between 0.19 and 1.20 and still gives the same PM cycle.

Another advantage, is that it is easy to detect different patterns in the plot, which may have impact on the choice of the optimal replacement time. One example is that the line can be very close to the curve at several time points, but only touches the curve at one time point. At all these time points, the cost for maintenance will be approximately the same. Some time points may be more appropriate from a planning point of view, and some time points may be more appropriate if PM or emergency repairs are preferred. Then the time point that is most appropriate can be chosen, even if that point is not the one indicated as optimal. Another example is when the line touches the plot at a very large (or small) time point, which may give a clearly unrealistic estimate. This may occur in the case of outliers, in-correct data or misclassified data. By studying the line and the plot, other possible time points can be detected which will give a more realistic estimate of the optimal replacement time.

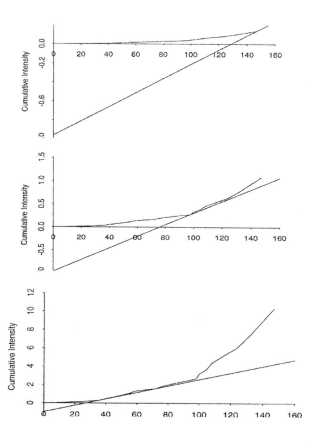

Figure 14.9   The optimal planned maintenance cycle length in hours.

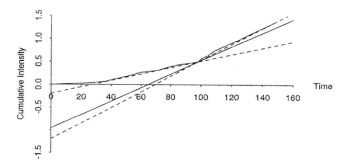

Figure 14.10   The plot is based on failure times only.

13.   CONCLUSIONS

The PHM should be used to find out whether the failure characteristic of a system depends on the failure times only or on the MPV also. In models from proportional hazards, the importance of the MPV in explaining the failure charecteristics can be identified without making any specific assumptions about the distribution of the baseline hazard rate. A large number of diagnostic techniques has been developed for testing the goodness-of-fit of such models. Censored data and tied failure times are easily accommodated in the PHM. Therefore, models from proportional hazards appear to be a better choice model among the different models. However, selection of a model should be done solely on the basis of objectives of the analysis, type of data available and the failure mechanisms of the system. Once the relative importance of the MPV are estimated, either graphical, numerical, or analytical methods can be used for maintenance scheduling based on the MPV. Graphical methods such as the TTT-plot or the cumulative intensity plot can be used. Among the advantages of using the graphical approach are that the failure process is visualised, the uncertainty in the costs involved can be dealt with, and that different failure patterns can be detected.

The proposed approach can be used for repairable as well as non-repairable systems. The advantage of this approach over the traditional approach considering only the failure times, is that we know in which way the MPV can vary. By keeping the value of the MPV in the interval, we can not only optimise the maintenance cost, but also the production cost (e.g. we know how much load we can set on the bucket elevator in the example illustrated) Application of such approaches will enhance the prediction capability in condition monitoring. However, there are some limitations and further research work needs to be done. A common drawback of the suggested method is that the value of the MPV is assumed to be constant or at least be within specified limits. This approach will be difficult to use when there is interaction among monitored parameter.

References

[1] Aalen, O. O. A Model for Nonparametric Regression Analysis of Counting Processes. *Lecture Notes in Statistics*, 2 Springer, New York, 1980, 1-25.

[2] Aalen, O. O. A Linear Regression Model for the Analysis of Life Times. *Statistics in Medicine*, **8** (1989) 907-925.

[3] Ansell, J. I. and Philips, M. J. Practical Problems in the Analysis of Reliability Data. *Applied Statistics*, **38** (2) (1989) 205-247.

[4] Ansell, J. I. Practical Reliability Data Analysis. *Reliability Engineering and System Safety*, **28** (1990) 337-356.

[5] Barlow, R. E. and Proschan, F. Statistical Theory of Reliability and Life Testing. *To Begin With*, Silver Spring, 1981.

[6] Barlow, R. E. and Campo, R. Total Time on Test Process and Applications to Failure Data Analysis. *In Reliability and Fault Tree Analysis*, Edited by R. E. Barlow, J. Fussel and N. D. Singpurwalla, SIAM, Philadelphia (1975) 451-481.

[7] Bergman, B. Some Graphical Methods for Maintenance Planning. *Annual Reliability and Maintainability Symposium*, (1977) 467-471.

[8] Bergman, B. & Klefsj, B. A Graphical Method Applicable to Age Replacement Problems. *IEEE Transaction on Reliability*, **R-31** (1982) 478-481.

[9] Bendell, A. and Walls, L. A. Exploring Reliability Data. *Quality and Reliability Engineering International*, **1** (1985) 37-51.

[10] Bendell, A., Wightman, D. W. and Walker, E. V. Applying Proportional Hazards Modelling in Reliability. *Reliability Engineering and System Safety*, **34** (1991) 35-53.

[11] Bennet, S. Log-logistic Regression Model for Survival Data Analysis. *Applied Statistics*, **32**(2) (1983) 165-171.

[12] Bennet, S. Analysis of Survival Data by Proportional Odds Model. *Statistics in Medicine*, **2** (1983) 273-277.

[13] Breslow, N. Covariance Analysis of Censored Survival Data. *Biometrics*, **30** (1974) 89-99.

[14] Brown, M. & Proschan, F. Imperfect Repair. *Journal of Applied Probability*, **20** (1983) 851-859.

[15] Cho, D. I. and Parlar, M. A Survey of Maintenance Models for Multi-Unit Systems. *European Journal of Operational Research*, **51** (1991) 1-23.

[16] Ciampi, A. and Etezadi-Amoli, J. A General Model for Testing the Proportional Hazards and the Accelerated Failure Time Hypothesis in the Analysis of Censored Survival Data with Covariates. *Communication in Statistics-Theory and Methods*, **14**(3) (1985) 651-667.

[17] Cox, D. R. Regression Models and Life-Tables. *Journal of Royal Statistical Society*, **B34** (1972) 187-220.

[18] Cox, D. R. The Statistical Analysis of Dependencies in Point Process. *In Symposium on point Processes*, Edited by P. A. W. Lewis, 1972, 55-66, New York.

[19] Cox, D. R. Partial Likelihood , *Biometrika,***62** (1975) 188-190.

[20] Cox, D. R. & Lewis, P. A. W. *The Statistical Analysis of Series of Events.* Chapman & Hall, England, 1966.

[21] Cox, D. R. and Oakes, D. *Analysis of Survival Data. Chapman and Hall,* New York, 1984.

[22] Crowder, M. J., Kimber, A. C., Smith, R. L. and Sweeting, T. J. *Statistical Analysis of Reliability Data.* Chapman & Hall, New York, 1991.

[23] Dohi, T., Kaio, N. and Osaki, S. Solution Procedure for a repair-limit problem using the TTT concept. *IMA Journal of Mathematics Applied in Business and Industry,* **6** (1) (1995) 101-111.

[24] Etezadi-Amoli, J. and Ciampi, A. Extended Hazard Regression for Censored Survival Data with Covariates: A Spline Approximation for the Baseline Hazard Function. *Biometrics,* 43 (1987) 181-192.

[25] Guo, R. and Love, C. E. Statistical Analysis of an Age Model for Imperfectly Repaired Systems. *Quality and Reliability Engineering International,* **8** (1992) 133-146.

[26] Hyland, A. and Rausand, M. *System Reliability Theory: Models and Statistical Methods.* John Wiley & Sons, New York, 1994.

[27] Jardine, A. K. S. anderson, P. M. and Mann, D. S. Application of the Weibull Proportional Hazards Model to Aircraft and Marine Engine Failure Data. *Quality and Reliability Engineering International* **3** (1987) 77-82.

[28] Kaplan, E. L. and Meier, P. Nonparametric Estimation from Incomplete Observations. *Journal of American Statistical Association,* **53** (1958) 457-481.

[29] Kalbfleisch, J. D. and Prentice, R. L. *The Statistical Analysis of Failure Time Data.* John Wiley & Sons, New York, 1980.

[30] Klefsj, B. TTT-transforms-A Useful Tool when Analysing Different Reliability Problems. *Reliability Engneering,* **4** (1986) 231-241.

[31] Kumar, D. Proportional Hazards Modelling of Repairable Systems. *Quality and Reliability Engineering International,* **11** (1995) 361-369.

[32] Kumar, D. and Klefsj, B. Proportional Hazards Model: A Review. *Reliability Engineering and System Safety,* **44** (1994) 177-188.

[33] Kumar, D. and Westberg, U. Proportional Hazards Modelling of Time -Dependent Covariates Using a Linear Regression: A Case Study. *IEEE Transactions on Reliability,* **45** (3) (1996) 386-392.

[34] Kumar, D. and Westberg, U. Some Reliability Models for Analysing the Effects of Operating Conditions. *International Journal of Reliability, Quality and Safety Engineering*, **4**(2), (1997) 133-148.

[35] Kumar, D. and Westberg, U. Maintenance scheduling under Age Replacement Policy using Proportional Hazards Model and TTT-plotting. *European Journal of Operational Research*, **99**(3), (1997) 507-515.

[36] Kumar U, Klefsj, B. and Granholm, S. Reliability Investigation for a Fleet of Load Haul Dump Machines in a Swedish Mine. *Reliability Engineering and System Safety*, **26** (1989) 341-369.

[37] Lancaster, T. *The Econometric Analysis of Transition Data*. Cambridge University Press, UK, 1990.

[38] Lawless, J. F. *Statistical Models and Methods for Lifetime Data*. John Wiley & Sons, New York, 1982.

[39] Lawless, J. F. Statistical Methods in Reliability. *Technometrics*, **25**(4) (1983) 305-335.

[40] Lawless, J. F. A Note on Lifetime Regression Models. *Biometrika*, **73**(2) (1986) 509-512.

[41] Lawless, J. F. Regression Methods for Poisson Process Data. *Journal of American Statistical Association*, **82** (1987) 805-815.

[42] Louis, A. T. Nonparametric Analysis of an Accelerated Failure Time Model . *Biometrika*, **68**(2) (1981) 381-90.

[43] Love, C. E. and Guo, R. Using Proportional Hazard Modelling in Plant Maintenance. *Quality and Reliability Engineering International*, **7** (1991) 7-17.

[44] McCullagh, P. Regression Models for Ordinal Data (with discussion). *Journal of Royal Statistical Society*, **B42** (1980) 109-142.

[45] McCullagh, P. and Neder, J. A. *Generalized Linear Models*. Chapman and Hall. London, 1989.

[46] Nelson, W. Hazard Plotting for Incomplete Failure Data. *Journal of Quality Technology*, **1** (1969) 27-52.

[47] Nelson, W. *Accelerated Testing: Statistical Models, Test Plans, and Data Analyses*. Wiley & Sons, New York, 1990.

[48] Newby, M. Accelerated Failure Time Models for Reliability Data Analysis. *Reliability Engineering and System Safety*, **20** (1988) 187-197.

[49] Newby, M. A Critical look at Some Point Process Models for Repairable Systems. *IMA Journal of Mathematics Applied in Business and Industry*, **4** (1993) 375-394.

[50] Newby, M. Why no Additive Hazards Model. *IEEE Transaction on Reliability*, **43** (3) (1994) 484-488.

[51] Pijnenburg, M. Additive Hazards Models in Repairable systems Reliability. *Reliability Engineering and System Safety*, **31** (1991) 369 - 390.

[52] Pike, M. C. A Suggested Method of Analysis of a Certain Class of Experiments in Carcinogenesis. *Biometrics*, **22** (1966) 142-161.

[53] Prentice, R. L, Williams, B. J. and Peterson, A. V., On the Regression Analysis of Multivariate Failure Data. *Biometrika*, **68**(2) (1981) 373-379.

[54] Ridder, G. The Non-parametric Identification of Generalized Accelerated Failure Time Models . *Review of Economic Statistic*, **57** (1990) 167-182.

[55] Smith, R. L. Weibull regression Models for Reliability Data. *Reliability Engineering and System Safety*, **34** (1991) 35-57.

[56] Solomon, P. J. Effect of Misconception in the Regression Analysis of Survival Data. *Biometrika*, **71** (1984) 291- 308. Amendment, **73** (1986) 245.

[57] Valdez-Flores, C. and Feldman, R. M. A Survey of Preventive Maintenance Models for Stochastically Deteriorating Single-unit System. *Naval Research Logistics*, **36** (1989) 419-446.

[58] Vaupel, J. W., Manton, K. G. and Stallard, E. The Impact of Heterogeneity in Individual Frailty on the Dynamics of Mortality. *Demography*, **16** (1979) 439-454.

V

INTEGRATED MAINTENANCE, PRODUCTION AND QUALITY MODELS

Chapter 15

A GENERAL EMQ MODEL WITH MACHINE BREAKDOWNS AND TWO TYPES OF FAILURE

V. Makis, X. Jiang and E. Tse

Department of Mechanical and Industrial Engineering,
University of Toronto, 5 King's College Road,
Toronto, Ontario, Canada M5S 3G8

firstauthor@myuniv.edu

Abstract   In this paper, we consider a production-inventory control problem for an Economic Manufacturing Quantity (EMQ) model with machine breakdowns and planned preventive replacement (overhaul) of the production unit. Two types of failure are considered. When type I failure occurs (major failure), the unit is replaced by a new one and the interrupted lot is aborted. Starting a new lot after a replacement or after a complete production run incurs a setup cost. Type II failures (minor failures) are corrected by minimal repairs which take negligible time. After performing a minimal repair, production can be resumed immediately at a cost lower than the setup cost. It is assumed that both preventive and failure replacement times are random and the demand that cannot be satisfied during these replacement times is lost.

    The objective is to determine the lot sizing and preventive replacement policy minimizing the long-run expected average cost per unit time. The structure of the optimal policy is found by formulating and analyzing the problem in the framework of the semi-Markov decision processes. It is shown that if the replacement times are negligible, the optimal preventive replacement is the age replacement and the optimum lot size is a function of the age of the production unit. The EMQ model with lost sales is studied under the assumption of a constant production lot size. The formula for the expected average cost is obtained in the general case and analyzed under certain assumptions. A special case is discussed and numerical results are provided.

Keywords: : Production-inventory control, preventive replacement, EMQ model, optimal lot sizing , minimal repair

1.   INTRODUCTION

The classical EMQ model assumes constant production and demand rate, infinite horizon, linear holding cost, failure-free production facility and a production process that is always "in-control", producing good quality items. The model is widely used in industry, although some of the assumptions are not very realistic and are rarely met in practice.

Recent research in this area has focused on EMQ models with an unreliable production facility (joint production and maintenance control) and on the models with production processes subject to random deterioration (production and quality control).

Maintenance issues in EMQ models have been investigated recently by Posner and Berg (1989), Groenevelt et al. (1992a,b) and Makis (1998a).

Posner and Berg obtained some important system characteristics related to machine utilization and service level to customers under the assumptions of a constant production rate, exponential failure and repair time distributions and compound Poisson demand process. Groenevelt et al. studied two extensions of the classical EMQ model. In their first paper (1992a), they assumed that the production unit is maintained at failure time or after completing a production run, whichever occurs first. It was assumed that the maintenance operation brings the machine to the "as good as new" state. In their second paper (1992b), they considered an EMQ model with exponentially distributed failure times, generally distributed repair times and safety stocks used to meet a prescribed service level. In both papers, the objective was to find the optimum lot size.

Makis (1998a) investigated an EMQ model with lost sales , machine breakdowns and minimal repairs. He proved that the optimal preventive replacement is an age replacement and the optimal lot size is generally a function of the operating age of the machine.

For an EMQ model with lost sales and a constant lot size, he showed that a good approximation to the optimal policy can be found by solving two nonlinear equations.

Considerable attention has also been paid recently in the research literature to EMQ models with production processes that may go randomly out of control, resulting in producing a high proportion of items which are defective or of substandard quality. The effect of process deterioration on the optimal production time was first investigated by Rosenblatt and Lee (1986) and Porteus (1986). In both papers, they assumed that the production process is subject to random deterioration and no process monitoring was considered. They found that the optimal lot size in their models is smaller than that in the classical

EmQ models. Lee and Rosenblatt (1987) and Lee and Park (1991) considered the situation where the state of the process can be monitored through inspections. They assumed exponentially distributed in-control periods and perfect inspections. Lee and Rosenblatt (1987) considered the same cost of defectives regardless of whether the nonconforming item is reworked or sold and Lee and Park (1991) considered a warranty cost incurred by selling defective items. The objective in both papers was the joint determination of the optimal lot size and the number of inspections in a production cycle.

Makis (1998b) studied the problem of the joint determination of the production quantity and inspection schedule for an EMQ model with imperfect inspections. In the exponential case, he found the structure of the optimal inspection policy using Lagrange's method and showed that the optimal inspection times can be found by solving a nonlinear equation. Makis and Fung (1995 and 1998) studied the effect of machine failures and preventive maintenance on the optimal lot size and inspection schedule. They found that preventive maintenance considerably reduces the dependency of the lot size on the frequency of machine failures and, hence, stabilizes also the inspection schedule.

Rahim and Banerjee (1993) investigated the problem of determining jointly the optimal design parameters of $\overline{X}$ control chart and a preventive replacement time for a production system with an increasing failure rate. Rahim (1994) considered the joint determination of production quantity, inspection schedule, and control chart design. He chose the lengths of the sampling intervals in such a way to make the cumulative hazards over different intervals equal.

In this paper, we consider a general lost sales EMQ model with machine breakdowns, two types of failure , and planned replacements. The description of the model and the notation are given in Section 2. In Section 3, we present the structure of the optimal production / maintenance policy under the assumption of an instantaneous replacement of the production unit. In the subsequent sections, we study a model with nonzero replacement times controlled by a simpler policy with a constant production lot size and preventive replacement after completing n production runs. The derivation of the average cost function for this model and a detailed analysis of the exponential case are in Section 4. In Section 5, we provide some numerical results and perform sensitivity analysis for a special case. Finally, in Section 6 we extend the results obtained by Goevenelt et al. (1992a) for a model with preventive replacement after the first production run.

2.   MODEL DESCRIPTION

We consider a production-inventory model with machine breakdowns, two types of machine failure and planned preventive replacement . When a failure occurs, it is of type I (major failure) with probability $s(t)$ and of type II (minor failure) with probability $q(t)$, where t denotes the number of operating hours since the last renewal. When type I failure occurs, the machine is replaced by a new one (overhauled) at a cost $c_u(\lambda_u)$, where λ_u^{-1} is the mean failure replacement time, and the interrupted lot is aborted. The duration of the failure replacement is a random variable T_u with distribution function $F_u(t)$. Production will resume when all available inventory is depleted. Starting a new lot after a replacement or after a complete production run incurs a positive setup cost S. On the other hand, when type II failure occurs, minimal repair is performed at a cost $c_i(t)$ if it is the i-th minimal repair since the last replacement. Production can be resumed immediately at a cost R lower than the setup cost S. It is assumed that the planned preventive replacement (overhaul) can be initiated only in the idle period after completing a production run at a cost of $c_p(\lambda_p)$ where λ_p^{-1} is the mean planned replacement time and the duration of a planned replacement is a random variable T_p with distribution function $F_p(t)$. Demand that cannot be satisfied during the planned or unplanned replacement is lost and there is a penalty of \$$L$ for each unit of lost demand. It is assumed that the machine cannot fail in an idle period. Other basic assumptions include a deterministic demand at a constant rate D, continuous finite production rate P and inventory holding cost proportional to the on-hand inventory.

3.   THE STRUCTURE OF THE OPTIMAL POLICY

In this section, we establish the form of the optimal policy for a special case $T_p = 0, T_u = 0$. First, we derive the optimality equation and show the existence of a solution to this equation. We will need the following results.

Lemma 1. *(Block et al. 1985). If no planned replacements are considered, the survival function of Y is given by*

$$\overline{F}(y) = exp\{-\int_0^y s(t)r(t)dt\} \tag{15.1}$$

Lemma 2. *(Savits 1988). Let $M(t)$ be the number of minimal repairs in the operating period $[0,t]$. Then $M(t)$ is a nonhomogeneous Poisson process with intensity $q(t)r(t)$ and $E[M(t)] = \int_0^t q(u)r(u)du$.*

Table 15.1   Notation Summary for the Modified EMQ model

T	time to failure of the system
$r(t)$	failure rate of T
T_p	duration of a planned replacement
$F_p(t)$	distribution function of T_p
T_u	duration of an unplanned (failure) replacement
$F_u(t)$	distribution function of T_u
Y	random variable denoting the time (number of operating hours) between successive major failures if no planned replacements are considered
$F(y)$	distribution function of Y
$s(t)$	probability of type I failure
$q(t)$	probability of type II failure $(1 - s(t))$
λ_p^{-1}	mean value of the planned replacement time
λ_u^{-1}	mean value of the unplanned replacement time
$c_i(t)$	cost of the i-th minimal repair after operating for t hours (\$)
$c_p(\lambda_p)$	cost of the preventive replacement (\$)
$c_u(\lambda_u)$	cost of the unplanned replacement (\$)
v	cost of operating the system (\$/hour)
h	inventory holding cost (\$/hour/unit)
D	demand rate (unit/hour)
P	production rate (unit/hour)
Q	production lot size (unit)
R	production resumption cost after minimal repair (\$)
S	production setup cost (\$)
L	lost-sales cost (\$/unit of lost demands)

Lemma 3. *(Block et al. 1988). Let $\{M(t), t \geq 0\}$ be a nonhomogeneous Poisson process with intensity $\lambda(t), t \geq 0$. Denote the successive arrival times by $S_1, S_2, \ldots$. Assume that at time S_i a cost $c_i(S_i)$ is incurred. If $C(t)$ is the total cost incurred in $[0, t]$, then*

$$E[C(t)] = \int_0^t E[c_{M(z)+1}(z)]\lambda(z)dz \qquad (15.2)$$

The decisions are made at the instants when the inventory level reaches zero. At each decision epoch, it is decided whether to replace (overhaul) the production unit and the lot size for the next production run is selected.

Thus, the state is fully described by the age of the unit and the action space is given by $A = \{0, 1\} \times [0, \infty)$ where 0 denotes the continuation with the old production unit, 1 indicates preventive replacement and the lot size $Q \in [0, \infty)$. We will make the following additional assumptions:

i) The conditional distribution function

$$\overline{F}_t(s) = P(Y > t + s | Y > t)$$

is nonincreasing in t.

ii) The conditional probability of a minor failure, $q(t)$, given that a failure occurred at time t is a constant, q.

iii) The failure rate $r(t)$ and the minimal repair cost $c(t)$ are continuous and nondecreasing in t.

The assumptions i) - iii) are quite realistic and are satisfied for most deteriorating production systems encountered in practice.

The existence of the optimal policy is established by formulating the problem in the framework of semi-Markov decision processes and verifying the sufficient conditions presented by Kurano (1985). The results are summarized in the following theorem.

Theorem 1. *Under the assumptions i) -iii), there exist a bounded function $w(t)$ defined on $[0, +\infty)$ and a constant $g^* > 0$ satisfying the following optimality equation:*

$$
\begin{aligned}
w(t) = \ & min\{c_p(\lambda_p) + min_{Q \geq 0}\{S + \int_0^{Q/P}[h(P - D)Ps/D \\
& + (c(s) + R)qr(s) + v - g^*P/D]\overline{F}(s)ds + w(Q/P)\overline{F}(Q/P) \\
& + w(0)F(Q/P) + c_u(\lambda_u)F(Q/P)\}, min_{Q \geq 0}\{S + \int_0^{Q/P}[h(P \\
& - D)Ps/D + (c(t + s) + R)qr(t + s) + v - g^*P/D]\overline{F}_t(s)ds \\
& + w(t + Q/P)\overline{F}_t(Q/P) + w(0)F_t(Q/P) + c_u(\lambda_u)F_t(Q/P)\}\}
\end{aligned}
$$

$$(15.3)$$

The next lemma is useful for establishing the monotonicity of function $w(t)$.

Lemma 4. *Define for $g > 0$,*

$$
\begin{aligned}
W(t, Q(t), g, w) = \ & min_{Q \geq 0}\{\int_0^{Q/P}[h(P - D)Ps/D \\
& + (c(t + s) + R)qr(t + s) + v - gP/D]\overline{F}_t(s)ds \\
& - \overline{F}_t(Q/P)(c_u(\lambda_u) + w(0) - w(t + Q/P))\} \\
\equiv \ & min_{Q \geq 0}\{W(t, Q, g, w)\}
\end{aligned}
$$

$$(15.4)$$

where $Q(t)$ minimizes the right side of (15.4). Then, under assumptions i) and iii), $W(t, Q(t), g, w)$ is nondecreasing in t for any nonnegative, nondecreasing function $w(t)$ such that $w(t) \leq c_u(\lambda_u) + w(0)$ for all t.

The next theorem determines the structure of the optimal policy.

Theorem 2. *Let assumptions i) – iii) be satisfied and let $(w(t), g^*)$ be a solution to the optimality equation (15.3). Define*

$$t^* = inf\{t \geq 0 : w(t) = c_p(\lambda_p) + w(0)\}. \tag{15.5}$$

Function $w(t)$ is nondecreasing in t and the optimal production maintenance policy is described as follows.

If the age t of the production unit at a decision epoch is less than t^* , continue production with the old unit and choose the lot size $Q(t)$ minimizing $W(t, Q, g^*, w)$ defined by (15.4). If $t \geq t^*$, replace (overhaul) the production unit and choose the lot size $Q(0)$ minimizing $W(0, Q, g^*, w)$.

Though the structure of the optimal policy determined by Theorem 2 is not complex, the computation of function $w(t)$ and the optimal expected average cost $g*$ is not easy. Since the optimal lot size $Q(t)$ is a function of the age of the production unit, the policy is not easily implementable.

For practical applications, it seems worthwhile to investigate the behavior of the average cost function and find the optimal policy in a class of simpler policies with a constant lot size and preventive replacementafter completing n production runs. Such policies will be investigated in the subsequent sections.

4. AN EMQ MODEL WITH LOST SALES

Here we will provide a detailed analysis of an EMQ model with lost sales controlled by a policy with a constant production lot size and preventive replacement (overhaul) of the production unit after completing n production runs. First, we will summarize the assumptions made in this section.

4.1 Model Assumptions. An EMQ model with machine breakdowns and planned preventive replacement is considered.

1. Description of the failure mechanism and the maintenance actions

i) Two types of failure are considered.

ii) Type I failure (a major failure) requires a replacement of the machine. The replacement takes T_u time units.

iii) When type II failure occurs (a minor failure), the system is restored (in a negligible time) to its operating condition just prior to failure (minimal repair) and the production continues.

iv) The system is overhauled (returned to the "as new" condition) after completing n production runs since the last replacement. The preventive replacement (overhaul) takes T_p time units.

2. The cost structure

i) The failure replacement cost and the preventive replacement cost are $c_u(\lambda_u)$ and $c_p(\lambda_p)$, respectively, where λ_u^{-1} is the mean value of T_u and λ_p^{-1} is the mean value of T_p.

ii) A setup cost S is incurred each time a new production run begins and a cost R is incurred each time a production run is interrupted due to a minor failure.

iii) The other cost components include the operating cost per unit time v, the inventory holding h per unit time and per unit of inventory, and the lost sales cost L per unit of lost demand.

The objective is to determine the optimal values of the lot size and of the preventive replacement time by minimizing the long-run expected average cost per unit time. The inventory process for this model can be seen as a regenerative process. The process regenerates itself at the beginning of a new production run following a planned preventive replacement or an unplanned replacement. By using the well-known renewal reward theorem, the long-run expected average cost per unit time is the ratio of the expected cost per renewal cycle and the expected duration of a cycle.

Assume that the system is replaced after a type I failure or after n production runs, whichever occurs first. The cycle is defined as the length of time between two successive replacements. The total cost of cycle is the summation of the following cost components:

$$
\begin{aligned}
E[Cost\,of\,cycle] \;=\; & E(Setup\;cost) + E(Holding\;cost) + \\
& E(Resumption + maintenance + operating\;cost) \\
& + E(Lost\;sales\;cost).
\end{aligned}
$$

(15.6)

The cost components in (15.6) can be found by conditioning on Y, the length of time to a major failure. We have:

$$
E(Setup\;cost) \;=\; \int_0^{nQ/P} E(Setup\;cost|Y=y)dF(y) + \int_{nQ/P}^{\infty} E(Setup\;cost|Y=y)dF(y)
$$

(15.7)

Obviously, for $y < nQ/P$,

$$
E(Setup\;cost|Y=y) = S(\lceil yP/Q \rceil)
$$

(15.8)

and for $y > nQ/P$,

$$E(Setup\ cost|Y = y) = nS \tag{15.9}$$

where $\lceil x \rceil$ is the smallest integer greater than or equal to x.
From (15.7)-(15.9), we obtain

$$
\begin{aligned}
E(Setup\ cost) &= nS\overline{F}(nQ/P) + \int_0^{nQ/P} S(\lceil yP/Q \rceil)dF(y) \\
&= nS\overline{F}(nQ/P) + S\sum_{i=0}^{n-1}(i+1)(\overline{F}(iQ/P) - \\
&\quad \overline{F}((i+1)Q/P)) \\
&= S\sum_{i=0}^{n-1}\overline{F}(iQ/P)
\end{aligned}
\tag{15.10}
$$

Similarly, we have for the expected inventory holding cost in a cycle,

$$
\begin{aligned}
E(Holding\ cost) &= hn\frac{(P-D)Q^2}{2PD}\overline{F}(nQ/P) + h\int_0^{nQ/P}(\frac{(P-D)Q^2}{2PD}\lfloor yP/Q \rfloor \\
&\quad +\frac{(P-D)P}{2D}(y - Q/P\lfloor yP/Q \rfloor)^2)dF(y) \\
&= \frac{h(P-D)P}{D}\sum_{i=0}^{n-1}\int_{iQ/P}^{(i+1)Q/P}(y - iQ/P)\overline{F}(y)dy
\end{aligned}
\tag{15.11}
$$

where $\lfloor x \rfloor$ is the integer part of x, $\lfloor x \rfloor \le x < \lfloor x \rfloor + 1$.

By applying Lemma 2 and Lemma 3, we obtain expressions for the expected resumption + maintenance + operating cost:

$$
\begin{aligned}
&E(Resumption+\ maintenance\ +operating\ cost) = \\
&\int_0^{nQ/P}[\int_0^y(R + E[c_{M(u)+1}(u)])q(u)r(u)du + c_u(\lambda_u) + vy]dF(y) \\
&+\overline{F}(nQ/P)[\int_0^{nQ/P}(R + E[c_{M(u)+1}(u)])q(u)r(u)du + c_p(\lambda_p) + vnQ/P]
\end{aligned}
\tag{15.12}
$$

The expected lost sales cost in a cycle is determined by

$$
\begin{aligned}
E(Lost\ sales\ cost) &= \int_0^{nQ/P} E(Lost\ sales\ cost \mid Y = y)dF(y) + \\
&\quad \overline{F}(nQ/P)E(Lost\ sales\ cost \mid Y \ge nQ/P) \\
&= LD\int_0^{nQ/P}\int_{t_2}^\infty(t - t_2)dF_u(t)dF(y) + \\
&\quad LD\overline{F}(nQ/P)\int_{Q/D-Q/P}^\infty(t - Q/D + Q/P)dF_p(t)
\end{aligned}
\tag{15.13}
$$

where $t_1 = y - Q/P\lfloor yP/Q \rfloor$ and $t_2 = (P - D)t_1/D)$.

From (15.8)-(15.13) we obtain, after some manipulations, the following expression for the expected cycle cost:

$$
\begin{aligned}
E[\text{Cost of cycle}] \;=\; & S\sum_{i=0}^{n-1}\overline{F}(iQ/P) + c_p(\lambda_p)\overline{F}(nQ/P) \\
& + c_u(\lambda_u)F(nQ/P) \\
& + \int_0^{nQ/P}[v + (R + E[c_{M(y)+1}(y)])q(y)r(y)]\overline{F}(y)dy \\
& + \frac{h(P-D)P}{D}\sum_{i=0}^{n-1}\int_{iQ/P}^{(i+1)Q/P}(y - iQ/P)\overline{F}(y)dy \\
& + LD\overline{F}(nQ/P)\int_{Q/D-Q/P}^{\infty}\overline{F}_p(y)dy \\
& + LD\sum_{i=0}^{n-1}\int_{iQ/P}^{(i+1)Q/P}(\int_{(P-D)/D(y-iQ/P)}^{\infty}\overline{F}_u(t)dt)dF(y)
\end{aligned}
$$

$$(15.14)$$

The expression for the expected length of cycle is obtained by conditioning on the length of time to a major failure. We have:

$$
\begin{aligned}
E[\text{Cycle length}] \;=\; & P/D\int_0^{nQ/P}\overline{F}(y)dy \\
& + \overline{F}(nQ/P)\int_{(Q/D-Q/P)}^{\infty}\overline{F}_p(t)dt \\
& + \sum_{i=0}^{n-1}\int_{iQ/P}^{(i+1)Q/P}(\int_{(P-D)/D(y-iQ/P)}^{\infty}\overline{F}_u(t)dt)dF(y)
\end{aligned}
$$

$$(15.15)$$

The long-run expected average cost per unit time is then given by

$$
\overline{C}(Q, n) = \frac{E[\text{Cost per cycle}]}{E[\text{Cycle length}]}
\tag{15.16}
$$

The optimal lot size Q^* and the optimal number of production runs before a preventive replacement n^* can be found by minimizing $\overline{C}(Q, n)$. In the general case, the optimization can be done numerically by exploring the surface of $\overline{C}(Q, n)$ and finding the global minimum.

To get more insight into the model and better understand how maintenance and lot-sizing decisions are interrelated, we will investigate in the next section the model with the exponentially distributed failure times and two types of failure.

4.2   The Exponential Case.
Here we assume that the preventive and failure replacement times are equal to zero ($T_p = 0, T_u = 0$), the repair cost is a function of the operating time t and does not depend on the number of repairs before t ($c_n(t) = c(t)$), $q(t) = q, s(t) = s = 1-q$, and the time to failure is exponentially distributed with parameter λ_1.

Denote $\lambda = s\lambda_1$. Then, $\overline{F}(t) = e^{-\lambda t}$ and $\overline{F}_p(t) = \overline{F}_u(t) = 0$. From (15.14), we have for the expected cost of cycle

$$
\begin{aligned}
E[\textit{Cost of cycle}] \;=\; & S\sum_{i=0}^{n-1} e^{-\lambda i Q/P} + c_p\lambda_p) \\
& + (c_u(\lambda_u) - c_p(\lambda_p))(1 - e^{-\lambda n Q/P}) \\
& + \int_0^{nQ/P} [v + (R + c(y))q/s\lambda] e^{-\lambda y} dy \\
& + \frac{h(P-D)P}{D} \sum_{i=0}^{n-1} \int_{iQ/P}^{(i+1)Q/P} (y - iQ/P) e^{-\lambda y} dy \\
=\; & \frac{S(1-e^{-n\lambda Q/P})}{(1-e^{-\lambda Q/P})} + c_p(\lambda_p) + q/s \int_0^{nQ/P} c(y)) \lambda e^{-\lambda y} dy \\
& + (1 - e^{-n\lambda Q/P})(c_u(\lambda_u) - c_p(\lambda_p) + v/\lambda + Rq/s) \\
& + \frac{h(P-D)P}{D} \\
& + \frac{1-e^{-n\lambda Q/P}}{1-e^{-\lambda Q/P}}(1/\lambda^2 - e^{-\lambda Q/P}(1/\lambda^2 + Q/(\lambda P)))
\end{aligned}
$$

$$(15.17)$$

and from (15.15), the expected cycle length is given by

$$E[\textit{Cycle length}] = \frac{P}{\lambda D}(1 - e^{-n\lambda Q/P}) \qquad (15.18)$$

From (15.17) and (15.18), we have for the expected average cost per unit time:

$$\overline{C}(Q,n) = \frac{c_1 - c_2 Q e^{-\lambda Q/P}}{1 - e^{-\lambda Q/P}} + \frac{c_3 + c_4 \int_0^{nQ/P} c(y)\lambda e^{-\lambda y} dy}{1 - e^{-n\lambda Q/P}} + K \quad (15.19)$$

where

$$
\begin{aligned}
c_1 =\; & S\lambda D/P, c_2 = h(P-D)/P, c_3 = c_p(\lambda_p)\lambda D/P, \\
c_4 =\; & \frac{\lambda q D}{sP}, K = \frac{\lambda D}{P}[c_u(\lambda_u) - c_p(\lambda_p) + v/\lambda + Rq/s] + \frac{h(P-D)}{\lambda}.
\end{aligned}
$$

$$(15.20)$$

If we put $z = nQ/P$, the cost function in (15.19) has the form

$$C(Q,z) = \frac{c_1 - c_2 Q e^{-\lambda Q/P}}{1 - e^{-\lambda Q/P}} + \frac{c_3 + c_4 \int_0^z c(y)\lambda e^{-\lambda y} dy}{1 - e^{-\lambda z}} + K. \quad (15.21)$$

In the next theorem, we will show that if we treat n as a continuous variable, the optimal solution is unique and can be obtained by solving two nonlinear equations.

Theorem 3. *Assume that $c(t)$ is an increasing function. Then $C(Q,z)$ has a global minimum and the optimal values of Q^* and z^* can be found as a unique solution to the following two equations:*

$$c_4 \int_0^z (c(z) - c(y))\lambda e^{-\lambda y} dy = c_3 \qquad (15.22)$$

$$c_2(\lambda Q/P + e^{-\lambda Q/P} - 1) = \lambda/Pc_1. \qquad (15.23)$$

We can see that the equations in (15.22) and (15.23) can be solved separately. If (Q^*, z^*) is the solution, then an approximation to the optimum number of production runs before a preventive replacement can be obtained from

$$n^* = \lfloor Pz^*/q^* \rfloor. \tag{15.24}$$

In the next Lemma, we will show how to find the optimum number of production cycles before a preventive replacement for a given lot size.

Lemma 5. *Assume that $c(t)$ is increasing. For any $Q > 0$, let $n(Q)$ be determined by*

$$n(Q) = inf\{n \geq 1 : \int_0^{nQ/P} (c(nQ/P) - c(y))\lambda e^{-\lambda y} dy \geq c_3/c_4\} \tag{15.25}$$

Then, the optimum number of production runs before a preventive replacement , $n_o(Q)$ is given by

$$\overline{C}(Q, n_o(Q)) = min\{\overline{C}(Q, n(Q)), \overline{C}(Q, n(Q) - 1)\} \tag{15.26}$$

if $n(Q) > 1$ and finite, and $n_o(Q) = 1$ if $n(Q) = 1$.

If $c(t) = c, \overline{C}(Q, n + 1) < \overline{C}(Q, n)$, i.e., no preventive replacement should be planned.

To summarize, if the objective is to find the optimal preventive replacement policy for a given lot size, one can use the result in Lemma 5. The optimum number of production runs before a preventive replacement is determined by (15.25) and (15.26).

A good approximation to the joint optimal lot sizing and replacement policy can be obtained quickly by finding the solution Q^* to the equation (15.23) and then applying Lemma 5. The optimal policy can always be found numerically by identifying the global minimum of $\overline{C}(Q, n)$.

5.   NUMERICAL EXAMPLE

Assume that $s(t) = 0, q(t) = 1$, i.e., the time to failure and the duration of the preventive replacement are exponentially distributed with parameters λ and λ_p, respectively. Next, consider linear minimal repair cost $c_i(t) = at + b$ and linear preventive replacement cost $c_p(\lambda_p) = c\lambda_p$. From (15.17) - (15.19), the expected long-run expected average cost per unit time is given by

$$\overline{C}(Q, n) = \frac{E[Cyclecost]}{E[Cyclelength]} \tag{15.27}$$

where

$$E[Cyclecycle] = \frac{nQ}{D} + \frac{1}{\lambda_p}e^{-\lambda_p(\frac{Q}{D} - \frac{Q}{P})}$$

$$E[Cyclelength] = nS + \frac{n\lambda RQ}{P} + c\lambda_p + \frac{nh(P-D)Q^2}{2PD} + \frac{vnQ}{P}$$
$$+ \frac{a\lambda n^2 Q^2}{2P^2} + \frac{b\lambda nQ}{P} + \frac{LD}{\lambda_p}e^{-\lambda_p(\frac{Q}{D} - \frac{Q}{P})}$$

The optimal values of Q and n have been found by exploring the surface of $\overline{C}(Q,n)$ numerically and identifying the global minimum. Some numerical results for this model are discussed below. Table 15.2 summarizes the numerical data used to generate Figures 15.1 - 15.6.

Table 15.2   Numerical Data for Figures 15.3 - 15.8.

Figure:	15.3	15.4	15.5	15.6	15.7	15.8
λ	0.2	0.2	0.2	0.2	set	set
λ_p	set	set	varies	varies	0.3	0.3
v	50	50	50	50	50	50
h	75	75	75	75	75	75
D	30	30	30	30	30	30
P	35	35	35	35	35	35
R	400	400	400	400	400	400
S	500	500	500	500	500	500
L	80	80	80	80	80	80
a	100	100	100	100	100	100
b	50	50	50	50	50	50
c	5000	5000	set	set	5000	5000
n	optimal	varies	optimal	optimal	optimal	varies
Q	varies	optimal	optimal	optimal	varies	optimal

Note: "varies": parameter is on the abscissa axis of the graph,

"set" : graph is given for a set of values of the parameter,

"optimal" : optimal value that is computed for a particular set of input data.

Figure 15.1 shows the total cost as a function of the lot size Q for the preventive replacement rate $\lambda_p = 0.1, 0.5$ and 1.0 and Figure 15.2 shows the total cost as a function of the number of production runs before a preventive replacement for the preventive replacement rate $\lambda_p = 0.1, 0.5$ and 1.0. It is interesting to observe in both figures that the total cost function shifts downwards when λ_p changes from 0.1 to 0.5. Then it shifts upwards when λ_p goes from 0.5 to 1.0. This can be explained by the fact that when the replacement rate increases, the lost-sales cost will decrease and the preventive replacement cost will increase.

Therefore, with increasing λ_p, when the decrease in the lost-sales cost is more significant than the increase in preventive replacement cost, the total cost function will shift downwards. Then as the replacement rate continues to increase, the decrease in the lost-sales cost will become less significant than the increase in the preventive replacement cost and the total cost function will move upwards.

Figures 15.3 and 15.4 depict the optimal lot size Q^* and the optimal number of production runs before a preventive replacement n^*, respectively, as a function of the preventive replacement rate λ_p for $c = 1000, 5000$, and 10000. From both figures, the optimal values Q^* and n^* both first decrease and then increase as λ_p gets larger. This behavior is again mainly caused by the inverse relationship between the preventive replacement cost and the lost-sales cost. From Figures 15.3 and 15.4, we can also see that the increase in c (constant in the preventive replacement cost) will cause both the optimal lot size $Q*$ and the optimal number of production runs before a preventive replacement n^* to shift upward.

Figures 15.5 and 15.6 depict the total cost as a function of the lot size Q and of the number of production runs before a preventive replacement , respectively, for $\lambda = 0.1, 0.3, 0.6$ and 1.0. From both figures, we can see that as λ increases, the total cost function shifts upward which is due to the increase in repair spending. It is interesting to observe in 15.5 that the optimum lot size remains relatively stable whereas n^* in Figure 15.6 decreases considerably with the increasing failure rate λ.

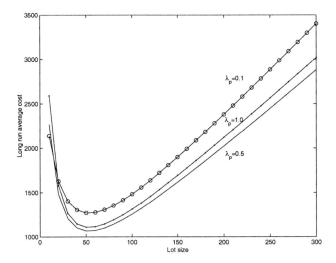

Figure 15.1 Long run average cost as a function of the lot size for several values of preventive replacement rate λ_p.

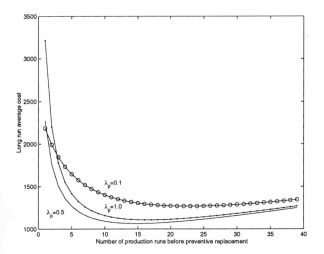

Figure 15.2 Long run average cost as a function of the number of production runs before a preventive replacement for several values of preventive replacement rate λ_p.

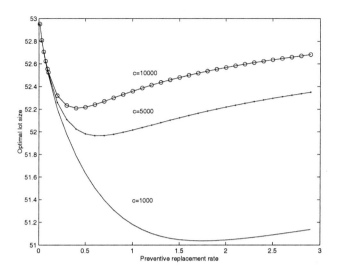

Figure 15.3   Optimal lot size as a function of the preventive replacement rate λ_p.

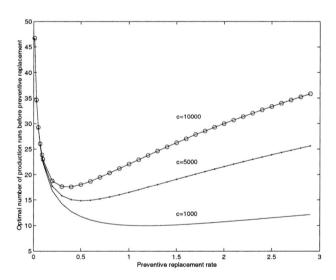

Figure 15.4   Optimal number of production runs before a preventive replacement n^* as a function of the preventive replacement rate λ_p.

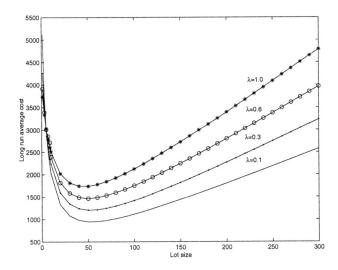

Figure 15.5 Long run average cost as a function of the lot size for several values of the failure rate λ.

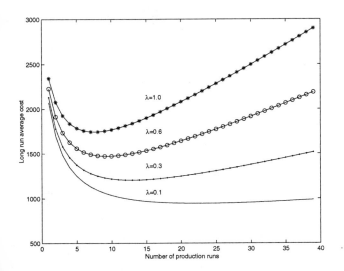

Figure 15.6 Long run average cost as a function of the number of production runs before a preventive replacement for several values of the failure rate λ.

6.   MODEL WITH PREVENTIVE REPLACEMENT AFTER THE FIRST PRODUCTION RUN

For this model, we assume that $s(t) = 1, q(t) = 0, c_p = 0, v = 0, n = 1, T_u = 0$, and $T_p = 0$, i.e., the failure and preventive replacement times are negligible. This is the model considered by Groenevelt et al. (1992a). They assumed that at each machine failure, some maintenance actions are performed and these actions will restore the system to the "as good as new" condition. They also assumed that after each production run, the system will be restored to the "as good as new" condition (i.e. n = 1). The only parameter of interest for this model is the optimal lot size Q^*.

It follows from the assumptions in the model that $\overline{F}_u(t) = 0$ and $\overline{F}_p(t) = 0$. Then from (15.14), (15.15), and (15.16), we have the following formula for the expected average cost:

$$\overline{C}(Q) = \frac{S + c_u F(Q/P) + h(P - D)/DP \int_0^{Q/P} t\overline{F}(t)dt}{P/D \int_0^{Q/P} \overline{F}(t)dt} \tag{15.28}$$

It is not difficult to see that the formula in (15.28) can be expressed as

$$\overline{C}(Q) = \frac{S + c_u F(\frac{Q}{P}) + 0.5h(P - D)\{(\frac{Q}{P})^2(1 - F(\frac{Q}{P}))\} + \int_0^{Q/P} t^2 f(t)dt}{\frac{Q}{P} \int_0^{\frac{Q}{P}} tf(t)dt + Q/D(1 - F(\frac{Q}{P}))} \tag{15.29}$$

where $f(t)$ is the density function of Y.

This formula was obtained by Groenevelt et al. (1992a). They analyzed the cost function in (15.29) for the exponential failure distribution and found that the optimal lot size in their model is always greater than the optimal lot size in the classical EMQ model. In the next theorem, we will show that if the failure rate is nondecreasing, the optimal lot size can be found as the unique solution to a nonlinear equation. The result is obtained by analyzing the cost function in (15.29) for a general failure distribution.

Theorem 4. *Assume that $r(t)$ is nondecreasing. Then the cost function $\overline{C}(Q)$ in (15.29) has the global minimum at Q^* which can be obtained as the unique solution to the equation*

$$c_u \int_0^{Q/P} (r(\frac{Q}{P}) - r(t))\overline{F}(t)dt + h(P - D)\frac{P}{D} \int_0^{Q/P} (Q/P - t)\overline{F}(t)dt = S \tag{15.30}$$

7. CONCLUSIONS

In this paper, we have studied a general EMQ model with machine breakdowns and repairs, two types of failure , lost sales and planned preventive replacement . The structure of the optimal production/maintenance policy has been found in a special case when the replacements are instantaneous. It has been shown that the optimal preventive replacement policy is of a control limit type and the optimal lot size is a function of the age of the production unit. The model with lost sales has been analyzed under the assumption of a constant production lot size and preventive replacement in an idle period after completing n production runs. An explicit formula for the long-run expected average cost per unit time has been obtained in the general case under certain assumptions.

We have found that both the failure rate and the replacement rate have some effects on the optimal lot size and the optimal preventive replacement time, and the optimal lot size also depends on the maintenance policy being used. Numerical results indicate that when applying preventive maintenance the optimal lot size remains relatively stable whereas the optimal number of production runs before preventive replacement decreases considerably with an increasing failure rate.

References

[1] Block, H.W., Borges W.S., and Savits T.H. Age-Dependent Minimal Repair. *Journal of Applied Probability*, **22** (1985) 370-385.

[2] Block, H.W., Borges W.S., and Savits T.H. A General Age Replacement Model with Minimal Repair. *Naval Research Logistics*, **35** (1988) 365-372.

[3] Groenevelt, H., Pintelon L., and Seidmann A. Production Lot Sizing with Machine Breakdowns. *Management Science*, **38** (1992a) 104-123.

[4] Groenevelt, H., Pintelon A. , and Seidmann A. Production Batching with Machine Breakdowns and Safety Stocks. *Operations Research*, **40** (1992b) 959-971.

[5] Kurano, M. Semi-Markov Decision Processes and Their Applications in Replacement Models. *Journal of the Operations Research Society of Japan*, **40** (1985) 959-971.

[6] Lee, H.L. and Rosenblatt M.J. Simultaneous Determination of Production Cycle and Inspection Schedules in a Production System. *Management Science* **33** (1987) 1125-1136.

[7] Lee, J.S. and Park K.S. Joint Determination of Production Cycle and Inspection Intervals in a Deteriorating Production System. *Journal of the Operational Research Society* **42**, (1991) 775-783.

[8] Makis, V. Optimal Lot-Sizing/ Preventive Replacement Policy for an EMQ model with Minimal Repair. *International Journal of Logistics: Research and Applications*, **1** (1998a) 173-180.

[9] Makis, V. Optimal Lot Sizing and Inspection Policy for an EMQ model with Imperfect Inspections. *Naval Research Logistics*, **45** (1998b) 165-186.

[10] Makis, V. and Fung J. Optimal Preventive Replacement , Lot Sizing and Inspection Policy for a Deteriorating Production System. *Journal of Quality in Maintenance Engineering*, **1(4)** (1995) 41-55.

[11] Makis, V. and Fung J. An EMQ model with Inspections and Random Machine Failures. *Journal of the Operational Research Society*, **49** (1998) 66-76.

[12] Porteus, E.L. Optimal Lot Sizing , Process Quality Improvement and Setup Cost Reduction. *Operations Research*, **34** (1986) 137-144.

[13] Posner, M.J.M. and Berg M. Analysis of a Production-Inventory System with Unreliable Production Facility. *Operations Research Letters*, **8** (1989) 339-345.

[14] Rahim, M.A. and Banerjee P.K. A Generalized Model for the Economic Design of $\overline{X}$ Control Charts for Production Systems with Increasing Failure Rate and Early Replacement. *Naval Research Logistics* **40** (1993) 787-809.

[15] Rahim, M.A. Joint Determination of Production Quantity, Inspection Schedule and Control Chart Design. *IIE Transactions*, **26-6** (1994) 2-11.

[16] Rosenblatt, M.J. and Lee H.J. Economic Production Cycles with Imperfect Production Processes. *IIE Transactions*, **18** (1986) 48-55.

[17] Savits, T.H. Some Multivariate Distribution Derived from Nonfatal Shock Model. *Journal of Applied Probability*, **25** (1988) 382-390.

Chapter 16

STOCHASTIC MANUFACTURING SYSTEMS: PRODUCTION AND MAINTENANCE CONTROL
1

E. K. Boukas

Mechanical Engineering Department, École Polytechnique de Montréal, P.O. Box 6079, station "Centre-ville", Montréal, Québec, Canada H3C 3A7

boukas@meca.polymtl.ca

Q. Zhang

Department of Mathematics, University of Georgia, Athens, GA 30602, USA

qingz@math.uga.edu

Abstract   : This chapter deals with stochastic manufacturing systems. It focuses on continuous-flow models that were extensively studied by researchers in this field. Various types of models that incorporate production, corrective maintenance and preventive maintenance are described. Related numerical methods are given.

Keywords: Manufacturing systems, Production control, Maintenance, Dynamic programming.

1.   INTRODUCTION

Manufacturing systems are complex systems and are in general very hard to model and control. In the last two decades a number of techniques have been proposed. From among them we quote the continuous flow model which has attracted a lot of researchers from the control and operations research communities. The first model was proposed by Olsder and Suri [66]. They stated the production control problem based on the formalism of the class of jump linear systems. In their formulation, they modeled the capacity of the production system by a continuous-time Markov process with a finite discrete state space. The Production planning has been formulated as a stochastic optimization problem. This contribution has been in some sense the starting point of a new direction of research in manufacturing systems. Many extensions of this model have been proposed,which we can gather into two groups. The first one contains all the models that concentrate only on production control. Among these models, we quote the contribution of Kimemia and Gershwin [57], Akella and Kumar [1], Bielecki and Kumar [8]. The other group contains all the models that combine production and maintenance control in an optimization stochastic control problem. This direction has been initiated by Boukas and his coauthors [13]. The purpose of maintenance is to restore a failed or deteriorating system to the normal operating state. From among the types of maintenance, we quote preventive maintenance and corrective maintenance . In preventive maintenance, the system is periodically inspected, some components are replaced, lubrication is checked and adjustments are made before the system fails. The Corrective maintenance is used once the system fails. While the aim of preventive maintenance is to increase the reliability and to prolong the life of the system by overcoming the effects of aging, fatigue and wear, the goal of corrective maintenance is to bring the system from the failed state to the operating state as quickly as possible to increase its availability. A complete review of the literature on this direction of research up to 1994 can be found in Sethi and Zhang [75] and Gershwin [43].

The aim of this chapter is to describe various contributions in this direction of research. We will give all the details of the different models of the two groups and put considerable emphasis on the practical meaning.

The rest of the paper is organized as follows. In section 2, the production control of manufacturing systems is formulated and the solution is given. In section 3, corrective maintenance is added to the model of the problem of the previous section and a new optimization problem that can determine simultaneously the production and the corrective maintenance policies is formulated and the solution is given. In section 4, an optimization stochastic control problem that determines the production and the preventive maintenance strategies is formulated and solved numerically. In sections 5, numerical methods are described and some examples are given to show the usefulness of some of our models.

2.   PRODUCTION CONTROL PROBLEM

In this section we will propose a certain number of models that can be used to compute the optimal control planning production for manufacturing systems with failure prone machines. We will consider a simple system to show the models.

2.1   Model 1.   To state clearly the production control planning problem, let us consider a production system that consists of one machine, producing one item. Let the state of the inventory be described by the continuous-time variable, $x(t) \in \mathbb{R}$. Let us also assume that the state of this machine is described by a continuous-time Markov process $\{\xi(t), t \geq 0\}$ that takes values in a finite space $S = \{1, 2\}$. For the rest of this section, $\xi(t) = 1$ means that the machine is up and it can produce parts and $\xi(t) = 2$ means that the machine is down and no part can be produced. The dynamics of the state of the machine is completely defined by the following transition probabilities:

$$P\left[\xi(t + \Delta t) = \beta | \xi(t) = \alpha\right] = \begin{cases} q_{\alpha\beta}\Delta t + o(\Delta t) & \text{when } \beta \neq \alpha \\ 1 + q_{\alpha\beta}\Delta t + o(\Delta t) & \text{otherwise} \end{cases} \tag{16.1}$$

with $\lim_{\Delta t \to 0} \frac{o(\Delta)}{\Delta t}$.

The inventory level at time t is described by the following differential equation:

$$\dot{x}(t) = u_p(t) - d, x(0) = x_0 \tag{16.2}$$

where d is the demand rate that we will assume to be constant for all the chapter and $u_p(t)$ is the production rate at which we produce the item at time t.

From the physical point of view the capacity of the machine is limited by the following constraints:

$$0 \leq u_p(t) \leq \bar{u}_p I_{\{\xi=1\}} \tag{16.3}$$

where $\bar{u}_p$ is the maximum production rate at which the machine can produce parts when it is up and $I_{\{\xi=1\}}$ is the indicator function that takes 1 when $\xi(t) = 1$ and 0 otherwise.

Remark 16.1 *Notice that the stock level, $x(t)$, can take positive and negative real values. Therefore, when the stock level is positive we will refer to surplus and when it is negative we will refer to backlog.*

The goal of the production planning is to produce parts in order to satisfy the given demand rate and also reduce the inventory level. This can be obtained by seeking a control law that minimizes the following cost function:

$$J_u(x_0, \xi_0) \qquad (16.4)$$

$$= \mathbb{E}\left[\int_0^\infty e^{-\rho t} h(x(t), \xi(t)) dt | x(0) = x_0, \xi(0) = \xi_0\right]$$

where $\rho > 0$ is the discount rate and $h(x(t), \xi(t))$ is defined as follows:

$$h(x(t), \xi(t)) = c^+ x^+(t) + c^- x^-(t) \qquad (16.5)$$

with c^+ and c^- are given positive constant and $x^+(t)$ and $x^-(t)$ are defined respectively as $x^+(t) = \max(0, x(t))$ when $x(t) \geq 0$ and $x^-(t) = \max(0, -x(t))$ when $x(t) \leq 0$.

Remark 16.2 *We will assume that the system is feasible. This means that its availability, which depends on the jump rates q_{12} and q_{21}, is higher and allows the production system to respond to the demand d.*

It is not difficult to show that the value function $v(x(t), \xi(t))$ is convex and continuously differentiable. Therefore, the corresponding optimality conditions are then given by the following dynamic programming equations:

$$\rho v(x(t), 1) = \min_{u_p(t)} [h(x(t), 1) + [u_p(t) - d] v_x(x(t), 1)$$

$$+ q_{12} [v(x(t), 2) - v(x(t), 1)]] \qquad (16.6)$$

$$\rho v(x(t), 2) = h(x(t), 2) + [-d] v_x(x(t), 2) + q_{21} [v(x(t), 1) - v(x(t), 2)] \qquad (16.7)$$

From these optimality conditions, it follows that the optimal control law is then given by the following expression:

$$u_p^\star(t) = \begin{cases} \bar{u}_p & \text{if } v_x(x(t), 1) < 0 \\ d & \text{if } v_x(x(t), 1) = 0 \\ 0 & \text{otherwise} \end{cases} \qquad (16.8)$$

Remark 16.3 *Notice that the choice of the control when $v_x(x(t), 1) = 0$ is arbitrary. To avoid the chattering problem, it is therefore necessary to choose $u_p(t) = d$ when $v_x(x(t), 1) = 0$. This means that once we reach* the minimum of the value function, $v(x(t), 1)$, it is necessary to keep the inventory at this level which is referred to in the literature as the hedging point.

Therefore, since the value function, $v(x(t), \xi(t))$, is convex and continuously differentiable in $x(t)$ for all $\xi(t)$ in S, it has a global minimum that we can denote by $x^\star$. This minimum can be computed from the solution of the previous optimality conditions (see Akella and Kumar [1].

It can shown that the hedging point is given by:

$$x^\star = max\left\{0, \frac{1}{\underline{a}} \log \left[\frac{c^+}{c^+ + c^-} \left[1 + \frac{\rho d}{\lambda d - (\rho + \mu + d\underline{a})(\bar{u}_p - d)} \right] \right] \right\} \tag{16.9}$$

where $\underline{a}$ is the only negative eigenvalue of the matrix A with the following expression:

$$A = \begin{bmatrix} \frac{\rho + \lambda}{\bar{u}_p - d} & \frac{-\lambda}{\bar{u}_p - d} \\ \frac{\mu}{d} & -\frac{\mu + \rho}{d} \end{bmatrix} \tag{16.10}$$

where the parameters ρ, c^+, c^-, $\lambda = q_{12}$, $\mu = q_{21}$.

Therefore, the final expression of the production policy becomes:

$$u_p^\star(t) = \begin{cases} \bar{u}_p & \text{if } x(t) < x^\star \\ d & \text{if } x(t) = x^\star \\ 0 & \text{otherwise} \end{cases} \tag{16.11}$$

Remark 16.4 *Notice that the hedging level, $x^\star$, is always greater than or equal to zero. Its value depends on the data of the manufacturing system and its feasibility to respond to the given demand rate d.*

Remark 16.5 *The expression of the optimal control policy can also be obtained in the following way. Since our goal consists of keeping the stock level at $x^\star$ when the machine is up, it follows that:*

- *if $x(t) < x^\star$, then $v_x(x(t), 1) < 0$ since the value function is convex. The only feasible control that can drive the stock level to $x^\star$ is $\bar{u}_p$.*

- *if $x(t) > x^\star$, then $v_x(x(t), 1) > 0$ since the value function is convex. The only feasible control that can drive the stock level to $x^\star$ is 0.*

- *if $x(t) = x^\star$, then $v_x(x(t), 1) = 0$ any control will be feasible but only the feasible control $\bar{u}_p^\star = d$ will keep the stock level at the hedging level $x^\star$ and prevent the chattering.*

Remark 16.6 *From the practical point of view this control law is easy to implement once we know the value of the hedging level $x^\star$. This value depends on the data of the manufacturing system that can be estimated easily.*

2.2   Model 2.

Let us consider a single product manufacturing system with stochastic production capacity modelled by a Markov process, $\alpha(t)$ and constant demand, d, for its production over time. Let the dynamics of $x(t)$ be the same as the previous ones and assume that the production has many modes, let say m modes, $\mathcal{M} = \{1, \ldots, m\}$.

Given an admissible $u(\cdot)$, we define

$$J(x, k, u(\cdot)) = \limsup_{T \to \infty} \frac{1}{T} \mathbb{E} \int_0^T (h(x(t)) + c(u(t))) \, dt. \qquad (16.12)$$

The objective is to choose $u(\cdot) \in \mathcal{A}(k)$ so as to minimize the cost functional $J(x, k, u(\cdot))$.

Assume the cost functions $h(\cdot)$ and $c(\cdot)$ to be smooth and convex functions. Moreover, the average capacity $\bar{\alpha} \equiv \sum_{i=0}^m i\nu_i > d$ and $d \notin \mathcal{M}$.

An admissible control $u(\cdot)$ is called *stable* if it satisfies the condition

$$\lim_{T \to \infty} \frac{E|x(T)|^{\kappa+1}}{T} = 0. \qquad (16.13)$$

The associated HJB equation takes the following form:

$$\lambda = F(k, W_x(x, k)) + h(x) + QW(x, \cdot)(k), \qquad (16.14)$$

where $F(k, r) = \inf_{0 \le u \le k}\{(u - d)r + c(u)\}$, λ is a constant and W is a real-valued function defined on $\mathbb{R} \times \mathcal{M}$.

Let $\mathcal{G}$ denote the family of real-valued functions $W(\cdot, \cdot)$ defined on $\mathbb{R} \times \mathcal{M}$ such that (i) $W(\cdot, k)$ is convex; (ii) $W(\cdot, k)$ is continuously differentiable; (iii) $W(\cdot, k)$ has polynomial growth. A solution to the HJB equation (16.14) is a pair (λ, W) with λ a constant and $W \in \mathcal{G}$. The function W is called a *potential function* for the control problem, if λ is the minimum long-run average cost.

Theorem 16.1 ([73]) (i) (λ^*, V) *is a viscosity solution to the HJB equation* (16.14). *Moreover, the constant* λ^* *is unique.*

(ii) *The function* $V(x, k)$ *is continuously differentiable in* x, *and* (λ^*, V) *is a classical solution to the HJB equation. Moreover,* $V(x, k)$ *is convex in* x *and*

$$|V(x, k)| \le C(1 + |x|^{\kappa+1}).$$

Theorem 16.2 ([73]) *Let* (λ, W) *be a solution to the HJB equation* (16.14). *Then*

(i) *If there is a control* $u^*(\cdot) \in \mathcal{A}(k)$ *such that*

$$F\left(\alpha(t), W_x(x^*(t), \alpha(t))\right) = (u^*(t) - d)W_x(x^*(t), \alpha(t)) + c(u^*(t)) \tag{16.15}$$

for a.e. $t \geq 0$ *with probability 1, where* $x^*(\cdot)$ *is the surplus process corresponding to the control* $u^*(\cdot)$, *and*

$$\lim_{T \to \infty} \frac{W(x^*(T), \alpha(T))}{T} = 0, \tag{16.16}$$

then

$$\lambda = J(x, k, u^*(\cdot)).$$

(ii) *For any* $u(\cdot) \in \mathcal{A}(k)$, *we have* $\lambda \leq J(x, k, u(\cdot))$, *i.e.,*

$$\limsup_{t \to \infty} E \int_0^t \left(h(x(t)) + c(u(t))\right) dt \geq \lambda.$$

(iii) *For any (stable) control policy* $u(\cdot) \in \mathcal{B}(k)$, *we have*

$$\liminf_{t \to \infty} \frac{1}{t} E \int_0^t \left(h(x(t)) + c(u(t))\right) dt \geq \lambda. \tag{16.17}$$

We define a control policy $u^*(\cdot, \cdot)$ via the relative function $V(\cdot, \cdot)$ as follows:

$$u^*(x, k) = \begin{cases} 0 & \text{if} \quad V_x(x, k) > -c_u(0), \\ (c_u)^{-1}\left(-V_x(x, k)\right) & \text{if} \quad -c_u(k) \leq V_x(x, k) \leq -c_u(0), \\ k & \text{if} \quad V_x(x, k) < -c_u(k), \end{cases} \tag{16.18}$$

if the function $c(\cdot)$ is strictly convex, or

$$u^*(x, k) = \begin{cases} 0 & \text{if} \quad V_x(x, k) > -c, \\ \min\{k, d\} & \text{if} \quad V_x(x, k) = -c, \\ k & \text{if} \quad V_x(x, k) < c, \end{cases} \tag{16.19}$$

if $c(u) = cu$. Therefore, the control policy $u^*(\cdot, \cdot)$ satisfies the condition (16.15).

Using the convexity of the function $V(\cdot, k)$, we can show that there are x_k, y_k, $-\infty < y_k < x_k < \infty$ such that

$$U(x) = (x_k, \infty) \quad \text{and} \quad L(k) = (-\infty, y_k).$$

The control policy $u^*(\cdot, \cdot)$ can be written as

$$u^*(x, k) = \begin{cases} 0 & x > x_k, \\ (c_u)^{-1}\left(-V_x(x, k)\right) & y_k \le x \le x_k, \\ k & x < y_k. \end{cases}$$

Theorem 16.3 ([73]) *The control policy $u^*(\cdot, \cdot)$, defined in (16.18) or (16.19) as the case may be, is optimal.*

If $c(u) = 0$, i.e., there is no production cost in the model, the optimal control policy can be chosen to be a hedging point policy , which has the following form: There are real numbers x_k, $k = 1, \ldots, m$, such that

$$u^*(x, k) = \begin{cases} 0 & x > x_k \\ \min\{k, d\} & x = x_k \\ k & x < x_k. \end{cases}$$

2.3   Model 3. Let us consider finite horizon production planning of stochastic manufacturing systems. Note that there are some distinct differences between the finite time and the infinite horizon formulations. For an infinite horizon formulation, the dynamics of the systems are essentially homogeneous, and therefore, the hedging point (or turnpike sets) consist of constants. If the system performance is evaluated over a finite time horizon the threshold levels are no longer constants, but "time dependent threshold curves". Therefore, the problem becomes much more complicated. One expects that the essence of the turnpike sets is that they produce at the maximum speed if the inventory level is below the turnpike, produce nothing if the inventory level is above the turnpike, and produce exactly enough to meet the demand if the inventory reaches the turnpike, should still work. Nevertheless, the inhomogeneous nature makes it very difficult to obtain explicit optimal solutions. To fulfill our goal of achieving the optimality, the turnpike sets must be smooth enough and be "traceable" by the trajectory of the system.

Let $x(t) \in \mathbb{R}$ denote the surplus (inventory/backlog) process and $u(t) \ge 0$ denote the rate of production planning of a manufacturing system. The product demand is assumed to be a constant and denoted by d. Then,

$$\dot{x}(t) = u(t) - d, \quad x(s) = x, \quad 0 \le s \le t \le T \tag{16.20}$$

where T is a finite horizon.

Let $\mathcal{M} = \{\alpha_1, \alpha_2\}$ ($\alpha_1 > \alpha_2 \ge 0$) denote the set of machine states and let $\alpha(t) \in \mathcal{M}$ denote the machine capacity process. If $\alpha(t) = \alpha_1$, it means the machine is in good condition with capacity α_1. If $\alpha(t) = \alpha_2$, the machine (or part of the machine) breaks down with a remaining capacity α_2. Assume that $\alpha_1 > d > \alpha_2$, i.e., the demand can be satisfied if the machine is in good condition and cannot be satisfied if the machine (or part of the machine) breaks down.

We consider the cost function $J(s, x, \alpha, u(\cdot))$ with $\alpha(s) = \alpha \in \mathcal{M}$ defined by

$$J(s, x, \alpha, u(\cdot)) = E \int_s^T e^{-\rho t} h(x(t)) dt, \qquad (16.21)$$

where $\rho \geq 0$ is the discount factor. Here ρ is allowed to be zero, since we are now considering a finite horizon problem. Our goal is to find a production plan $0 \leq u(t) \leq \alpha(t)$ as a function of the past $\alpha(\cdot)$ that minimizes $J(s, x, \alpha, u(\cdot))$.

Assumptions:

(A1) $h(x)$ is a convex function such that for positive constants C_h and k_h,

$$0 \leq h(x) \leq C_h(1 + |x|^{k_h}) \text{ and } h(x) > h(0) = 0 \text{ for all } x \neq 0.$$

Moreover, there exists a constant $c_h > 0$ such that

$$\frac{h_{x+}(x_2) - h_{x+}(x_1)}{x_2 - x_1} \geq c_h \text{ for all } -|\alpha_2 - d|T \leq x_1 < 0 \leq x_2 \leq |\alpha_2 - d|T,$$
$$(16.22)$$

where $h_{x+}(x)$ denotes the right-hand derivative of $h(x)$.

(A2) $\alpha(t) \in \mathcal{M}$ is a two state Markov chain governed by

$$L_\alpha f(\cdot)(i) = \begin{cases} \lambda(f(\alpha_2) - f(\alpha_1)) & \text{if } i = \alpha_1 \\ 0 & \text{if } i = \alpha_2. \end{cases} \qquad (16.23)$$

for any function f on $\mathcal{M}$. Here $\lambda \geq 0$ is the machine breakdown rate.

Examples. Let us give a few examples of $h(x)$ that satisfy Assumption (A1).

(1) $h(x) = x^2$;

(2) $h(x) = h^+ \max\{0, x\} + h^- \max\{0, -x\}$ where $h^+ > 0$ and $h^- > 0$ are constants. This cost function was employed in [1].

(3) $h(x)$ is convex and piecewise linear with $h(0) = 0$, $h_{x-}(0) < 0$, and $h_{x+}(0) > 0$.

Assumption (A2) is a condition on the machine capacity process $\alpha(t)$. It indicates that once the machine goes down it will never come up again. Such a situation occurs when the cost of repairing is very expensive, or no repair facilities are available. As a result, replacement is a better alternative than repair.

Definition 16.1. A control $u(\cdot) = \{u(t) : t \geq 0\}$ is *admissible* if $u(t)$ is an $\mathcal{F}_t = \sigma\{\alpha(s), s \leq t\}$ adapted measurable process and $0 \leq u(t) \leq \alpha(t)$ for all $0 \leq t \leq T$. $\mathcal{A}$ will denote the set of all admissible controls in the sequel.

Let $v(s, x, \alpha)$ denote the value function of the problem, i.e.,

$$v(s, x, \alpha) = \inf_{u(\cdot) \in \mathcal{A}} J(s, x, \alpha, u(\cdot)), \text{ for } \alpha \in \mathcal{M}.$$

It can be shown as in [104] that the value function $v(s, x, \alpha)$ is convex in x for each $s \in [0, T]$ and $\alpha \in \mathcal{M}$. Moreover, $v(s, x, \alpha) \in C([0, T], I\!R)$ is the only viscosity solution the following dynamic programming equations.

$$\begin{cases} 0 = -v_s(s, x, \alpha_1) + \sup_{0 \leq u \leq \alpha_1} [-(u - d)v_x(s, x, \alpha_1)] \\ \qquad - \exp(-\rho s)h(x) - \lambda(v(s, x, \alpha_2) - v(s, x, \alpha_1)), \qquad (16.24) \\ 0 = v(T, x, \alpha_1) \end{cases}$$

and

$$\begin{cases} 0 = -v_s(s, x, \alpha_2) + \sup_{0 \leq u \leq \alpha_2} [-(u - d)v_x(s, x, \alpha_2)] - \exp(-\rho s)h(x) \\ 0 = v(T, x, \alpha_2). \end{cases}$$

$$(16.25)$$

Next, we modify the turnpike definition given in [75] to incorporate the variation of the turnpike sets with the changes of time.

Definition 16.2. $\phi(s)$ and $\psi(s)$ are said to be the *turnpike sets* for $\alpha = \alpha_1$ and $\alpha = \alpha_2$, respectively, if for all $s \in [0, T]$,

$$v(s, \phi(s), \alpha_1) = \min_x v(s, x, \alpha_1)$$
$$\text{and} \quad v(s, \psi(s), \alpha_2) = \min_x v(s, x, \alpha_2), \text{ respectively.}$$

([104]) Let $\psi(s)$ be defined as follows:

$$\int_s^T e^{-\rho t} h_{x+}(\psi(s) + (\alpha_2 - d)(t - s))dt = 0. \qquad (16.26)$$

Then $\psi(s)$ is continuous, uniquely determined by (16.26) and satisfies
(a) $0 < \psi(s) < |\alpha_2 - d|(T - s)$ for $s \in [0, T)$ and $\psi(T) = 0$;
(b) $\psi(s)$ is monotone decreasing and absolute continuous. Moreover,

$$\dot{\psi}(s) + d > \alpha_2, \text{ a.e.}$$

Remark. The absolute continuity of $\psi(s)$ implies that it is differentiable almost everywhere in s. $\dot{\psi}(s) + d > \alpha_2$ says that if $x(t) \leq \psi(t)$ for some t_1, then $x(t)$ will stay below $\psi(t)$ for all $t_1 \leq t \leq T$.

Let

$$H(s, x) = h(x) + \lambda e^{\rho s} v(s, x, \alpha_2).$$

Then,

$$J(s, x, \alpha_1, u(\cdot)) = \int_s^T e^{-\lambda(t-s)} e^{-\rho t} H(t, x(t)) dt.$$

Note that $H(s, x)$ is convex in x for each s. We are to show that the turnpike set for $\alpha = \alpha_1$ is given by the minimizer of $H(s, x)$, i.e.,

$$H(s, \phi(s)) = \min_x H(s, x). \tag{16.27}$$

We need to mention an important property possessed by $\phi(s)$, which is described in the following definition.

([104]) Let $\phi(s)$ be the minimizer of $H(s, x)$. Then $\phi(s)$ is a single-valued function and satisfies:

(a) $0 \leq \phi(s) < \psi(s)$ for $s \in [0, T)$ and $\phi(T) = 0$;

(b) $\phi(s)$ is traceable, i.e., $\phi(s)$ is absolutely continuous on $[0, T]$ and

$$0 \leq \dot{\phi}(s) + d \leq d, \text{ a.e. in } s \in [0, T].$$

It can be seen easily that a traceable curve is always decreasing. If a function $\gamma(s)$ is traceable, then there exists a control $0 \leq u(s) = \dot{\gamma}(s) + d \leq d < \alpha_1$ such that the corresponding system trajectory $x(t)$ may stay on the curve $\gamma(s)$ after it reaches $\gamma(s)$.

Let

$$\tilde{H}(s, x) = h(x) + \lambda e^{\rho s} \int_s^T e^{-\rho t} h(x + (\alpha_2 - d)(t - s)) dt. \tag{16.28}$$

Note that $\phi(s) < \psi(s)$ and

$$v(s, x, \alpha_2) = \int_s^T e^{-\rho t} h(x + (\alpha_2 - d)(t - s)) dt \text{ if } x \leq \psi(s).$$

It follows that

$$q \min_x H(s, x) = \min_x \tilde{H}(s, x) \tag{16.29}$$

$$= h(\phi(s)) + \lambda e^{\rho s} \int_s^T e^{-\rho t} h(\phi(s) + (\alpha_2 - d)(t - s)) dt.$$

It can be shown that $\phi(s)$ is the only solution to (16.29).

Theorem 16.4 ([104]) *Let $\phi(s)$ and $\psi(s)$ be given as in (16.29) and (16.26), respectively. Then $\phi(s)$ and $\psi(s)$ are the turnpike sets for $\alpha = \alpha_1$ and $\alpha = \alpha_2$, respectively. Moreover, the feedback control $u^*(t) = u^*(t, x^*(t), \alpha(t))$ given below is optimal:*

$$u^*(t, x, \alpha_1) = \begin{cases} 0 & \text{if } x > \phi(t) \\ \dot{\phi}(t) + d & \text{if } x = \phi(t) \\ \alpha_1 & \text{if } x < \phi(t); \end{cases} \qquad (16.30)$$

$$u^*(t, x, \alpha_2) = \begin{cases} 0 & \text{if } x > \psi(t) \\ \alpha_2 & \text{if } x \leq \psi(t). \end{cases}$$

Moreover, it is easy to see that under the control policy $u^*(t) = u^*(t, x(t), \alpha(t))$, the ordinary differential equation

$$\dot{x}^*(t) = u^*(t, x^*(t), \alpha(t)) - d, \ x^*(s) = x$$

has a unique solution.

Using the control given in (16.30), the value function $v(s, x, \alpha_1)$ can be written as follows:

$v(s, x, \alpha_1) =$

$$\begin{cases} \displaystyle\int_s^T e^{-\lambda(t-s)}[e^{-\rho t}h(x - d(t-s)) + \lambda v(t, x - d(t-s), \alpha_2)]dt \\ \qquad\qquad \text{if } x \geq d(T-s) \\[2mm] \displaystyle\int_s^{s_1} e^{-\lambda(t-s)}[e^{-\rho t}h(x - d(t-s)) + \lambda v(t, x - d(t-s), \alpha_2)]dt \\ \quad + \displaystyle\int_{s_1}^T e^{-\lambda(t-s)}[e^{-\rho t}h(\phi(t)) + \lambda v(t, \phi(t), \alpha_2)]dt \\ \qquad\qquad \text{if } \phi(s) < x < d(T-s) \\[2mm] \displaystyle\int_s^T e^{-\lambda(t-s)}[e^{-\rho t}h(\phi(t))\lambda v(t, \phi(t), \alpha_2)]dt \\ \qquad\qquad \text{if } x = \phi(s) \\[2mm] \displaystyle\int_s^{s_2} e^{-\lambda(t-s)}[e^{-\rho t}h(x + (\alpha_1 - d)(t-s)) \\ \quad + \lambda v(t, x + (\alpha_1 - d)(t-s), \alpha_2)]dt \\ \quad + \displaystyle\int_{s_2}^T e^{-\lambda(t-s)}[e^{-\rho t}h(\phi(t)) + \lambda v(t, \phi(t), \alpha_2)]dt \\ \qquad\qquad \text{if } -(\alpha_1 - d)(T-s) < x < \phi(s) \\[2mm] \displaystyle\int_s^T e^{-\lambda(t-s)}[e^{-\rho t}h(x + (\alpha_1 - d)(t-s)) \\ \quad + \lambda v(t, x + (\alpha_1 - d)(t-s), \alpha_2)]dt \\ \qquad\qquad \text{if } x \leq -(\alpha_1 - d)(T-s), \end{cases}$$

where s_1 is the first time that $x - d(t - s)$ hits $\phi(t)$ and s_2 is the first time that $x + (\alpha_1 - d)(t - s)$ hits $\phi(t)$, respectively. Thus,

$$x - d(s_1 - s) = \phi(s_1) \text{ and } x + (\alpha_1 - d)(s_2 - s) = \phi(s_2), \text{ respectively.}$$

We can write the value function $v(s, x, \alpha_2)$ as follows:

$$v(s, x, \alpha_2) = \begin{cases} \int_{s_0}^{T} e^{-\rho t} h(x - d(t - s))dt & \text{if } x \geq d(T - s) \\[2mm] \int_{s}^{s_0} e^{-\rho t} h(x - d(t - s))dt \\[2mm] \quad + \int_{s_0}^{T} e^{-\rho t} h(x - d(s_0 - s) + (\alpha_2 - d)(t - s_0))dt \\[2mm] \qquad\qquad \text{if } \psi(s) < x < d(T - s) \\[2mm] \int_{s}^{T} e^{-\rho t} h(x + (\alpha_2 - d)(t - s))dt & \text{if } x \leq \psi(s) \end{cases}$$

where s_0 is the first time that $x - d(t - s)$ hits $\psi(t)$. Thus, $x - d(s_0 - s) = \psi(s_0)$.

. The cost function in this example is given by

$$h(x) = h^{+} \max\{0, x\} + h^{-} \max\{0, -x\}.$$

Then, (16.26) becomes

$$\int_{s}^{t_1} e^{-\rho t} h^{+} dt - \int_{t_1}^{T} e^{-\rho t} h^{-} dt = 0,$$

where t_1 is given by $\psi(s) + (\alpha_2 - d)(t_1 - s) = 0$. This yields

$$\psi(s) = \frac{\alpha_2 - d}{\rho} \log \frac{h^{+} + h^{-} e^{-\rho(T-s)}}{h^{+} + h^{-}}.$$

We now identify $\phi(s)$. Recall that $0 \leq \phi(s) < \psi(s) \leq |\alpha_2 - d|(T - s)$ and for $x < \psi(s)$,

$$v(s, x, \alpha_2) = \int_{s}^{T} e^{-\rho t} h(x + (\alpha_2 - d)(t - s))dt.$$

Moreover, for $0 \leq x \leq |\alpha_2 - d|(T - s)$,

$$\begin{aligned} v(s, x, \alpha_2) &= \exp(-\rho s) \int_{0}^{\frac{x}{|\alpha_2 - d|}} e^{-\rho t} h^{+}(x + (\alpha_2 - d)t)dt \\ &\quad - \int_{\frac{x}{|\alpha_2 - d|}}^{T-s} e^{-\rho t} h^{-}(x + (\alpha_2 - d)t)dt \end{aligned}$$

$$= \exp(-\rho s)h^+\rho^{-1}[x + (\alpha_2 - d)\rho^{-1}(1 - e^{-\rho x/|\alpha_2 - d|})]$$
$$+ h^-\rho^{-1}[(x + (\alpha_2 - d)(T - s)e^{-\rho(T-s)} +$$
$$|\alpha_2 - d|\rho^{-1}(e^{-\rho x/|\alpha_2 - d|} - e^{-\rho(T-s)})]$$

(16.31)

This together with (16.29) yields

$$\phi(s) = \max\left\{0, -\frac{|\alpha_2 - d|}{\rho}\log\frac{\rho h^+ + \lambda(h^+ + h^- e^{-\rho(T-s)})}{\lambda(h^+ + h^-)}\right\}. \quad (16.32)$$

Equivalently, $\phi(s)$ can also be written as:

$$\phi(s) = 0 \text{ for all } s \in [0, T] \text{ if } \lambda h^- - \rho h^+ \leq 0;$$

otherwise,

$$\phi(s) = \begin{cases} -\dfrac{|\alpha_2 - d|}{\rho}\log\dfrac{\rho h^+ + \lambda(h^+ + h^- e^{-\rho(T-s)})}{\lambda(h^+ + h^-)} \\ \qquad \text{if } s \leq T + \dfrac{1}{\rho}\log\dfrac{\lambda h^- - \rho h^+}{\lambda h^-} \\ 0 \qquad \text{if } s > T + \dfrac{1}{\rho}\log\dfrac{\lambda h^- - \rho h^+}{\lambda h^-}. \end{cases}$$

As $T \to \infty$, it is easily seen that

$$\phi(s) \to \max\left\{0, -\frac{|\alpha_2 - d|}{\rho}\log\frac{(\rho h^+ + \lambda h^+)}{\lambda(h^+ + h^-)}\right\},$$

which gives the same turnpike set as in [1] provided that the repair rate vanishes.

In this example, we are able to solve (16.26) and (16.29) to obtain explicitly the turnpike sets $\phi(s)$ and $\psi(s)$. It should be noted that such explicit turnpike sets are not available for general $h(\cdot)$. However, in many applications of manufacturing systems, $h(\cdot)$ appears to be piecewise linear or a linear combination of linear functions. Then (16.26) and (16.29) are solvable.

2.4   Model 4.

We consider a variation of the problem with constant demand. Assume that the system has a constant machine capacity $\alpha_0 > 0$ but a random demand rate $d(t)$. The system equation is described as follows:

$$\dot{x}(t) = u(t) - d(t), \quad x(s) = x, \quad 0 \leq s \leq t \leq T < \infty \quad (16.33)$$

with the production constraints $0 \le u(t) \le \alpha_0$.

Let $\mathcal{Z} = \{d_1, d_2\}$ denote the set of demand rates with $0 < d_1 < \alpha_0 < d_2$.

The cost function $J(s, x, d, u(\cdot))$ with $d(s) = d \in \mathcal{Z}$ is defined by

$$J(s, x, d, u(\cdot)) = E \int_s^T e^{-\rho t} h(x(t)) dt. \tag{16.34}$$

The objective of the problem is to find a production plan $0 \le u(t) \le \alpha_0$ as a function of the past $d(t)$ to minimize $J(s, x, d, u(\cdot))$.

(A1') Let Assumption (A1) be satisfied with (16.22) replaced by

$$\frac{h_{x+}(x_2) - h_{x+}(x_1)}{x_2 - x_1} \ge c_h \text{ for all } -|\alpha_0 - d_2|T \le x_1 < 0 \le x_2 \le |\alpha_0 - d_2|T,$$

with d_2 given below.

(A2') $d(t) \in \mathcal{Z}$ is also a two state Markov chain governed by

$$L_d f(\cdot)(i) = \begin{cases} \lambda'(f(d_2) - f(d_1)) & \text{if } i = d_1 \\ 0 & \text{if } i = d_2. \end{cases} \tag{16.35}$$

for any function f on $\mathcal{Z}$.

The corresponding value function is given by

$$v(s, x, d) = \inf_{u(\cdot) \in \mathcal{A}} J(s, x, d, u(\cdot)), \text{ for } d \in \mathcal{Z}.$$

We can show as in [104] that the value functions $v(s, x, d)$ are convex functions in x for each $s \in [0, T]$ and $d \in \mathcal{Z}$. Moreover, $v(s, x, d) \in C([0, T], \mathbb{R})$ are the only viscosity solutions the following dynamic programming equations.

$$\begin{cases} 0 = -v_s(s, x, d_1) + \sup_{0 \le u \le \alpha_0} [-(u - d_1) v_x(s, x, d_1)] \\ \quad - \exp(-\rho s) h(x) - \lambda'(v(s, x, d_2) - v(s, x, d_1)), \\ 0 = v(T, x, d_1) \end{cases} \tag{16.36}$$

and

$$\begin{cases} 0 = -v_s(s, x, d_2) + \sup_{0 \le u \le \alpha_0} [-(u - d_2) v_x(s, x, d_2)] - \exp(-\rho s) h(x) \\ 0 = v(T, x, d_2). \end{cases}$$
$$\tag{16.37}$$

Definition 16.3. $\phi(s)$ and $\psi(s)$ are said to be *turnpike sets* if

$$v(s, \phi(s), d_1) = \min_x v(s, x, d_1)$$
$$\text{and} \quad v(s, \psi(s), d_2) = \min_x v(s, x, d_2).$$

Let

$$\hat{H}(s, x) = h(x) + \lambda \int_s^T e^{-\rho t} h(x + (\alpha_0 - d_2)(t - s)) dt.$$

Then $\psi(s)$ and $\phi(s)$ are determined by:

$$\int_s^T e^{-\rho t} h_{x+}(\psi(s) + (\alpha_0 - d_2)(t - s)) dt = 0 \qquad (16.38)$$

and

$$\hat{H}(s, \phi(s)) = \min_x \hat{H}(s, x). \qquad (16.39)$$

([104]) $\phi(s)$ and $\psi(s)$ are single-valued absolute continuous functions. They satisfy the following properties:
  (a) $0 < \psi(s) < |\alpha_0 - d_2|(T - s)$ for $s \in [0, T)$ and $\psi(T) = 0$;
  (b) $\psi(s)$ is monotone decreasing and $\psi(s) + d_2 > \alpha_0$, a.e.;
  (c) $0 \leq \phi(s) < \psi(s)$ for $s \in [0, T)$ and $\phi(T) = 0$;
  (d) $\phi(s)$ is monotone decreasing and $\phi(s) + d_1 \geq \alpha_0 - d_2 + d_1$, a.e.
  Notice that by (4) of the above lemma, a sufficient condition for $\phi(s)$ to be traceable (i.e., $0 \leq \dot{\phi}(s) + d_1 \leq \alpha_0$) is $d_1 \geq d_2 - \alpha_0$.

Theorem 16.5 ([104]) *Suppose that* (A1'), (A2') *are satisfied, and* $d_1 \geq d_2 - \alpha_0$. *Let* $u^*(t, x(t), d)$ *be defined as follows:*

$$u^*(t, x, d_1) = \begin{cases} 0 & \text{if } x > \phi(t) \\ \dot{\phi}(t) + d_1 & \text{if } x = \phi(t) \\ \alpha_0 & \text{if } x < \phi(t), \end{cases} \qquad (16.40)$$

$$u^*(t, x, d_2) = \begin{cases} 0 & \text{if } x > \psi(t) \\ \alpha_0 & \text{if } x \leq \psi(t). \end{cases}$$

Then under the control $u^*(t) = u^*(t, x(t), d(t))$, *the equation*

$$\dot{x}^*(t) = u^*(t, x^*(t), d(t)) - d(t), \quad x_0^* = x$$

has a unique solution. Therefore, the control $u^*(t)$ *is optimal.*

. We consider the cost function

$$h(x) = h^+ \max\{0, x\} + h^- \max\{0, -x\}.$$

Then,

$$\psi(s) = \frac{\alpha_0 - d_2}{\rho} \log \frac{h^+ + h^- e^{-\rho(T-s)}}{h^+ + h^-}.$$

$$\phi(s) = 0 \text{ if } \lambda h^- - \rho h^+ \le 0.$$

If $\lambda h^- - \rho h^+ > 0$, then

$$\phi(s) = \begin{cases} -\dfrac{|\alpha_0 - d_2|}{\rho} \log \dfrac{\rho h^+ + \lambda(h^+ + h^- e^{-\rho(T-s)})}{\lambda(h^+ + h^-)} \\ \qquad \text{if } s \le T + \dfrac{1}{\rho} \log \dfrac{\lambda h^- - \rho h^+}{\lambda h^-} \\ 0 \qquad \text{if } s > T + \dfrac{1}{\rho} \log \dfrac{\lambda h^- - \rho h^+}{\lambda h^-}. \end{cases}$$

The assumption $d_1 \ge d_2 - \alpha_0$ in the above theorem is a relatively conservative one. In the previous example, this condition can be relaxed to

$$d_1 \ge |\alpha_0 - d_2| \frac{\lambda h^- - \rho h^+}{\lambda(h^+ + h^-)}. \tag{16.41}$$

Actually, (16.41) is also necessary for the traceability of $\phi(s)$ in this example. If (16.41) fails. $\phi(s)$ will no longer be the turnpike of the problem since it is not traceable on $[0, T]$. Let $\tilde{\phi}(s)$ denote the turnpike set for $\alpha = \alpha_1$ defined as $v(s, \tilde{\phi}(s), d_1) = \min_x v(s, x, d_1)$. Note that $0 \le \tilde{\phi}(s) \le \psi(s)$. It can be shown that $h(x) + \lambda v(s, x, d_2)$ is strictly convex on $[0, \psi(s)]$, which implies that $v(s, x, d_1)$ is also strictly convex on $[0, \psi(s)]$ (see [75] for details). Therefore, $\tilde{\phi}(s)$ is a continuous function. Intuitively, the optimal control $u^*(t, x, \alpha_1)$ should be given as in (16.40) with $\tilde{\phi}(s)$ in place of $\phi(s)$ provided that $\tilde{\phi}(s)$ is traceable. Let

$$s_0 = T + \frac{1}{\rho} \log \frac{\lambda h^- - \rho h^+}{\lambda h^-}.$$

Then, $0 \le \dot{\tilde{\phi}}(s) + d_1 \le \alpha_0$ for $s \ge s_0$. This implies $\tilde{\phi}(s) = \phi(s)$ for $s \ge s_0$. However, if $0 > s_0 > T$ and d_2 is large enough, the traceability property of $\tilde{\phi}(s)$ will not hold, which makes the problem very complicated; explicit optimal solution is very difficult to obtain.

Remark 16.7 *This section presents a certain number of models that determines the production planning when the machine is subject to random breakdowns or when the demand is random. In all the cases we developed here, we were able to get the closed-form solution. For the multi-machine multi-part manufacturing system, the extension is direct, but it is in general impossible to get a closed form solution for the production planning problem. A suboptimal solution for such an optimization problem can be obtained using numerical methods (see Boukas [11] or [75]).*

3.   PRODUCTION AND CORRECTIVE MAINTENANCE CONTROL

For the case of a one machine one part manufacturing system, the previous model can be modified to handle the corrective maintenance. Our way of doing corrective maintenance is different from the ones developed in the literature. Our technique consists of controlling the jump rate from the failure state to the operational state. The idea consists in some sense of choosing the company that can assure the shortest mean time to repair, that is the inverse of the jump rate from the failure state to the operational one. Let $u_r(t)$ denote this jump rate. In this case, the probability of the transition rates matrix is given by:

$$Q(u_r(t)) = \left[\begin{array}{cc} q_{11} & q_{12} \\ u_r(t) & -u_r(t) \end{array} \right] \tag{16.42}$$

The dynamics of the stock level remains the same as in the previous section.

Remark 16.8 *The control variable $u_r(t)$ has to satisfy the following conditions:*

$$0 < \underline{u}_r \leq u_r(t) \leq \bar{u}_r \tag{16.43}$$

where $\underline{u}_r$ and $\bar{u}_r$ are known positive constants. The choice of $\underline{u}_r > 0$ is due to the fact that we don't want the machine in failure mode to remain for ever in this state. The lower constraint and the upper constraint will give respectively the longest and the shortest mean times to repair.

The goal of the production and corrective maintenance planning is twofold:

- produce parts when the machine is up in order to satisfy the given demand rate and also to reduce the inventory level;

- bring the machine to the operational state when it is in failure mode in the shortest time when it is needed.

This goal can be reached by seeking a control law that minimizes the following cost function:

$$J_u(x_0, \xi_0) \tag{16.44}$$
$$= \mathbb{E}\left[\int_0^\infty e^{-\rho t} \left[c^+ x^+(t) + c^- x^-(t) + c(\xi(t)) \right] dt \mid x(0) = x_0, \xi(0) = \xi_0 \right]$$

where $\rho > 0$ is the discount rate and $c(\xi(t))$ is equal to c_p when $\xi(t) = 1$ (c_p a positive given real number) or equal to $c_r u_r(t)$ when $\xi(t) = 2$ (c_r is a given positive given real number). The rest of the variables are similar to the ones defined previously.

The optimal control law of our optimization problem has two components. The first one represents the production rate u_p^* and the second one represents the corrective maintenance rate u_r^*, i.e $u^*(t) = (u_p^*(t), u_r^*(t))$. Based on the structure of the optimization problem and its Hamilton Jacobi Bellman equations, one can show that the optimal control law is given by the following theorem:

Theorem 16.6 *([11]) The optimal policy for the production rate u_p^* and the corrective maintenance rate u_r^* are given respectively by:*

$$u_p^*(t) = \begin{cases} \bar{u}_p & \text{if } x < x^* \\ d & \text{if } x = x^* \\ 0 & \text{if } x > x^* \end{cases} \qquad (16.45)$$

$$u_r^*(t) = \begin{cases} \bar{u}_r & \text{if } \left[c_r + v(x,1) - v(x,2) \right] \leq 0 \\ \underline{u}_r & \text{otherwise} \end{cases} \qquad (16.46)$$

where x^* is the hedging point defined by Eq. (16.9).

Remark 16.9 *A certain number of properties for the value function, $v(x, \alpha)$, can be shown. See Boukas [11].*

3.1 Closed Form Solution of the HJB Equation.

Let (u_p, u_r) be the optimal control law. Then the dynamic programming equations become:

$$\rho v(x,1) = c^+ x^+ + c^- x^- + c_p + \dot{v}(x,1)(u_p - d) + \lambda \left[v(x,2) - v(x,1) \right] \qquad (16.47)$$

$$\rho v(x,2) = c^+ x^+ + c^- x^- + \dot{v}(x,2)(-d) + u_r \left[c_r + v(x,1) - v(x,2) \right] \qquad (16.48)$$

where $\lambda = q_{12}$.

Let $V(x)$ be defined by:

$$V(x) = \begin{bmatrix} v(x,1) \\ v(x,2) \end{bmatrix} \qquad (16.49)$$

For a given control, the dynamic programming equations can be rewritten as:

$$\dot{V}(x) = \begin{bmatrix} \frac{\rho+\lambda}{u_p-d} & -\frac{\lambda}{u_p-d} \\ \frac{u_r}{d} & -\frac{\rho+u_r}{d} \end{bmatrix} V(x) + \begin{bmatrix} -\frac{1}{u_p-d} \\ \frac{1}{d} \end{bmatrix} \left(c^+ x^+ + c^- x^- \right)$$

$$+ \begin{bmatrix} -\frac{c_p}{u_p-d} \\ \frac{c_r u_r}{d} \end{bmatrix} \qquad (16.50)$$

which can be in turn rewritten in the following generic form:

$$\dot{V}(x) = A(u_p, u_r)V(x) + B(u_p, u_r)U(x) + C(u_p, u_r) \qquad (16.51)$$

where the matrices $A(u_p, u_r)$, $B(u_p, u_r)$ and $C(u_p, u_r)$ are dependent on the control law.

Let $F(u_p, u_r)$ be a constant matrix when the control law is fixed and defined by:

$$C(u_p, u_r) = A(u_p, u_r)F(u_p, u_r) \qquad (16.52)$$

Remark 16.10 *Since the matrix $A(u_p, u_r)$ is not singular, the matrix $F(u_p, u_r)$ exists and it is unique for a fixed control law. Its expression is given by:*

$$F(u_p, u_r) = \left[\begin{array}{c} -\dfrac{\lambda c_r u_r + c_p(\rho + u_r)}{\rho(\rho + \lambda + u_r)} \\[2ex] -\dfrac{\rho + \lambda}{\lambda}\left[\dfrac{\lambda c_r u_r + c_p(\rho + u_r)}{\rho(\rho + \lambda + u_r)} \right] + \dfrac{c_p}{\lambda} \end{array} \right] \qquad (16.53)$$

By using the following change of variable:

$$Z = V + F(u_p, u_r) \qquad (16.54)$$

we get:

$$\dot{Z} = A(u_p, u_r)Z + B(u_p, u_r)U, \quad z_0 = V(x_0) + F(u_p, u_r). \qquad (16.55)$$

which has the following closed form solution:

$$Z(x) = \Phi_u(x)z_0 + \int_{z_0}^{x} \Phi_u(x - y)B(u_p, u_r)U(y)dy \qquad (16.56)$$

where $\Phi_u(x)$ is the transition matrix with the following expression:

$$\Phi_u(x) = e^{A(u_p, u_r)x} \qquad (16.57)$$

To simplfy the expression of the HJB equation in each region, we will use the following:

$$C_r(x, c_r) = c_r + v(x, 1) - v(x, 2) \qquad (16.58)$$

The HJB equation can be rewritten as follows in the different regions of the stock level:

- Region R_0: $(x > x^\star \geq 0)$

$$\dot{V}(x) = \begin{cases} \begin{bmatrix} -\frac{\rho+\lambda}{d} & \frac{\lambda}{d} \\ \frac{u_r}{d} & -\frac{\rho+u_r}{d} \end{bmatrix} V(x) + \begin{bmatrix} \frac{c^+}{d} \\ \frac{c^+}{d} \end{bmatrix} x + \begin{bmatrix} \frac{c_p}{d} \\ \frac{c_r u_r}{d} \end{bmatrix} \\ \qquad\qquad \text{if } C_r(x,c_r) > 0 \\[2em] \begin{bmatrix} -\frac{\rho+\lambda}{d} & \frac{\lambda}{d} \\ \frac{\bar{u}_r}{d} & -\frac{\rho+\bar{u}_r}{d} \end{bmatrix} V(x) + \begin{bmatrix} \frac{c^+}{d} \\ \frac{c^+}{d} \end{bmatrix} x + \begin{bmatrix} \frac{c_p}{d} \\ \frac{c_r \bar{u}_r}{d} \end{bmatrix} \\ \qquad\qquad \text{otherwise} \end{cases}$$

$$(16.59)$$

- Hedging point line: $(x = x^\star)$

$$v(x,1) = \frac{\lambda}{\lambda+\rho} v(x,0) + \frac{c^+}{\lambda+\rho} x + \frac{c_p}{\lambda+\rho} \qquad (16.60)$$

$$\dot{v}(x,2) = \begin{cases} -\frac{\rho}{d}\left[1 + \frac{u_r}{\rho+\lambda} \right] v(x,0) + \frac{c^+}{d}\left[1 + \frac{u_r}{\rho+\lambda} \right] x \\ +\frac{u_r}{d}\left[c_r + \frac{c_p}{\rho+\lambda} \right] \qquad \text{if } C_r(x,c_r) > 0 \\[1.5em] -\frac{\rho}{d}\left[1 + \frac{\bar{u}_r}{\rho+\lambda} \right] v(x,0) + \frac{c^+}{d}\left[1 + \frac{\bar{u}_r}{\rho+\lambda} \right] x \\ +\frac{\bar{u}_r}{d}\left[c_r + \frac{c_p}{\rho+\lambda} \right] \qquad \text{otherwise} \end{cases}$$

$$(16.61)$$

- Region R_1: $(0 \leq x \leq x^\star)$

$$\dot{V}(x) = \begin{cases} \begin{bmatrix} \frac{\rho+\lambda}{\bar{u}_p-d} & -\frac{\lambda}{\bar{u}_p-d} \\ \frac{u_r}{d} & -\frac{\rho+u_r}{d} \end{bmatrix} V(x) + \begin{bmatrix} -\frac{c^+}{\bar{u}_p-d} \\ \frac{c^+}{d} \end{bmatrix} x + \begin{bmatrix} -\frac{c_p}{\bar{u}_p-d} \\ \frac{c_r u_r}{d} \end{bmatrix} \\ \qquad\qquad \text{if } C_r(x,c_r) > 0 \\[2em] \begin{bmatrix} \frac{\rho+\lambda}{\bar{u}_p-d} & -\frac{\lambda}{\bar{u}_p-d} \\ \frac{\bar{u}_r}{d} & -\frac{\rho+\bar{u}_r}{d} \end{bmatrix} V(x) + \begin{bmatrix} -\frac{c^+}{\bar{u}_p-d} \\ \frac{c^+}{d} \end{bmatrix} x + \begin{bmatrix} -\frac{c_p}{\bar{u}_p-d} \\ \frac{c_r \bar{u}_r}{d} \end{bmatrix} \\ \qquad\qquad \text{otherwise} \end{cases}$$

$$(16.62)$$

- Region R_2: $(x \leq 0)$

$$\dot{V}(x) = \begin{cases} \begin{bmatrix} \frac{\rho+\lambda}{\bar{u}_p-d} & -\frac{\lambda}{\bar{u}_p-d} \\ \frac{u_r}{d} & -\frac{\rho+u_r}{d} \end{bmatrix} V(x) + \begin{bmatrix} -\frac{c^-}{\bar{u}_p-d} \\ \frac{c^-}{d} \end{bmatrix} x + \begin{bmatrix} -\frac{c_p}{\bar{u}_p-d} \\ \frac{c_r u_r}{d} \end{bmatrix} \\ \qquad\qquad \text{if } C_r(x,c_r) > 0 \\[2em] \begin{bmatrix} \frac{\rho+\lambda}{\bar{u}_p-d} & -\frac{\lambda}{\bar{u}_p-d} \\ \frac{\bar{u}_r}{d} & -\frac{\rho+\bar{u}_r}{d} \end{bmatrix} V(x) + \begin{bmatrix} -\frac{c^-}{\bar{u}_p-d} \\ \frac{c^-}{d} \end{bmatrix} x + \begin{bmatrix} -\frac{c_p}{\bar{u}_p-d} \\ \frac{c_r \bar{u}_r}{d} \end{bmatrix} \\ \qquad\qquad \text{otherwise} \end{cases}$$

$$(16.63)$$

By defining $x^\star$ as follows:

$$x^\star = max\left\{0, \frac{1}{\underline{a}} \log\left[\frac{c^+}{c^+ + c^-}\left[1 + \frac{\rho d}{\lambda d - (\rho + u_r + d\underline{a})(\bar{u}_p - d)}\right]\right]\right\}$$
(16.64)

we can show that the solutions of the above cases are as follows:

$$V(x) = \begin{cases} e^{A(0,u_r)(x-x^\star)}\left[\,A^{-2}(0,u_r)b_1(0) - A^{-1}(0,u_r)b_1(0)\lambda^{-1}\,\right] \\ \quad -A^{-1}(0,u_r)b_1(0)x - A^{-2}(0,u_r)b_1(0) \qquad \text{if } x \geq x^\star \\[2mm] e^{A(\bar{u}_p,u_r)(x-x^\star)}\left[\,A^{-2}(\bar{u}_p,u_r)b_1(0) - A^{-1}(\bar{u}_p,u_r)b_1(0)\lambda^{-1}\,\right] \\ \quad -A^{-1}(\bar{u}_p,u_r)b_1(0)x - A^{-2}(\bar{u}_p,u_r)b_1(0) \qquad \text{if } 0 \leq x \leq x^\star \\[2mm] e^{A(\bar{u}_p,u_r)x}\left\{e^{-A(\bar{u}_p,u_r)x^\star}[A^{-2}(\bar{u}_p,u_r)b_2(0)\right. \\ \quad \left. -A^{-1}(\bar{u}_p,u_r)b_2(0)\lambda^{-1}] - A^{-2}(\bar{u}_p,u_r)b_2(0)(c^+ + c^-)\right\} \\ \quad +A^{-1}(\bar{u}_p,u_r)b_2(0)x + A^{-2}(\bar{u}_p,u_r)b_2(0)c^- \qquad \text{if } x \leq 0 \end{cases}$$
(16.65)

where the matrices $A(u_p, u_r)$, $b_1(u_p)$, $b_2(u_p)$ and $c(u_p, u_r)$ are defined as follows:

$$A(\bar{u}_p, u_r) = \begin{bmatrix} \frac{\rho+\lambda}{\bar{u}_p - d} & \frac{-\lambda}{\bar{u}_p - d} \\ \frac{\mu}{d} & -\frac{\mu+\rho}{d} \end{bmatrix}$$
(16.66)

$$b_1(u_p) = \begin{bmatrix} -\frac{c^+}{u_p - d} \\ \frac{c^+}{d} \end{bmatrix}$$
(16.67)

$$b_2(u_p) = \begin{bmatrix} -\frac{c^-}{u_p - d} \\ \frac{c^-}{d} \end{bmatrix}$$
(16.68)

$$c(u_p, u_r) = \begin{bmatrix} -\frac{c_p}{u_p - d} \\ \frac{c_r u_r}{d} \end{bmatrix}$$
(16.69)

The goal of the rest of this section is to provide a comparison between the policy obtained by Akella and Kumar and the one we developed in this paper. We show that the fact of using the corrective maintenance will assure a lower hedging level.

3.2   Analytic Results.

Theorem 16.7 *([11]) The negative eigenvalue, $\underline{a}_2$, corresponding to $\bar{u}_r$ is less or equal to the one, $\underline{a}_1$ corresponding to $\underline{u}_r$.*

Let us now return to the expression of the hedging level as given in the expression (16.9). The important case is the one with

$$\frac{\rho d}{\lambda d - (\rho + \mu + d\underline{a})\,(\bar{u}_p - d)} \leq 0 \qquad (16.70)$$

where $\underline{a}$ is the negative eigenvalue of the matrix $A(u_p, u_r)$.

This requires the following relation:

$$\lambda d - (\rho + \mu + d\underline{a})\,(\bar{u}_p - d) \leq 0 \qquad \forall \mu \in [\underline{u}_r, \bar{u}_r] \qquad (16.71)$$

Theorem 16.8 *([11]) The hedging level, $x^{\star}_{\bar{u}_r}$, corresponding to $\bar{u}_r$ is less than or equal to the one, $x^{\star}_{\underline{u}_r}$ corresponding to $\underline{u}_r$ if $(\bar{u}_r - \underline{u}_r) > d(\underline{a}_1 - \underline{a}_2)$.*

Remark 16.11 *In Akella and Kumar since we do not control the jump rate u_r which is all the time equal to $\underline{u}_r$, the result is that their hedging point level is greater than the one obtained in this paper.*

Remark 16.12 *This model can be generalized to the case of multi-machine multi-part. The same results will remain valid but unfortunately the closed-form solution can not be obtained and the only way to approximate the optimal solution is the numerical methods .*

4.   PRODUCTION AND PREVENTIVE MAINTENANCE CONTROL

The model of section two can be modified to include preventive maintenance. The action of preventive maintenance we are considering here is different from the ones already reported in the literature. Our idea consists of what we did in the previous section to control the jump rate from the operational state to the preventive state. In this section, we will model the state of the machine by a continuous-time controlled Markov process with three modes. The state space of the process $\{\xi(t), t \geq 0\}$ will be $\mathcal{S} = \{1, 2, 3\}$. $\xi(t) = 1$ and $\xi(t) = 2$ will keep the same meanings as in the previous section and $\xi(t) = 3$ means that the machine is under preventive maintenance. In this mode, the machine can't produce parts and the operators will change some critical parts of the machine to prevent breakdowns. The sojourn time of the machine in this mode is shorter than the one in mode 2, and therefore, by doing preventive maintenance when it is required and permitted, we increase the reliability of the machine and consequently a greater demand rate can be satisfied.

To make our model more realistic, we will consider that we have complete access to the age of the machine which will be measured from the last intervention (a repair, or a preventive maintenance action). The aging of the machine is then considered as a function of time and the use of the machine. This can be described by the following differential equation:

$$\dot{a}(t) = f(u_p), a(0) = a_0 \qquad (16.72)$$

where $f(.)$ is a known function.

Remark 16.13 *For instance, we can consider that function $f(.)$ is defined by the following expression:*

$$f(u_p(t)) = c_0 + c_1 u_p(t) \qquad (16.73)$$

where c_0 and c_1 are positive constant real numbers.

When $u_p(t)$ is equal to zero for all t, this expression means that the age will increase with time and when $u_p(t)$ is not equal to zero, which means that the machine is used to produce parts, it will give the age of the machine at time t. The more we use the machine in operational mode, the higher the age will be.

The complete dynamics of the production system is then given by the following differential system of equations:

$$\dot{x}(t) = u_p(t) - d, x(0) = x_0 \qquad (16.74)$$

$$\dot{a}(t) = f(u_p), a(0) = a_0 \qquad (16.75)$$

The transition matrix between the different modes is given by:

$$Q(u_m(t)) = \begin{bmatrix} q_{11} & q_{12}(a(t)) & u_m(t) \\ q_{21} & q_{22} & 0 \\ q_{31} & 0 & q_{33} \end{bmatrix} \qquad (16.76)$$

where $u_m(t)$ is the jump rate from operational mode to preventive mode and represents a control varaiable.

The goal of this section is to produce parts in order to satisfy the given demand rate, d, and keep the stock level close to zero. A sequence of preventive maintenance actions are also to be planned. This goal can be reached by seeking a control law that minimizes the following cost function:

$$J_{u_p, u_m}(x_0, \xi_0) = \mathbb{E}\left[\int_0^\infty e^{-\rho t} \left[c^+ x^+(t) + c^- x^-(t) + c(\xi(t))\right] dt \big| x(0) = x_0,\right.$$

$$\left. \xi(0) = \xi_0\right] \tag{16.77}$$

where $\rho > 0$ is the discount rate and $c(\xi(t))$ is equal to c_p when $\xi(t) = 1$ (c_p a positive given real number (production cost)) or equal to c_r when $\xi(t) = 2$ (c_r is a given positive given real number (repair cost)) or finally equal to c_m when $\xi(t) = 3$ (c_m is a given positive real number (preventive maintenance cost)). The rest of the variables are similar to the ones defined previously.

Let $H(.)$ be a real-valued function defined by:

$$H(x, a, \alpha, v(x, a, .), r_1, r_2, u_p, u_m) = (u_p - d)r_1 + f(u_p, \alpha)r_2$$
$$+ c^+ x^+ + c^- x^- + c(\alpha) + \sum_{\beta \in \mathcal{S}} q_{\beta\alpha}\left(v(x, a, \beta) - v(x, a, \alpha)\right)$$

Based on the previous results, the HJB equation is then given by the following:

$$\rho v(x, a, \alpha) \tag{16.78}$$
$$= \min_{u_p, u_m} H(x, a, \alpha, v(x, a, .), v_x(x, a, \alpha), v_a(x, a, \alpha), u_p, u_m), \forall \alpha \in \mathcal{S}$$

The following theorems give some interesting results for this stochastic optimization control problem.

Theorem 16.9 *([20]) The value function $v(x(t), \xi(t))$ satisfies the following:*

i. $\leq v(x, a, \alpha) \leq K\left(1 + |x|^k + |a|^k\right)$;

ii. $|v(x, a, \alpha) - v(x', a', \alpha)| \leq K\left(1|x|^k + |x'|^k + |a|^k + |a'|^k\right)(|x - x'| + |a - a'|)$

iii. $v(x, a, \alpha)$ *is the only viscosity solution to the Hamilton-Jacobi-Isaac (HJI) (16.78).*

Theorem 16.10 *([20, 11]) Let $w(x, a, \alpha)$ denote a differentiable solution of the HJI (16.78) such that:*

$$0 \leq w(x, a, \alpha) \leq K\left(1 + |x|^k + |a|^k\right)$$

Then:

i. $w(x, a, \alpha) \leq \hat{J}(x, a, \alpha, u_p(.), u_m(.))\forall(u_p(.), u_m(.)), \in \mathcal{A}$

ii. If $(u_p^\star(x, a, \alpha), u_m^\star(x, a, \alpha))$ is an admissible feedback control such that:

$$\begin{aligned}
\dot{x}^\star(t) &= u_p^\star(t) - d, x^\star(0) = x_0 \\
\dot{a}^\star(t) &= f(u_p^\star), a^\star(0) = a_0
\end{aligned}$$

and

$$\min_{u_p, u_m} H(x, a, \alpha, v(x, a, .), v_x(x, a, \alpha), v_a(x, a, \alpha), u_p, u_m) =$$

$$\min_{u_p, u_m} H(x, a, \alpha, v(x, a, .), v_x(x, a, \alpha), v_a(x, a, \alpha), u_p^\star, u_m^\star)$$

Thus $(u_p^\star(.), u_m^\star(.))$ is optimal, i.e. $\hat{J}(x, a, \alpha, u_p^\star, u_m^\star) = w(x, a, \alpha) = v(x, a, \alpha)$

Obtaining the closed form solution for the production and maintenance planning problem even in the case of one machine one part is impossible. Therefore, the only way to solve this problem is the use of computation techniques (see Boukas [11] or Kushner and Dupuis [60]).

Based on our experience with the numerical techniques, the optimal solution for this simple problem has the following characteristics:

- the production rate is still a bang-bang control, but in this case the hedging level, that fixes the switching, is not constant but depends on the age;

- the preventive maintenance rate is also a bang-bang control that depends on the age and the stock levels. The following rules can be used for practical use:

  - no preventive maintenance action is taken when the stock level is negative for all the values of the age.

  - a preventive maintenance action is taken when stock level is positive and the machine is old.

5.   NUMERICAL METHODS AND EXAMPLES

In order to solve the optimal control problem and implement the control policies in practice, it is necessary to resort to numerical methods because an analytic solution is difficult to obtain in general. In this section, we discuss related computational methods. We begin with a finite difference method for solving dynamic programming equations. Then we will discuss a hierarchical approach and stochastic approximation for dealing with large scale systems.

5.1   Finite Difference Approach..
Let us consider the control problem in which $x(t)$ is the state, $u(t)$ is the control, and $\alpha(t)$ is a finite-state Markov chain generated by matrix Q. The system equation is given by

$$\dot{x}(t) = f(x(t), u(t), \alpha(t)), x(0) = x,$$

where f is a function that takes the form $u - d$ as in Model 1. The objective is to choose an admissible control $u(t) \in \Gamma$, a compact subset of $\mathbb{R}^n$ to minimize a discounted cost function

$$J(x, \alpha, u(\cdot)) = E \int_0^\infty e^{-\rho t} G(x(t), u(t), \alpha(t)) dt.$$

For simplicity, we only consider the 1-dimensional case. Let $v(x, \alpha) = \inf_{u(\cdot)} J(x, u(\cdot), \alpha)$ denote the value function. Then, the associated HJB equation is

$$\rho v(x, \alpha) = \min_{u \in \Gamma} \left\{ f(x, u, \alpha) v_x(x, \alpha) + G(x, u, \alpha) \right\} + Qv(x, \cdot)(\alpha).$$
$$(16.79)$$

In view of the verification theorem (such as in Theorem 2.2), in order to find an optimal control for the problem, the dynamic programming approach requires a solution to the associated HJB equation. However, a closed-form solution of the corresponding HJB equation is difficult to obtain. It is necessary to develop numerical algorithms to resolve the problem. Here, we use Kushner's finite difference method for stochastic controls. The idea is to use an approximation method for the partial derivatives of the value function $v(x, \alpha)$ within a finite grid of the state vector x and a finite grid for the control vector, which transforms the original optimization problem to an auxiliary discounted Markov decision process. Such transformation allows us to apply the well-known techniques, such as a successive approximation or policy improvement, to solve the HJB equations and then the original optimization problems.

Let $\Delta x > 0$ denote the length of the finite difference interval of the variables x. Using this finite difference interval, approximate the value function $v(x, \alpha)$ by a sequence of functions $v^\Delta(x, \alpha)$ and the partial derivatives $v_x(x, \alpha)$ by

$$\begin{cases} \dfrac{1}{\Delta x}(v^\Delta(x + \Delta x, \alpha) - v^\Delta(x, \alpha)), & \text{if } f(x, u, \alpha) \geq 0, \\[2ex] \dfrac{1}{\Delta x}(v^\Delta(x, \alpha) - v^\Delta(x - \Delta x, \alpha)), & \text{if } f(x, u, \alpha) < 0. \end{cases}$$

This leads to

$$f(x, u, \alpha)v_x(x, \alpha) \doteq \frac{|f(x, u, \alpha)|}{\Delta x}v^\Delta(x + \Delta x, \alpha)I_{\{f(x,u,\alpha) \geq 0\}}$$
$$+ \frac{|f(x, u, \alpha)|}{\Delta x}v^\Delta(x - \Delta x, \alpha)I_{\{f(x,u,\alpha) < 0\}}$$
$$- \frac{|f(x, u, \alpha)|}{\Delta x}v^\Delta(x, \alpha).$$

Using these approximations, we can "rewrite" the HJB equation (16.79) in terms of $v^\Delta(x, \alpha)$ as

$$v^\Delta(x, \alpha) = \min_{u \in \Gamma}\left(\rho + |q_{\alpha\alpha}| + \frac{|f(x, u, \alpha)|}{\Delta x}\right)^{-1}\left\{\frac{|f(x, u, \alpha)|}{\Delta x}\right.$$
$$\left(v^\Delta(x + \Delta x, \alpha)I_{\{f(x,u,\alpha) \geq 0\}} + v^\Delta(x - \Delta x, \alpha)I_{\{f(x,u,\alpha) < 0\}}\right)$$
$$\left. + G(x, u, \alpha) + \sum_{\beta \neq \alpha}q_{\alpha\beta}v^\Delta(x, \beta)\right\}.$$

$$(16.80)$$

The next theorem shows that $v^\Delta(x, \alpha)$ converges to $v(x, \alpha)$ as the step size Δx goes to zero.

Theorem 16.11. *Suppose that $v^\Delta(x, \alpha)$ is a solution to (16.80) and*

$$0 \leq v^\Delta(x, \alpha) \leq K(1 + |x|^k),$$

for some constants $K > 0$ and $k > 0$. Then

$$\lim_{\Delta x \to 0}v^\Delta(x, \alpha) = v(x, \alpha). \qquad (16.81)$$

5.2   Approximation Methods..   The finite difference work when the dimension of the underlying system is not very large. For large systems, the computational burden could be huge and make it impossible to obtain any meaningful solution. In this subsection, we briefly discuss other alternatives for large systems.

An Hierarchical Approach.

An important method in dealing with the optimization of large, complex systems is that of the hierarchical control approach. The basic idea is to reduce the overall complex problem into manageable approximate problems or subproblems, to solve these problems, and to construct a solution of the original problem from the solutions of these simpler problems.

To illustrate the idea, we consider Model 1 in which the jump rates of $\alpha(.)$ are of order $1/\varepsilon$. Then as ε gets smaller and smaller, $\alpha(t)$ jumps more and more rapidly. This gives rise to a limiting problem, which is simpler to solve than the given problem. This limiting problem is obtained by replacing the stochastic machine availability process by the average total capacity of machines and by appropriately modifying the objective function. From its optimal control, one constructs an asymptotically optimal control of the original, more complex, problem. We refer to the book by Sethi and Zhang [75] for a comprehensive treatment of hierarchical controls of manufacturing systems.

A Stochastic Approximation Approach.

As seen in Model 1, the optimal control policy is of the hedging type, i.e., the corresponding feedback control is characterized by a few threshold constants. Such control policies are natural for many manufacturing systems. One may use a stochastic approximation approach by focusing on the class of hedging policies and treat the corresponding cost as a function of the hedging points. As a result, the optimal control problem is converted to an optimization problem, such that the main task reduces to finding the optimal or nearly optimal hedging points. We refer to the book by Kushner and Yin [61] for more details in this connection.

Note that the stochastic approximation algorithms can be used for the limited problem arising in the hierarchical control framework as well as for other systems not having the hierarchical structure.

A Numerical Example

To show the usefulness of our results and prove their importance, let us consider a numerical example that consists of one machine producing one part as considered by Akella and Kumar. Let the data used in simulation be given by the table (16.1).

c^+	c^-	c_r	λ	μ	ρ	d	$\bar{u}_p$	$\bar{u}_r$	$\underline{u}_r$
1	5	10	0.01	0.015	0.06	0.20	0.27	0.15	0.25

Table 16.1 Simulation data

A Matlab program has been developed to find the solution of the DP equation of this problem. The results are plotted in Figures 16.1, 16.2, 16.3 and 16.4. Figures 16.1 and 16.2 represent respectively the evolution of the value function in function of the stock level in mode 1 and mode 2. Figures 16.3 and 16.4 represent respectively the evolution of the production rate and the corrective maintenance rate in function of the stock level. The resolution of the DP equations gives the same results as the ones given by previous theorems and confirms effectively that the hedging point obtained here is less than the one obtained without corrective maintenance .

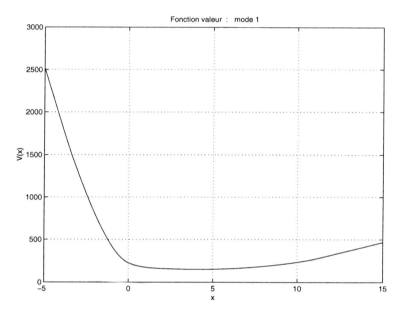

Figure 16.1 Evolution of the value function versus the stock level in mode 1 when the corrective maintenance is considered

From this simulation results, we conclude that the value function is convex in the two modes and the controls in the two modes are in the same form as it was shown in theory. These simulation results show also that by using the model we propose, we can reduce the hedging level and therefore decrease the operating cost of the production system.

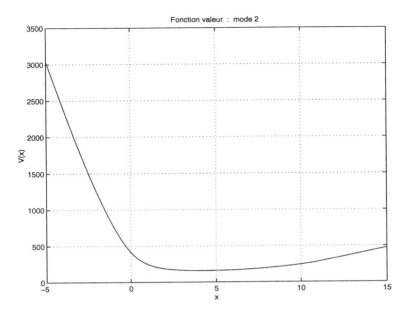

Figure 16.2   Evolution of the value function versus the stock level in mode 2 when the corrective maintenance is considered

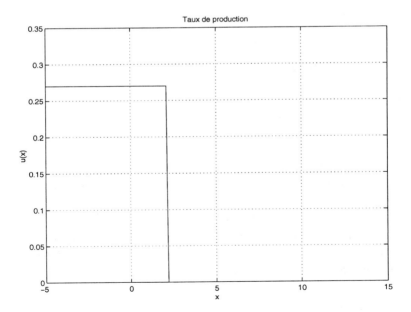

Figure 16.3   Evolution of the production rate versus the stock level when the corrective maintenance is considered

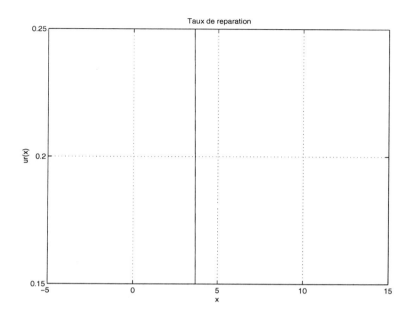

Figure 16.4 Evolution of the corrective maintenance rate versus the stock level when the corrective maintenance is considered

6.   CONCLUSION

This chapter presented a certain number of models on production planning and production and maintenance (corrective and preventive) planning. In order to make clear the concept of production planning and production and maintenance planning, we dealt only with the case of one machine one part. The proposed models can be extended to the more general cases easily. The only problem we will face will be difficulty in getting a clsoed-form solution for instance in the case of the production planning. The proposed numerical methods can then be used to approximate the optimal strategies of production and maintenance. Our models can be extended to handle the production and maintenance planning for production systems with deteriorating items.

Notes

1. This work is supported by the Natural Sciences and Engineering Research Council of Canada under Grants OGP0036444, the US Air Force Grant F30602-99-2-0548 and the Office of Naval Research Grant N00014-96-1-0263.

References

[1] Akella, R., and Kumar, P. R. Optimal Control of Production Rate in a Failure prone manufacturing system , *IEEE Transaction on Automatic Control*, **AC–31, No. 2** (1986) 106-126.

[2] Algoet, P. H. Flow Balance Equations for the Steady-State Distribution of a flexible Manufacturing System, *IEEE Transaction on Automatic Control*, **34-8** (1988) 917-921.

[3] Auger, P. *Dynamics and Thermodynamics in Hierarchically Organized Systems*, Pergamon Press, Oxford, England, 1989.

[4] Basar, T. and Bernhard, P. H^∞ - *Optimal Control and Related Minimax Design Problems*, Birkhauser, Boston, 1991.

[5] Barron, E.N. and Jensen *Total risk aversion, stochastic optimal control, and differential games*, Appl. Math. Optim., **19** (1989) 313-327.

[6] Bensoussan, A. and Nagai, H. An Ergodic Control Problem Arising from the Principal Eigenfunction of an Elliptic Operator, *J. Math. Soc. Japan*, **43** (1991) 49-65.

[7] Bertsekas, D. P. *dynamic programming : Deterministic and Stochastic Models*, Prentice Hall, Inc, 1987.

[8] Bielecki, T. and Kumar, P. R. Optimality of Zero-Inventory Policies for Unreliable Manufacturing Systems, *Operation Research*, **36-4** (1988) 532-541.

[9] Boukas, E. K. Hedging Point Policy improvement, *Journal of Optimization Theory and Applications*, **97-1** 1998.

[10] Boukas, E. K. Control of Systems with Controlled Jump Markov Disturbances, *Control Theory and Advanced Technology, 9-2* (1993) 577-595.

[11] Boukas, E. K. Numerical Methods for HJB Equations of Optimization Problems for Piecewise Deterministic Systems, *Optimal Control Applications & Methods, 16* (1995) 41-58.

[12] Boukas, E. K. and Haurie, A. optimality conditions for continuous time systems with controlled jump Markov disturbances: Application to production And maintenance scheduling, Proceeding INRIA 8 th International Conference Springer Verlag on *Analysis and Optimization of Systems* , Antibes June 1988.

[13] Boukas, E. K. *Commande optimale stochastique appliquée aux systèmes de production*, Ph. D Thesis, Ecole Polytechnique de Montréal, Université de Montréal, 1987.

[14] Boukas, E. K. and Haurie, A. Manufacturing Flow Control and Preventive Maintenance : a stochastic control approach, *IEEE Transaction on Automatic Control*, **35-9** (1990) 1024-1031.

[15] Boukas, E. K. and Kenne, J. P. Maintenance and production control of manufacturing system with setups, *Lectures in Applied Mathematics*, **33** (1997) 55-70.

[16] Boukas, E. K. and Liu, Z. K. Jump linear Quadratic Regulator with Controlled Jump Rates, *IEEE Transaction on Automatic Control*, to appear, 1999.

[17] Boukas, E. K. and Liu, Z. K. Production and Maintenance Control for Manufacturing System, *Proceeding of 38th IEEE CDC*, Dec. 1999.

[18] Boukas, E. K. and Liu, Z.K. Production and Corrective Maintenance Control for Flexible Manufacturing System, Submitted for publication, 1999.

[19] Boukas E. K., and Yang, H. Manufacturing Flow Control and Preventive Maintenance : A Stochastic Control Approach, *IEEE Transaction on Automatic Control*, **41-6** (1996) 881-885.

[20] Boukas, E. K., Zhang, Q. and Yin, G. On Robust Design for a Class of Failure Prone Manufacturing System , in *Recent advances in control and optimization of Manufacturing Systems*, Yin and Zhang ed. Lecture Notes in control and information sciences 214, Springer-Verlag, London, 1996.

[21] Boukas, E. K., Zhu, Q. and Zhang, Q. Piecewise Deterministic Markov Process Model for Flexible Manufacturing Systems with Preventive Maintenance . *Journal of Optimization Theory and Application* , **81-2** (1994) 258-275.

[22] Bremaud, P., Malhame, R. P., and Massoulie, L. A manufacturing System with General Failure Process: stability and IPA of hedging control policies, *IEEE Transaction on Automatic Control*, **42-2** (1997) 155-170.

[23] Buzacott, J. A. and Shanthikumar, J. G. *Stochastic Models of Manufacturing Systems*, Prentice-Hall, Englewood Cliffs, NJ, 1993.

[24] Connolly, S., Dallery, Y., and Gershwin, S. B. A Real-Time Policy for Performing Setup Changes in a Manufacturing System, *Proceeding of the 31st IEEE Conference on Decision and Control*, Dec, Tucson, AZ, 1992.

[25] Caramanis, M. and Liberopoulos, G. Perturbation Analysis for the Design of Flexible Manufacturing System Flow controllers, *Operations Research*, **40** (1992) 1107-1125.

[26] Caramanis, M. and Sharifnia, A. Optimal Manufacturing Flow Controller Design, *Int. J. Flex. Manuf. Syst.* **3** (1991) 321-336.

[27] Costa, O. L. V. Impulse Control of Piecewise-Deterministic Processes via Linear Programming, *IEEE Transaction on Automatic Control*, **36-3** (1991) 371-375.

[28] Costa, O. L. V. Average Impulse Control of Piecewise Deterministic Processes. *IMA Journal of Mathematical Control and Information*, **6-4** (1989) 379-397.

[29] Costa, O. L. V. and Davis, M. H. A. Impulse Control of Piecewise-Deterministic Processes. *Mathematics of Control, Signals, and Systems*, **2-3** (1989) 187-206.

[30] Courtois, P.J. *Decomposability: Queueing and Computer System Applications*, Academic Press, New York, 1977.

[31] Dassios, A. and Embrechts, P., Martingales and insurance risk. *Commun. Statist. - Stochastic Models*, **5-2** (1989) 181-217.

[32] Davis, M. H. A. Piecewise Deterministic Markov Processes: a General Class of Non-diffusion Stochastic Models. *Journal of Royal Staistical Society*, **46-3** (1984) 353-388.

[33] Davis, M. H. A. Control of Piecewise-Deterministic Processes via Discrete-Time Dynamic Programming, in *Stochastic Differential Systems*, 140-150, Helmes K. and Christopeit, Kohlman, Springer-Verlag, 1986.

[34] Davis, M. H. A. *Markov Modeling and Optimization* , Chapman and Hall, 1993.

[35] Davis, M. H. A., Dempster, M. A. H. Sethi, S. P. and Vermes, D. Optimal Capacity Expansion under Uncertainty. *Advance in Applied Probability*, **19** (1987) 156-176.

[36] Dempster, M. A. H. Optimal Control of Piecewise Deterministic Markov Processes. In *Applied Stochastic Analysis, 303-325* , Gordon and Breach, New York, 1991.

[37] Duncan, T. E., Pasik-Duncan, B. and Zhang, Q. Adaptive Control of Stochastic Manufacturing System with Hidden Markovian Demands and Small Noise, *IEEE Transaction on Automatic Control*, **44-2** (1999) 427-431.

[38] Fleming, W.H. and McEneaney, W.M. Risk Sensitive Control on an Infinite Horizon. *SIAM J. Control Optim.*, **33** (1995) 1881-1921.

[39] Fleming, W. H., and Soner, H. M. *Controlled Markov Processes and Viscosity Solutions*, Springer-Verlag, New York, 1992.

[40] Fleming, W.H., Sethi, S.P. and Soner, H.M. An optimal stochastic production Planning Problem with Random Fluctuating Demand, *SIAM J. Control Optim.*, **25** (1987) 1494-1502.

[41] Fleming, W.H. and Zhang, Q. Risk-sensitive Production Planning in a Stochastic Manufacturing System, *SIAM Journal on Control and Optimization*, **36** (1998) 1147-1170.

[42] Gatarek, D. Optimality Conditions for Impulsive Control of piecewise-deterministic processes. *Mathematics of Control, Signals, and Systems*, **5-2** (1992) 217-232.

[43] Gershwin, S. B. *Manufacturing Systems Engineering*, Prentice Hall, Englewood Cliffs, 1993.

[44] Gershwin, S. B. Hierarchical Flow Control: A framework for Scheduling and Planning Discrete Events in Manufacturing Systems, *Proceeding of the IEEE, special issue on discrete event dynamic systems*, **77-1** (1989) 195-209.

[45] Gershwin, S. B., Caramanis, M, and Murray, P. Simulation Experience with a Hierarchical Scheduling Policy for a Simple Manufacturing system, *Proceeding of the 27th IEEE Conference on Decision and Control*, Dec., Austtin, TX, 1941-1849, 1988.

[46] Godbole, D.N. and Lygeros, J. (preprint) Hierarchical Hybrid Control: A case Study.

[47] Gonzales, R. and Roffman, E. On the Deterministic control problems: an approximation procedure for the Optimal Cost I. The stationary problem, *SIAM Control and Optimization* **23-2** (1985) 242-266.

[48] Haurie, A. and L'Ecuyer, P. Approximation and Bounds in Discrete Event Dynamic Programming , *IEEE Transaction on Automatic Control*, **31-3** (1986) 227-235.

[49] Haurie, A. and van Delft, Ch. Turnpike Properties for a Class of Piecewise Deterministic Control Systems arising in Manufacturing Flow Control, *Annals of O. R.*, **29** (1991) 351-373.

[50] Hillier, F.S. and Lieberman, G.J. *Introduction to Operations Research*, McGraw-Hill, New York, 1989.

[51] Hirata,, H. Modeling and Analysis of Ecological Systems: the large-scale system viewpoint, *Int. J. Systems Science*, **18** (1987) 1839-1855.

[52] Hu, J. and Caramanis, M. Near Optimal Setup Scheduling for Flexible Manufacturing Systems, *Proc. of the Third RPI International Conference on Computer Integrated Manufacturing*, Troy, NY, May 20-22, 1992.

[53] Hu, J. Q., and Caramanis, M. Dynamic set-up Scheduling of Flexible Manufacturing Systems: Design and Stability of Near Optimal General Round Robin Policies, *Discrete Event Systems, IMA volumes in mathematics and applications series*, Kumar, P.R. and Varaiya, P. (eds.), Springer-Verlag, 1994.

[54] Hu J. Q., Vakili, P. and Yu, G. X. Optimality of Hedging Point Policies in the Production Control of Failure Prone Manufacturing Systems, *IEEE Transaction on Automatic Control*, **39-9** (1994) 1875-1880.

[55] Hu, J. Q. and Xiang, D. Structural Properties of Optimal Flow Control for Failure Prone Production System, *IEEE Transaction on Automatic Control*, **39-4** (1994) 640-642.

[56] Jiang, J. and Sethi, S.P. A State Aggregation Approach to Manufacturing Systems having Machines States with Weak and Strong Interactions, *Operations Research*, **39** (1991) 970-978.

[57] Kimemia, J. and Gershwin, S. B. An Algorithm for Computer Control of a Flexible Manufacturing System, *IIE Transactions*, **15-4** (1982) 353–362.

[58] Koktovic, P. Application of Singular Perturbation Techniques to Control Problems, *SIAM Review*, **26** (1984) 501-550.

[59] Kushner, H. J. *Probability Methods for Approximation in Stochastic Control and for Elliptic Equations*, Academic Press, New York, 1977.

[60] Kushner, H. J. and Dupuis, P. G. *numerical methods for Stochastic Control Problems in Continuous Time*, Springer-Verlag, New York, 1992.

[61] Kushner, H. J. and Yin, G., *Stochastic Approximation Algorithms and Applications*, Springer-Verlag, New York, 1997.

[62] Lehoczky, J., Sethi, S.P., Soner, H.M. and Taksar, M. An Asymptotic Analysis of Hierarchical Control of Manufacturing Systems under Uncertainty, *Mathematics of Operations Research*, **16** (1992) 596-608.

[63] Lenhart M. S. and Liao, Y. C. Switching Control of Piecewise-Deterministic Processes. *Journal of Optimization Theory and Applications*, **59-1** (1988) 99-115.

[64] Liberopoulos, G. and Caramanis, M. Production control of Manufacturing System with Production Rate-Dependent Failure Rates, *IEEE Transaction on Automatic Control*, **39-4** (1994) 889-895.

[65] Menaldi, J. L. Some Estimates for Finite Difference Approximations. *SIAM Journal on Control and Optimization*, **27** (1989) 579-607.

[66] Olsder, G. J. and Suri, R. Time Optimal of Parts-Routing in a Manufacturing system with Failure prone Machines. *Proc. 19th IEEE. Conference on Decision and Control* Alburquerque, New Mexico (1980) 722-727.

[67] Phillips, R. G. and Koktovic, P. A Singular Perturbation Approach to Modelling and Control of Markov Chains, *IEEE Trans. Automatic Control*, **AC-26** (1981) 1087-1094.

[68] Presman, E., Sethi, S.P. and Zhang, Q. Optimal Feedback Production Planning in a Stochastic N-Machine Flowshop, *Automatica*, **31** (1995) 1325-1332.

[69] Rishel, R. Control of Systems with Jump Markov Disturbances, *IEEE Transactions on Automatic Control* , **20** (1975) 241-244.

[70] Ross, S. M. *Applied Probability Methods with Optimization Applications*, Holden-day, San Francisco, 1970.

[71] Sethi, S. P., Taksar, M. I., and Zhang Q. Optimal Production Planning in a Stochastic Manufacturing System with Long-Run Average Cost. *Journal of Optimization Theory and Applications,* **92-1** (1997) 161-188.

[72] Sethi, S. P., Taksar, M. I., and Zhang Q. Capacity and Production Decisions in Stochastic Manufacturing Systems: An asymptotic optimal Hierarchical Approach, *Prod. & Oper. Mgmt.*, **1** (1992) 367-392.

[73] Sethi, S. P., Suo, W., Taksar, M. I., and Zhang Q. Optimal Production Planning in a Stochastic Manufacturing System with Long-Run Average Cost, *Journal of Optimization Theory and Applications,* **92** (1997) 161-188.

[74] Sethi, S.P., Zhang, H. and Zhang, Q. Hierarchical Production Planning in a Stochastic Manufacturing System with Long-Run Average Cost, *Journal of Mathematical Analysis and Applications,* **214** (1997) 151-172.

[75] Sethi, S. P., and Zhang, Q. *Hierarchical Decision Making in Stochastic Manufacturing Systems.* Birkhauser, Boston, 1994.

[76] Sethi, S.P. and Zhang, Q. Multilevel Hierarchical Decision Making in Stochastic Marketing-Production Systems , *SIAM J. Control and Optim.*, **33** (1995) 528-553.

[77] Sethi, S.P. and Zhang, Q. Hierarchical Production Planning in Dynamic Stochastic Manufacturing Systems: Asymptotic Optimality and Error Bounds, *J. Math. Anal. and Appl.*, **181** (1994) 285-319.

[78] Sethi, S.P., Zhang, Q., and Zhou, X.Y. Hierarchical Controls in Stochastic Manufacturing Systems with Machines in Tandem, *Stochastics and Stochastics Reports*, **41** (1992) 89-118.

[79] Sethi, S.P., Zhang, Q., and Zhou, X.Y. Hierarchical Controls in Stochastic Manufacturing Systems with Convex Costs, *J. Opt. Theory and Appl.*, **80** (1994) 303-321.

[80] Sethi, S.P. and Zhou, X.Y. Stochastic Dynamic Job Shops and Hierarchical Production Planning, *IEEE Trans. Auto. Contr.*, **39** (1994) 2061-2076.

[81] Sharifnia, A. Production Control of a Manufacturing System with Multiple Machine States', *IEEE Transaction on Automatic Control*, **33-7** (1988) 620-625.

[82] Sharifnia, A., Caramanis, M., and Gershwin, S. B. Dynamic Setup Scheduling and Flow Control in Manufacturing Systems, *Discrete*

Event Dynamic Systems: Theory and Applications, **1** (1991) 149-175.

[83] Siljak, D.D. *Large-Scale Dynamic Systems*, North-Holland, New York, 1978.

[84] Simon, H.A. The Architecture of Complexity, *Proc. of the American Philosophical Society*, **106** (1962) 467-482. reprinted as Chapter 7 in Simon, H.A. *The Sciences of the Artificial*, 2nd Ed., The MIT Press, Cambridge, MA (1981).

[85] Singh, M.G. *Dynamical Hierarchical Control*, Elsevier, rev., 1982.

[86] Smith, N. J. and Sage, A. P. An Introduction to Hierarchical Systems Theory, *Computers and Electrical Engineering*, **1** (1973) 55-72.

[87] Soner, H. M. Optimal control with state-space constraint II. *SIAM Journal on Control and Optimization*, **24-6** (1986) 1110-1122.

[88] Stadtler, H. *Hierarchische Produktionsplanung bei Losweiser Fertigung*, Physica-Verlag, Heidelberg, Germany, 1988.

[89] Switalski, M. *Hierarchische Produktionsplanung*, Physica-Verlag, Heidelberg, Germany, 1989.

[90] Tu, F. S., Song, D. P. and Lou, S. X. C. Preventive Hedging Point Control Policy and its Realization, *Preprint of the XII World IFAC congress*, July 18-23, Sydney, Australia, **5** (1993) 13-16.

[91] Sworder, D. D. and Robinson V. G. Feedback Regulators for Jump Parameter Systems with State and Control dependent Transition Rates. *IEEE Transaction on Automatic Control*, **18** (1974) 355-359.

[92] Veatch, M. H. and Caramanis, M. C. Optimal Average Cost Manufacturing Mlow Montrollers: Convexity and Differentiability, *IEEE Transaction on Automatic Control*, **44-4** (1999) 779-783.

[93] Viswanadham, N. and Narahari, Y. *Performance Modeling of Automated Manufacturing Systems*, Prentice Hall, Englewood Cliffs, 1992.

[94] Wonham, W. M. Random Differential Equations in Control Theory. in *Probabilistic Methods in Applied Mathematics*, **2** A. T. Bharucha-reid, Ed. New York: Academic 1971.

[95] Whittle, P. *Risk-Sensitive Optimal Control*, Wiley, New York, 1990.

[96] Xie, X.L. Hierarchical Production Control of a Flexible Manufacturing System, *Applied Stochastic Models and Data Analysis*, **7** (1991) 343-360.

[97] Yan, H. and Zhang, Q. A numerical Method in Optimal Production and Setup Scheduling of Stochastic Manufacturing System, *IEEE Transaction on Automatic Control*, **42-10** (1997) 1452-1555.

[98] Yin, G. and Zhang, Q. *Continuous-Time Markov Chains and Applications: A Singular Perturbation Approach*, Springer-Verlag, 1998.

[99] Yin, G., and Zhang, Q. *Recent Advances in Control and Optimization of Manufacturing System,* Lecture notes in control and information sciences 214, Springer-Verlag, London, 1996.

[100] Yu, X. Z. and Song, W. Z., Further properties of optimal hedging points in a class of manufacturing system, *IEEE Transaction on Automatic Control,* **44-2** (1999) 379-382.

[101] Zhang, Q. Risk Sensitive Production Planning of Stochastic Manufacturing Systems: A Singular Perturbation Approach, *SIAM J. Control Optim.,* **33** (1995) 498-527.

[102] Zhang, Q. Nonlinear Filtering and Control of a Switching Diffusion with Small Observation Noise, *SIAM J. Control Optim.,* **36** (1998) 1738-1768.

[103] Zhang, Q. Optimal Filtering of Discrete-Time Hybrid Systems, *J. Optim. Theory Appl.,* **100** (1999) 123-144.

[104] Zhang, Q. and Yin, G. Turnpike Sets in Stochastic Manufacturing Systems with Finite Time Horizon, *Stochastics Stochastic Rep.* **51** (1994) 11-40, .

VI

MAINTENANCE & NEW TECHNOLOGIES

Chapter 17

JIT AND MAINTENANCE

G. Waeyenbergh, L. Pintelon, and L. Gelders

Center for Industrial Management

Celestijnenlaan 300A

B-3001 Heverlee-Leuven

Belgium

geert.waeyenbergh@cib.kuleuven.ac.be

Liliane.Pintelon@cib.kuleuven.ac.be

Ludo.Gelders@cib.kuleuven.ac.be

Abstract   Due to the emphasis on cost-reduction and customer service, JIT (Just-in-Time) has become a very popular concept in logistic control systems. In the current competitive environment with short lead times and on-time deliveries, maintenance management also plays an important role in the optimization of business processes. Both manufacturing and service companies have realized that production , logistics and maintenance, can not be managed as separated functions. Boundaries are to be crossed in order to gain competitiveness through the complete package of operating functions. As a consequence borders between functional departments have disappeared over the past decade. During the last few years more and more emphasis has been put on company wide integration of maintenance into other business functions and on the contribution of maintenance to overall performance. We do no longer talk about maintenance contributing to life cycle costs, but rather to life cycle profit. The link between maintenance and performance is especially important in a JIT environment. Maintenance is a vital component here for achieving increased internal capability, which in turn leads to improved product quality and stronger market penetration. The TPM concept has integrated maintenance into machine design and into the production and quality improvement processes of the current organizations. In this chapter we describe the consequences of the JIT-philosophy on maintenance.

Keywords: Just-In-Time (JIT), Total Productive Maintenance (TPM), Total Quality Management (TQM), Reliability Centered Maintenance (RCM).

1.   INTRODUCTION

In the past few decades, production and operations managers have been confronted with a variety of concepts like MRP (Manufacturing Resource Planning), OPT (Optimized Production Technology) and JIT (Just-In-Time). Due to the emphasis on cost-reduction and customer service, JIT has become one of the most popular philosophies in logistic systems. In the current competitive environment of short lead times and on-time deliveries, maintenance management plays an ever more important strategic role in the optimization of business processes. In this contribution we highlight the consequences of the JIT-philosophy on maintenance.

2.   THE JIT-PHILOSOPHY

Devastated by war, Japanese firms, where JIT is said to have been developed, didn't have the cash to invest in excess inventories. Thus, lean production was born from necessity. But inventories are not only expensive; they also mask underlying problems. A popular story here is to compare a company with a ship on a river. The river represents the production process and the water level in the river the level of inventory. When the water level is high, the water will cover the rocks (i.e. problems like breakdowns, material shortage, etc.). Likewise, when the inventory levels are high, problems are masked. So, the ship will be able to sail, but when the water level decreases, the ship may hit the rocks. However, when the water level is low, the rocks are evident and maybe they can be removed. The same simple rule is in force in a company. When the level of inventory is low, problems will become visible and can be solved. Especially in a JIT system, where items are moved through the system in small batches, 100% inspection will be feasible. Seen in this light, JIT can be easily incorporated into an overall quality control strategy like Total Quality Management (TQM).

Although originally intended as a means of moving materials through a plant, JIT has evolved into much more. The core of the JIT philosophy is to eliminate all kinds of waste (in Japanese 'muda') including waste resulting from defective parts, machine changeovers, machine breakdowns, etc...

Although JIT focuses at first glance on the material flow, the relationship with equipment status and maintenance policy is straightforward. Just-in-Time is difficult when unscheduled breakdowns occur. If a breakdown or any other disturbance occurs, workers' attention must be immediately focussed on the productive process. Besides the awareness of the system, workers should be empowered to stop the flow of production and to undertake corrective action.

3. MAINTENANCE MEETS PRODUCTION

Not so long ago, price and quality were the most important competitive factors, and maintaining large inventories of finished and unfinished products could ensure customer satisfaction. Such a strategy would be uneconomical now, because of the rapid technological changes and smaller profit margins, which force companies to run with lower inventory levels. Moreover, customers expect an ever-higher product quality, products with a higher degree of customization, quick response to orders, fast and reliable product delivery and a good after-sales service, all against reasonable costs. Due to this increasing pressure, manufacturing firms are forced to improve continuously in several domains.

It is clear that not only the production system, but also its maintenance plays an important role. In this respect, we will define two important concepts, i.e. the 'reliability' and the 'availability ' of a production system.

3.1 Reliability. In literature, the reliability of a production system is often expressed in terms of the probability that the system will operate satisfactorily (i.e. without failures) for at least a certain period of time. Quantitatively, it can be described in two ways, depending on the point of view:

- the definition can be translated strictly mathematically on the basis of the statistical description. This description concerns only the failure time and is used especially in theoretical studies of reliability and in optimizing maintenance policies.

- a practical approach can be adopted by looking at the failure time as well as the corresponding repair time. This technique is applicable to repairable systems and is used especially in availability studies.

As reliability concerns the full life of the equipment, the costs incurred during this life will also be affected, and it is therefore useful to study the cost implications in detail (life cycle costing).

The reliability $R(t)$ is therefore the probability that the life T will exceed t. Expressed by means of the cumulative failure distribution function $F(t)$,

$$R(t) = Prob(T > t) = 1 - F(t) = 1 - \int_0^t f(t)dt$$

where f(t) is the failure distribution function. The determination of reliability is closely related to the failure rate. The failure rate h(t) is a measure of the number of failures occurring per time unit:

$$h(t) = \frac{f(t)}{R(t)}$$

The life of a non-repairable system can be divided into three different stages, which are often represented by means of the bathtub curve concept (figure 17.1), comprising

- the running-in period: initially, the system displays a high but decreasing failure rate (DFR);

- the useful life period: the system now exhibits a relatively constant failure rate (CFR), with the failures occurring at random and therefore unpredictably;

- the wear period: this is characterized by a strong increase in the failure rate (IFR).

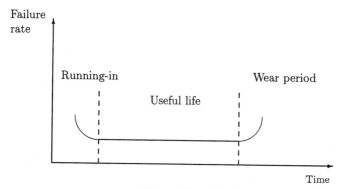

Figure 17.1   The bathtub curve.

3.2   Availability. In a similar way, the availability of a production system is usually defined as the probability that the system will be operating satisfactorily at an arbitrary point in time, or equivalently as the long run average fraction of time that the system is operational. The availability (A) is often expressed in relation to breakdowns and maintenance activities and is closely related to the downtime D (as a fraction of total available time).

The availability can be estimated by

$$A = 1 - D = \frac{total\ operational\ time}{total\ operational\ time + total\ downtime}$$
$$= \frac{MTBF}{MTBF + MTTR} \tag{17.1}$$

where MTTR is the Mean Time To Repair (more generally, the mean time that the process is inoperable when it is down for maintenance or because of a breakdown) and MTBF is the Mean Time Between Failures (more generally, the mean operating time between one downtime and the next, where each downtime can be due to maintenance or a breakdown).

But not only reliability and availability are important. Variability also matters. A production system should be able to produce at a more or less constant rate, so the variability should be low. The variability is often expressed in terms of an interval availability distribution.

Efficient utilization of the available resources has always been essential to keep the operating costs at a competitive level. Customer and producer, however, have different objectives here. For the customer, the unit cost (i.e. total cost divided by total volume) is important. The producer on the other hand, looks at the total cost. This implies that both cost reduction and volume enhancement are commercially worthy objectives. In this respect, not only the frequency, but also the timing of maintenance matters. In general, preventive maintenance involves lower cost and less time compared with corrective maintenance. But preventive maintenance has many other advantages:

- preventive maintenance can be carried out at more convenient times (e.g. during the night, in weekends, during holidays, during idle times because of withdrawn orders, while repairing machine failures, etc.), whereas corrective maintenance usually requires interruption of the production process.

- preventive maintenance on different components can often be grouped into maintenance packages (clustering) to reduce set-up times and/or costs, whereas this is somewhat more complicated in the case of corrective maintenance.

- preventive maintenance allows for the reduction or elimination of intermediate buffers and safety stocks, which are usually maintained to keep production going if one or more machines have failed. Also the number of spare parts stocked can often be reduced significantly, since the majority of repairs and replacements can be planned in advance.

Cost and quality have always been critical factors. During the last decade, however, speed has become a factor of critical importance too. The rapid development rates of new products, together with the need for fast and on-time delivery, are some of the pillars of manufacturing strategies like JIT. Responsive delivery, without inefficient excess inventory, requires short cycle times, reliable processes, and effective integration of disparate functions (e.g. maintenance and production). In this respect, there is a perspective of significant improvements in system efficiency, if the variabilities in the production processes can be reduced. Unscheduled downtimes due to random breakdowns are one of the largest and most disruptive sources of variability. Ideally, production systems should be able to produce at a constant rate, without service interruptions, while retaining a satisfactory production capacity. Therefore, a production system with frequent, predictable and short interruptions is to be preferred above one with infrequent, unpredictable and long interruptions, all other things being equal. In practice, performing preventive maintenance at regular intervals can reduce the variability of a production system.

Nowadays, it is often found cost-effective to leave the day-to-day or routine maintenance activities (e.g. lubrication, cleaning, monitoring) in the hands of machine operators, since they often know best when their equipment exhibits abnormal behavior.

4. FROM PREVENTIVE MAINTENANCE TO PRODUCTIVE MAINTENANCE

Equipment management evolved from preventive maintenance to productive maintenance . Equipment management is a cornerstone of JIT . In a sense productive maintenance is as JIT - about eliminating waste. The goal of productive maintenance is to take away the obstacles (the so-called six big losses , table 17.1) to reach higher equipment effectiveness. Productive maintenance ensures that a machine runs so well that it never breaks down, always operates at the designed speed or faster with no idling nor minor stoppages, never produces a defective product, and causes a minimum of start-up, set-up and adjustment losses. Therefore appropriate equipment diagnosis, early detection of deviations, equipment records, breakdown data and an appropriate preventive maintenance (PM) plan deserve special attention.

But productive maintenance was not fully successful at achieving zero breakdowns and zero defects as long as practised by the maintenance department. Based on small-group activities, Total Productive Maintenance (TPM) takes productive maintenance company wide, with the support and co-operation of management and employees at all levels. In this way TPM generates (and requires) a major change in organizational culture.

The idea of improving performance through learning is an essential component of TPM . In this respect it has a lot of affinity with the TQM concept.

5.   THE TPM-PHILOSOPHY

TPM is an approach to improve the performance of maintenance activities. TPM improves the effectiveness as well as the efficiency of maintenance. The effectiveness of a maintenance concept relates to the results achieved in terms of reliability and availability. In other words it is a measure of how well 'the goal' is reached. The efficiency of a maintenance concept concerns the resources used, i.e. costs of manpower, spares, energy, etc. Efficient means 'with the best use of resources'.

However, TPM goes much further than maintenance only. In literature, but also in practice, one speaks more and more about Total Productive Manufacturing or even Total Productive Management. So, TPM is more than just another maintenance concept, it is more like a management strategy. The whole organization is getting involved in it. In this section, we deal with the fundamentals of TPM. Finally, we make some recommendations on how to improve the TPM development and implementation program in a more practical way.

TPM finds its roots - like so many recent (after the 60's) management philosophies - in Japan (The story tells that 'TPM' was born at the Japanese firm Nippondenso). Just like TQM , TPM is grown out of the KAIZEN-philosophy. KAIZEN is a Japanese word and means 'continuous improvement'. However, TPM can be considered as an indispensable contribution to lean production , supporting JIT manufacturing and TQM. So, TPM is an essential pillar alongside JIT and TQM to companies seeking world class manufacturing status.

Talking about the fundamentals of TPM , one can distinguish two main approaches, the Japanese one and the Western one. Because of the Japanese background of TPM, we first look at TPM in the Japanese way, before going to the Western approach. Nakajima (1986, 1988 and 1989) (sometimes called the father of TPM) attempts to summarize TPM as: "Productive maintenance involving total participation in addition to maximizing equipment effectiveness and establishing a thorough system of PM". Here PM is a comprehensive planned maintenance system. A more complete definition recognized by Nakajima contains the following 5 points:

- It aims at getting the most efficient use of equipment (improve overall efficiency).

- It establishes a complete PM program encompassing maintenance prevention, preventive maintenance, and improvement related maintenance for the entire life cycle of the equipment.

- It is implemented on a team basis and it requires the participation of equipment designers, equipment operators, and maintenance department workers.

- It involves every employee from top management down to the workers on the floor.

- It promotes and implements PM based on autonomous small-group activities (motivational management).

In the western world, there are a lot of definitions of TPM , all very closely related to the Japanese one, but more suited to western manufacturing and culture. It is not our intention to give an overview of all the TPM definitions (an overview can be found in literature (Bamber, 1999)), but the two following definitions are very interesting. Edward Hartmann, who is recognized by Nakajima as the father of TPM in the US, provides a definition that is suggested as being more readily adopted by western companies. Hartmann (1992) states "Total Productive Maintenance permanently improves the overall effectiveness of equipment with the active involvement of operators." Wireman (1991) gives a broader definition of TPM: "It encompasses all departments including, maintenance, operations, facilities, design engineering, project engineering, instruction engineering, inventory and stores, purchasing, accounting finances and plant/site management."

6. THE PILLARS OF TPM

TPM wants to achieve an overall workshop improvement by developing optimal human-machine conditions. To achieve this, one can build a TPM model based on five pillars. These pillars are commonly accepted as fundamental to the TPM success. The implementation strategy of each pillar can, however, vary, because each pillar is often presented with differing approaches. The five pillars can be described by the following definitions:

- Individual equipment improvements to eliminate the six big losses
  .

- Autonomous maintenance.

- Planned preventive maintenance.

- Maintenance and operations skills training.

- Maintenance plan design and early equipment management .

6.1 Individual Equipment Improvements. The production and maintenance engineering teams should be ready to continuously improve their machines and themselves, in order to achieve zero failures and zero defects. To do this, first of all the six big losses have to be identified. The TPM model is based on the computation of the O.E.E. (Overall Equipment Effectiveness). This O.E.E. is computed by reducing the theoretical availability of 100

6.2 Autonomous Maintenance.. The operator should be involved in the maintenance of his or her own machine. Operators, who better understand their equipment, should be able to work more effectively and take more care of their equipment. Some very simple activities can contribute a lot to the operator-machine relationship, like cleaning, lubricating and inspecting. Also some simple redesign, like making the inner parts of the equipment visible by replacing the metal covers by Plexiglas-covers, can make the detection of a future failure much easier.

6.3 Planned Preventive Maintenance.. The maintenance teams should perform periodic inspections and preventive maintenance in order to improve the lifetime of the equipment. Often maintenance is carried out when the equipment is idle, for example during night shifts and weekends. Also spare parts control and breakdown analysis can help to achieve more efficient, cost-effective maintenance operations.

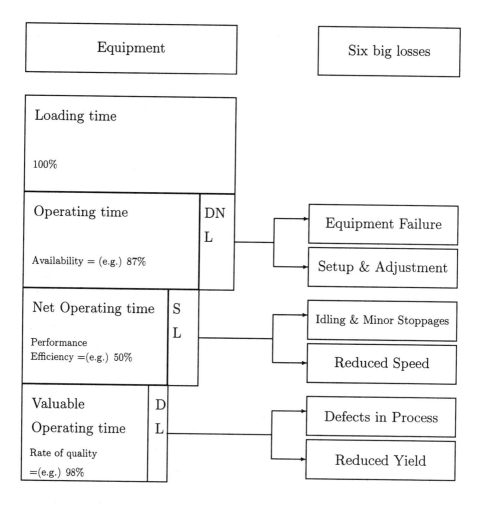

DN L = downtime losses,   S L = speed losses,   D L = defect losses.

OEE = Availability × Performance efficiency × Rate of quality
OEE = (e.g.)0.87 × 0.50 × 0.98 = 42.6%

Source: Nakajima(1986)

Figure 17.2 The six big losses.

Table 17.1   Definitions and examples of the six big losses.

Type of loss	Definition
Downtime losses Equipment failure	Whenever there is a loss of function. These failures have a relative low frequency and long duration.
Setup and adjustment	From the end of the production of one type until the standard quality of a new product is reached.
Speed losses Idling and minor stoppages	A minor stoppage occurs when production is interrupted by a temporary malfunction or when a machine is idling. These stoppages are short in duration, but can have a significant effect on capacity because of high frequency.
Reduced speed	Differences between actual speed and designed speed. Reasons for this can be e.g. that there are too many quality problems at the nominal speed, etc.
Quality losses Defects in process	Losses caused by defect products and reworking of defects.
Reduced yield	Quality problems that are typical to the starting of machines.

6.4   Maintenance and Operations Skills Training..

The operators and maintenance workers should get training in maintenance fundamentals, predictive technology, troubleshooting and diagnosis. This training will give the operators and maintenance workers higher skill levels. Also a lot of attention is paid to multi-skill training. A multi-skilled workforce is considered to be an important asset.

6.5   Maintenance Plan Design and Early Equipment Management..

The production-, design- and maintenance-engineers should try to enhance maintainability, operability, and reliability at lower Life Cycle Costs (LCC). Anticipating on production- and maintenance- problems during the design stages of the equipment can do this (Life Cycle Engineering, LCE). In this way, more reliable equipment that is easier to operate and maintain will be developed. The production will be much more stable right after the installation of the equipment.

7.   THE NEED FOR SUPPORTING MAINTENANCE CONCEPTS

Of course, TPM alone is not enough. It requires the aid of other concepts to build an effective and efficient maintenance plan for a specific situation. In this section we describe and compare a limited selection of concepts for the organization of maintenance. A maintenance concept is defined here as the set of maintenance interventions of various types (corrective, preventive, condition-based, opportunistic, design-based, etc.) and the general structure in which these interventions are foreseen. The effectiveness of a maintenance concept relates to the results achieved in terms of reliability and availability. The efficiency of a maintenance concept concerns the resources used, i.e. costs of manpower, spares, etc. This efficiency can be seen as the sum of elementary efficiency and combinatorial efficiency, where elementary efficiency is the decrease in failure costs due to the maintenance plan and the combinatorial efficiency is the decrease in cost due to doing different maintenance tasks simultaneously. Neither elementary efficiency, nor combinatorial efficiency can guarantee that the desired effectivity is achieved. Therefore, optimum maintenance concepts are expressing the best trade-off between efficiency and effectiveness.

In literature, quite a few maintenance concepts are described. These concepts can be very different. Some of these concepts are fairly simple (e.g. a single decision chart) and provide only rough guidelines for establishing a concept, while other concepts involve thorough and time-consuming analyses and provide a detailed concept.

Here we restrict ourselves to a few rather well known concepts, which are comparable in terms of scope and amount of work. Geraerds (1978), Gits (1994), Kelly (1997), Moubray (1997), Smith (1993), Blanchard (1992) and Anderson and Neri (1990) describe these concepts in full detail. Geraerds does not offer a very practical concept, but he was one of the first to describe the interactions between production and maintenance. Gits provides a useful framework for establishing a maintenance concept, taking into account the characteristics of the technical system and its organizational context. Moubray and Smith both describe an extension to RCM , the so-called RCM II. RCM stands for Reliability Centred Maintenance. This concept, originally designed for the aircraft industry, is described in Nowlan and Heap (1978). The concept by Blanchard relates to LCC (life cycle costing) and considers the spectrum of activities needed for the logistic support in designing a maintenance concept. Anderson and Neri describe a concept that can be used for the development of large, high-tech systems. They distinguish two different approaches. In the first approach, the maintenance plan is developed with the use of a Fault Three Analysis (FTA). In the second approach, the maintenance plan is developed with the use of a Failure Mode Effect and Criticality Analysis (FMECA). Kelly describes a complex, but systematic and complete Business Centred Maintenance (BCM) approach. It is a framework based on the identification of the business objectives, which are then translated into maintenance objectives.

These selected concepts are discussed in Table 17.2. For a more extensive discussion, we refer to Pintelon and Waeyenbergh (1999). The following criteria were used in comparing the concepts:

- Objectives: what is the goal of the concept? Which end result is to be expected?

- Input: what data are needed and how can they be obtained? What are the human resources required to build the concept?

- Basic maintenance policies considered. In maintenance, there are two basic interventions: Corrective maintenance (CM), which means the repair of machines after breakdown, and preventive maintenance (PM), which means doing maintenance to reduce failure probability or reduce failure impact. According to the way CM and PM are combined, different basic policies can be distinguished. With failure based maintenance (FBM), also called operate-to-failure (OTF), only CM is carried out. In the case of a constant failure rate, and/or low breakdown costs, this may be a good policy. In an opportunity based maintenance (OBM) policy one waits to maintain the system until the opportunity arises to repair some other more critical components. The decision whether or not OBM is suited for a given component depends on the expectation of its residual life, which in turn depends on the utilization. Fixed time maintenance (FTM), also called use-based maintenance (UBM), prescribes to carry out PM after a specified amount of time. CM is applied when necessary. FTM assumes that the failure behavior is predictable and that there is an increasing failure rate. This policy, which is a PM policy, assumes that PM is cheaper than CM. With condition based maintenance (CBM), PM is carried out each time the value of a given system parameter exceeds a predetermined value. As in the previous policy, PM is assumed to be cheaper then CM. CBM is gaining popularity due to the fact that the underlying techniques (e.g. vibration analysis, oil spectrometry,) become more widely available and at better prices. A design-out maintenance policy (DOM) wants to improve the design of the technical system in order to make maintenance easier or even eliminate it. Ergonomic and technical aspects are important here.

- Basic maintenance decision rules: rules used to decide which policy is most appropriate.

- Feedback: is there a possibility to validate/evaluate the obtained maintenance concept?

- Specific points of action for this concept, special characteristics.

- Degree of complexity and accuracy: is the concept well structured? Is the concept ready to use?

These concepts all have their merits and provide a structured way to establish a maintenance concept. They are very useful, although e.g. the underlying decision methodology is sometimes vaguely described (e.g. "is it technically feasible: yes/no", without any further specifications). There are many similarities between the concepts, but also some differences: in focus, in terminology, in the readiness-to-use of the end result. These concepts are all very interesting from the academic point of view. For practitioners, it may be difficult to decide which concept is best for their situation and/or it may be necessary to fine-tune the concept for a practical situation.

Table 17.2 Comparison of five maintenance design concepts.

Cri-teria	Blanchard (ILS/LSA)	Anderson and Neri	BCM (kelly)
C1	Acheiving maximal effectivity of the TS, with a minimum on lifetime costs.	Development of an MP that guarantees reliability, maintainability and safety of high-tech TS in efficient way.	Development of a framework for deciding the optimum life plans for the TS and designing maintenance schedules based on the business objectives.
C2	Same as RCM but with more extensive application domain. analysis) Extensive	All the available, historical data. (FTA- or FMECA-Extensive analysis. analysis.	Production process & planning, forecast, workload, life plan and & expected availability of the TS, spare parts. Functional decomposition. Extensive analysis.
C3	Same as RCM	UBM, CBM, FBM	FBM, UBM, CBM, DOM, OM, inspection.
C4	Same as RCM	Same as RCM	Production schedule, hidden failures, safety requirements, equipment, redundancy, etc.
C5	-	-	Possible
C6	Logistic support, i.e. number and level of personnel, spare parts, facilities, intervals, maintenance level, training	Structured, but no real plan	The concept springs from the identification of business objectives, which are then translated into maintenance objectives
C7	Concatenation of procedures, no predefined structure. Good accuracy.	Structured. Good accuracy	Structured, but very complex. Very acurate.

C1: Objectives

C2: Input: Data requirements & human resources

C3: Basic policies

C4: Decision rules

C5: Feedback

C6 : Points of action

C7: Complexity and accuracy

FMECA: Failure Mode Effect and Criticality Analysis

Table 17.2 Continued...

Cri-teria	Geraerds	Gits	RCM II (Moubray)
C1	Development of guidelines for the maintenance of a TS	Development of an MP with focus on safety and contunity of the production process.	Development of an MP that determines the maintenance requirements to keep a TS in its operating context.
C2	System behaviour, expectation, desires, restrictions.	Failure modes & effects. Technical decomposition. Extensive analysis.	Failure modes, effects and consequences. Functional decomposition. Exten-sive analysis.
C3	'wait and see', PM	FBM, UBM, CBM	CBM, DOM, discard, restoration, failure-finding tasks (for hidden failures)
C4	Failure rate, reliability, maintainability, and availability.	Hidden failures, normative failure weight, failure rate, failure predictive measurement.	Failure consequences operational, non-operational, (safety, hidden-function, environment)
C5	-	Possible	Possible
C6	Interaction of maintenance with production.	Continuity of the production process.	Achieving higher safety, environmental integrity, efficiency and longer useful life of expensive items.
C7	Only descriptive. Poor accuracy.	Structured, systematic approach. Eight step concept. Good accuracy.	Complex. Good accuracy.

PM: Preventive Maintenance   MP: Maintenance Plan
CBM: Condition Based Maintenance   FBM: Failure based Maintenance
TS: Technical System   UBM: Use Based Maintenance
DOM: Design-out Maintenance   OM: Operator Monitoring
FTA: Fault Tree Analysis

8.   THE IDEAL BASE TO START WITH TPM

How to start with TPM? What is the first step in TPM implementation? In order to answer this question, one should know that TPM is characterized by the 5 S's. This means 5 Japanese words (translated to English below) that describe the TPM-philosophy completely.

- Seiri : Organization, eliminate the unnecessary

- Seiton : Tidiness, establish permanent locations for the essential

- Seiso : Purity, find ways to keep things clean and inspect through cleaning

- Seiketsu : Cleanliness, make cleaning easy

- Shitsuke : Discipline, always want to improve

If applied in the correct way, an RCM analysis results in the following outcomes:

- maintenance schedules to be done by the maintenance department.

- revised operating procedures for the operators of the asset.

- a list of areas where one-off changes must be made to the design of the asset or the way in which it is operated to deal with situations where the asset cannot deliver the desired performance in its current configuration.

- the participants in the process learn a great deal about how the asset works

- there is a tendency to function better as teams.

These outcomes, together with the fact that RCM is effective in most maintenance cultures, makes RCM a good starting point for TPM. Training, insight in the production environment, teamwork and restoring equipment condition are the major factors in TPM. The first key for TPM is getting back to the basics of good equipment management, focusing on educating and training the most important asset any company has: its people!

RCM considers the safety and environmental implications of every failure mode before considering its effect on operations. This means that steps are taken to minimize all identifiable equipment-related safety and environmental hazards, if not eliminate them altogether. By integrating safety into the mainstream of maintenance decision-making, RCM also improves attitudes to safety.

RCM recognizes that all types of maintenance have some value, and provides rules for deciding which is most suitable in every situation. By doing so, it helps ensure that only the most effective forms of maintenance are chosen for each asset, and that suitable action is taken in cases where maintenance cannot help. This much more tightly focused maintenance effort leads to quantum jumps in the performance of existing assets where these are sought.

If RCM is correctly applied, it reduces the amount of routine work (in other words, maintenance tasks to be undertaken on a cyclic basis). On the other hand, if RCM is used to develop a new maintenance program, the resulting scheduled workload is much lower than if the program is developed by traditional methods. Expensive items have a longer useful life, due to a carefully focused emphasis on the use of on-condition maintenance techniques. A comprehensive database makes it possible to adapt to changing circumstances (such as changing shift patterns or new technology) without having to reconsider all maintenance policies from scratch. It also enables equipment users to demonstrate that their maintenance programs are built on rational foundations (the audit trail required by more and more regulations).

The information stored on RCM worksheets reduces the effects of staff turnover with its loss of experience and expertise. An RCM review of the maintenance requirements of each asset also provides a much clearer view of the skills required to maintain each asset, and for deciding what spares should be held in stock.

RCM delivers improved drawings and manuals, and greater motivation of individuals. This leads to greatly improved general understanding of the equipment in its operating context, together with wider "ownership" of maintenance problems and their solutions. It also means that solutions are more likely to endure.There is better teamwork as RCM provides a common, easily understood technical language for everyone who has anything to do with maintenance. This gives maintenance and operations people a better understanding of what maintenance can (and cannot) achieve and what must be done to achieve it.

As a result, RCM is an ideal way to develop maintenance programs for new (unique) assets, especially complex equipment for which no historical information is available. All of these issues are part of the mainstream of maintenance management, and many are already the target of improvement programs. A major feature of RCM is that it provides an effective step-by-step framework for tackling all of them at once, and for involving everyone who has anything to do with the equipment in the process.

Another concept worth mentioning is RBM (Reliability Based Maintenance). This concept can be seen as an intermediate form of TPM and RBM. RBM is a comprehensive strategy that helps businesses reduce costs while increasing profit, through the careful balancing of preventive, proactive and predictive maintenance. The concept includes points of RCM and TPM . For example, RBM allows the operators to inspect their equipment on a regular basis and to perform little maintenance jobs. In this way, the future behavior of their machinery can be predicted more easily. The difference with TPM is that in TPM the total company is much more involved.

9.   MODELING TOOLS FOR IMPROVED MAINTENANCE

Implementing TPM is not an easy task, but different techniques and modeling tools are available to assist maintenance. We mention just a few.

9.1   Line Balancing.
The perfect balance of the production line is a key factor for the success of JIT . A line-balancing problem is one in which a collection of tasks must be performed on each item. Furthermore, the tasks must be performed in a specified sequence. From production point of view, the quality of the balance is measured by the idle time at each station. Determining the best mix of cycle time (amount of time allotted to each station) and number of stations is an extremely difficult analytical problem. A very simple heuristic solution method from Helgeson and Birnie (1961) can help to balance the line quickly.

9.2   Scheduling Techniques.
As mentioned earlier, JIT is very difficult when unscheduled breakdowns occur. If a breakdown or any other disturbance occurs, workers' attention must be immediately focussed on the reparation of the continuity of the productive process. Besides the awareness of the system, operators should be empowered to stop the flow of production and to undertake corrective action. So, scheduling is an important aspect of operations control in both production and maintenance. There are many different types of scheduling problems faced by an organization. In order to achieve maximum availability of the production system, job scheduling, personnel scheduling and project scheduling are very important for JIT and TPM.

9.3   Scheduling Irregular Activities.
Time lost while equipment is broken down or operating at lower capacity will increase

costs and decrease productivity. Briefly said, it will increase wastes of different kinds and disturb the JIT -environment, which includes TPM

Having a maintenance program that has the ability to complete repairs in a timely fashion will be very important to the successful implementation of TPM. An interesting tool, developed by Taylor (1996), that can be used to develop such a maintenance program is a linear programming model to schedule irregular maintenance activities. This linear programming model has been developed to process the irregular orders promptly and effectively. The linear programming model has been developed to cost-effectively allocate labor crews to different prioritized work orders. The model gives consideration to different craft areas, such as electrical, mason, and so on, as well as to the prioritization of work orders. The benefit of such a model goes much further than it was initially developed for. By changing the constraints in the model, the program could be used to eliminate or significantly reduce the need for overtime among maintenance personnel while simultaneously improving on-time performance on work orders. As part of a TPM implementation effort, this model would allow workers to better understand how TPM will positively impact both maintenance jobs and maintenance performance.

9.4   Pareto Analysis.   A maintenance service always has many tasks to perform and often suffers from staff shortages. The Pareto analysis is one of the most valuable methods to evaluate the inventory position. The method is very suitable for getting a deeper understanding in the different categories of items in the warehouse, but it can also be used to decide which machines are 'fast-failing', 'normal-failing' or 'slow-failing'. So, it can be used to determine which machines need preventive maintenance , and which machines need priority. Altogether, operations should be organized so that the service is provided as efficiently as possible. The 'Pareto' or 'ABC' analysis is an aid to achieve this; it is based on a classification of failures in terms of cost so as to give an order of priority among the actions to be taken. An ABC-analysis is done in two steps. The method starts by listing the machines in order of decreasing cost (i.e. hours of downtime), with the number of failures of each machine, and forming the cumulative sums of the costs and the corresponding failures. These results are plotted in a graph with the cumulative costs on the y-axis and the cumulative sum of the failures on the x-axis (both in percentages). In the second step, the graph is divided in to three zones (A, B and C). (The rules of common sense apply here too: the subdivision is done in such a way that on one hand the number of failures are relevant for the machines in question, and on the other there are sufficient machines in each subgroup to allow statistically relevant conclusions.) Using this subdivided graph, the machines responsible for a given percentage of downtime are obtained.

- Zone A: about 20% of the failures account for about 80% of the costs; failures in this zone clearly must have priority. Effort should be directed to improve machines in this class, with adequate stocks of replacement parts ensured.

- Zone B: about 30% of the failures account for about 15% of the costs, and these form the second priority class. A lower level of maintenance will suffice for elements corresponding to zone B.

- Zone C: the remaining 50% of the failures account for the remaining 5% of the costs, and are the lowest-priority class. Elements corresponding to zone C require little or no preventive maintenance .

9.5   Activity Based Costing (ABC).

The core aim of the JIT philosophy is to eliminate all kinds of waste. Waste is cost, and cost is to be minimized. ABC is a method which may be used to measure and assign costs to produced products or rendered services. Traditional cost accounting often treats maintenance as a simple overhead cost without detailed analysis. ABC, in contrast, focuses on activities or processes of business operations and can analyze maintenance cost in greater detail. By making use of ABC, management can minimize non-value-added activities and strive toward continuous improvement and re-engineering of business operations.

9.6   Simulation Techniques.

In general, there are two ways to solve a problem, analytical or experimental. In reality, experiments can only be carried out with the aid of a model that imitates actual situations. This is what simulation is all about. Simulation requires building a model to study the behavior of a system. This means that a simulation study comprises three important components: the physical system, the mathematical-logical model and the solution strategy.

The mathematical-logical model is a description of the physical system being studied. The description usually entails a number of simplifications and idealizations, and a number of hypotheses must be postulated. Naturally, the relevant characteristics of the system must be retained; otherwise the model loses its practical value. This model can be solved mathematically (analytically or numerically), or simulation runs can be performed (if the mathematical model is too complex or when the dynamic behavior of the system must be followed in detail). The difference between a simulation solution (SIM) and a solution obtained by mathematical programming (MP) is explained in 17.3. In simulation, a number of alternative solutions are introduced which are then evaluated.

With the development and introduction of powerful computers, computer simulation has become a technique with a wide application domain. Today, several kinds of problems can be solved with the aid of computer simulation: industrial, social, scientific, economical, logistic, etc. Initially, heavy programming efforts were required, but the current software products allow rapid, modular program structuring. Three different types of simulation software can be distinguished:

- General languages: e.g. Fortran, C, Pascal, Basic,

- Simulation Languages: e.g. GPSS, AutoMOD, Simula,

- (simulation languages are general languages with special characteristics for simulation modeling)

- Simulators: e.g. XCell+, Taylor-II, Witness, Arena, (simulators are easy-to-use programs, which do not require advanced programming knowledge)

MATHEMATICAL PROGRAMMING

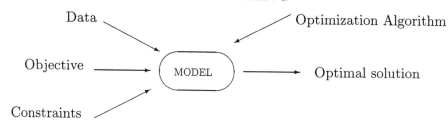

SIMULATION

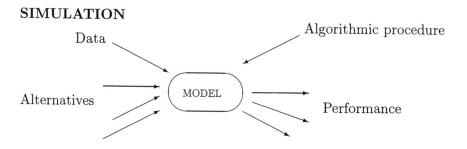

Figure 17.3 Problem solving approach with mathematical programming(MP) and simulation(SIM)

Philosophies like JIT can make us believe that the stochastic characters of processes are slowly becoming a part of the past. This is not correct. It is often uncertainty that disturbs a production process considerably. In order to gain insight in the behavior of the system as a whole, a simulation study can be set up, and models can be built based on the knowledge or assumptions regarding the behavior of parts of that system. Another obvious reason for switching to simulation is the limitation on experimentation with actual systems. For example:

- the disruption of operations: Production cannot allow Maintenance to experiment in order to determine the optimum mix of maintenance interventions;

- observation may disrupt the system: when studies are carried out on site, the employees feel they are being observed and the system won't function normally;

- experiments are expensive and/or time-consuming: setting up experiments (and carrying them out) in a manufacturing environment demands considerable planning. The testing period needed to ensure sufficient relevance of the results may be very long;

- conditions are difficult to maintain: in order to validate certain alternatives, the same starting conditions must prevail. If the set of jobs to be carried out differs from one day to the next, this is clearly impossible;

- some alternatives are so drastic that they cannot be performed as experiments. In some cases, the entire layout of the plant must be changed, e.g. when production and maintenance units are to be rationalized. Physical experiments are clearly impossible here.

Simulation can of course not eliminate uncertainties, but it can help to design the system in such a way that the sensitivity for uncertainties is minimal. Because of these characteristics, simulation is a very interesting tool for maintenance management. It can be used for the measurement of the effects of breakdowns, finding the optimal maintenance crew for the technical department, determining the effect of a certain maintenance policy, determining the optimal buffer capacity between the different machines in the production-line, etc.

In developing a simulation model, the analyst must select a conceptual framework within which the modeling will be done. This framework determines the "world view" within which the functional relations of the actual system will be modeled. The selection of this worldview determines the kind of simulation to be used.

Discrete simulation is used when the dependent system variables change discretely at specific moments (event times). The time variable may be continuous or discrete, depending on whether the changes can occur at any time or only at specific moments. The situation of a technician repairing a machine can be used as an example of a discrete simulation . The technician is busy with the machine until it is repaired. If the technician is already busy when another machine breaks down, the latter will have to queue and await its turn to be repaired. In order to construct a discrete simulation model of this system, the system status must be defined as well as the events that may change this status. In the example, the system status is determined completely by the status of the technician (busy or idle) and by the number of queuing machines. The system status can therefore be changed by the arrival of another break down or by a repair. A flow of machines and the associated repair times of them will therefore have to be generated. The object of the discrete simulation is to reproduce the activities of the system entities (i.e. machines, in this case) in order to understand the performance of the overall system. Besides the description of the system status and of these activities, a description of the process that makes the entities move through the system is needed. Depending on the simulation approach selected, the technique used is called event-oriented, process-oriented or activity-oriented.

In a continuous simulation model, the system status is described by dependent variables that may change continuously over time. Building such a model will entail the writing of analytical equations (often differential equations) for a set of status or condition variables in order to describe the dynamic behavior of the system over time. The simulation of distillation processes in the chemical industry is a typical example.

For some applications, it will be necessary to combine the two worldviews, e.g. when tankers are unloading in a port: arrivals and departures must be modeled discretely, whereas actual discharging (continuous change in the level of liquid) must be represented by a continuous model.

In the industrial environment, simulation studies have gained a great deal of interest and their potential contributions to process improvements are fully realized. At the same time, the capacities of software products have increased greatly and more realistic results can be obtained more quickly. However, evidently, the more complex the technique applied, the higher the development costs.

The different steps needed for a simulation study are summarized in 17.4.

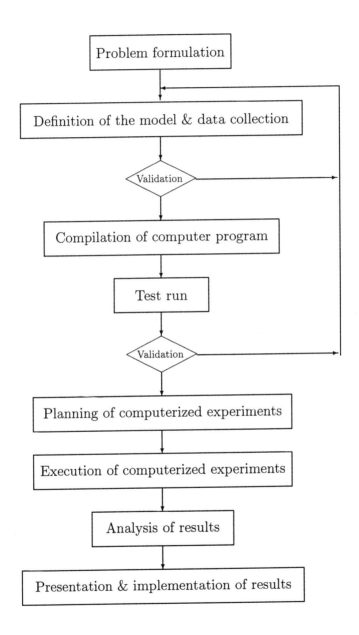

Figure 17.4 Steps of a simulation study.

9.7   Performance Measurement and Visual Management.

Performance measurement has the intention to examine the performances of the installation and personnel of a certain department, like for example Maintenance and Operations. Different systems of performance measurement for maintenance follow-up exist. In Table 17.3 a brief summary of the most commonly used performance indicator (PI) measuring systems with their advantages and drawbacks are given.

Table 17.3   Definitions and examples of the six big losses.

TYPE	ADVANTAGES	DRAWBACKS
(i) indicators		
Overall PI	popular (compact)	(excessive) aggregation
Set PI's	more complete	decision not always possible
Structured PI list	standardized	good follow-up difficult
(ii) benchmarks		
Checklists	quick understanding	superficial understanding
Surveys	OK, if available	use cautiously
(iii) graphs		
Diagram (bar, line, pie)	popular	can easily be manipulated
Potential graph	actual v. target performance	often subjective
radar plot	insight	limited number of PI's
(iv) extensive models		
Hibi method	overall	rigid, time consuming
Luck's method	fairly complete	complex
MMT	useful	implementation is critical
Balanced score card	integrated approach	implementation time

Source: Pintelon (1997)

All this, in combination with visual management techniques, can help to obtain an optimal motivation and information of the employees. Some of the basic factors in visual management are:

- Visual documentation

- Beaconing of the work area of the different teams

- Advertising the PI scores

- Advertising the progression (or regression)

For more information on this issue, we refer to specific literature about performance measurement systems and visual management.

10. ILLUSTRATIVE CASE: FAILURE ANALYSIS AND TPM IN A BREWERY

This study was carried out at the Centre for Industrial Management as a thesis project within a brewery that produces $1,3 \times 10^6$ Hl of beer a year. The study was prompted by the wish to improve the poor yield (60% f the rated capacity) of the filling line. This line comprises a depalletizing machine, an unpacker to remove the bottles from the trays, a washing plant, a filling section, a crown cork machine, a labeling machine and a palletizing machine. After an elementary failure analysis, it was decided to launch an improvement program based on TPM principles.

To understand the nature and causes of each type of loss, the available historical data were analyzed. This analysis was supplemented with additional measurements with the aid of new and more specific registration forms. A technical-organizational audit of the filling line was also done by means of observation and interviews. Table 17.4 gives more details of each type of loss and a typical example of each.

In order to illustrate the approach to the search for improvements, actions to reduce loss due to type 1 failures are discussed briefly below.

Failures of this sort can be divided into 2 classes: the sudden, completely unpredictable failures where a loss of function occurs immediately, and failures resulting from excessive wear. Before loss of function occurs (e.g. a production breakdown), a loss of function is usually observed in this latter case (production can still continue, but with much loss of quality). Here an adapted inspection policy is obviously indicated. The failures may be due to technical causes or to policy/organizational factors. Typical examples are the short life of certain spares or components in the first case and insufficient training of maintenance workers on the newer and more highly automated machines in the second.

As regards the maintenance approach to this sort of failure, it was clear that too little systematic preventive maintenance was being done. Better preventive maintenance can prevent much of the failures due to wear, and the causes of some of the sudden failures can be eliminated

Table 17.4 Definitions and examples of the six big losses.

Type of Loss	Example
Downtime losses due to	
(1) failures	palletizing plant motor breaks down
(2) setup and adjustment	avoid frequent setups by better planning
Loss of speed due to	
(3) Stoppages	dirty photocells on palletizing machines
(4) reduced speed	use of unadapted pallets (longer processing time for the same number of bottles)
Quality losses due to	
(5) Process defects	With some pallet types, trays get stuck in between depalletizer and unpacker
(6) reduced yields	poor preparation for morning shift by night shift (e.g. problems with the filling taps)

by removing broken components before other parts suffer consequential damage. On the basis of failure analyses (including ABC analyses of failure susceptibility of machine parts and the causes of failure) and interviews, preventive maintenance schemes were proposed. For a number of problem machines, explicit checklists were developed that must be completed during each inspection, so that important aspects cannot be neglected.

For the TPM approach, the scheme to be followed for 'zero breakdowns' was borrowed from the TPM philosophy. It has to be remarked that the operations and maintenance activities were not as strictly separated as is often the case in the West; for certain simple maintenance tasks, TPM relies on the operators. Such a plan is executed in four phases:

1. the stabilization of the MTBF by dealing with visible defects and the avoidance of accelerated wear by adequate cleaning and lubrication

2. prolonging of machine life by eliminating design weaknesses and ensuring that operators and maintenance personnel receive specific training to carry out their tasks properly

3. periodic repair of wear by means of preventive maintenance programs and learning to interpret warning signals that sometimes precede failures

4. analyzing the life cycle of the machine in greater detail by using specialized diagnostic techniques and technical analyses.

Furthermore, classical maintenance efforts are especially aimed at unforeseen failures and visible plant defects; TPM also attends to latent defects. Such defects may be physically concealed (e.g. inspections are too superficial, maintenance-unfriendly design, covered by dirt and dust) or psychologically invisible (e.g. underestimation of the problem, ignoring problem symptoms, forgetting to submit the maintenance request, etc). Maintenance-friendly design of new or modified machines also receives much more attention from TPM than is traditionally the case.

Evidently, a full TPM implementation cannot be realized within the very short term. Thorough planning and sufficient management support (approval of funds for its realization) is not only a step in the right direction, but a prerequisite for success.

11.   CONCLUSIONS

JIT , TPM and TQM are all related to each other, and should be considered as such. When one talks about JIT and maintenance, automatically TPM will be mentioned. But TPM is not just a maintenance concept; it is more like a management philosophy. It involves the total company, from top to bottom. This strength of TPM is at the same time its weakness. One of the problems with TPM is that it represents a major shift in the way an organization approaches the maintenance function. TPM involves the whole organization, because maintenance becomes the responsibility of everybody. Because of this typical characteristic, TPM can be interpreted as an attempt to make production employees do more work and to get by with fewer maintenance employees. Neither maintenance nor production will be very supportive if TPM is viewed in this light, and there will often be resistance from both production and maintenance personnel. One of the hardest aspects of implementing TPM is overcoming this resistance and making the personnel realize that TPM involves benefits for maintenance and production (and even for the quality of the product and the work environment). In this chapter, JIT and maintenance were discussed simultaneously. It became clear that the implementation of TPM is not an easy task that can be done in a short time. TPM is not a philosophy that can stand on his own. It needs the support of (an) other(s) concept(s) and methods, and what is more, it needs the support and the motivation of everyone involved. The whole company.

References

[1] Abdulnour, G., Dudek, R. A. and Smith, M.L. Effect of Maintenance Policies on the Just-in-Time production System. *International Journal of Production Research*, **33** (1995) 565-583.

[2] Anderson, R.T., Neri, L. Reliability Centred Maintenance: Management and Engineering Methods. *Elsevier Applied Sciences*, London, 1990.

[3] Bamber, C. J. Sharp, J. M., Hides, M. T. Factors Affecting Successful Implementation of Total Productive Maintenance: A UK Manufacturing Case Study Perspective. *Journal of Quality in Maintenance Engineering (to a ear)*, 1999.

[4] Blanchard, B.S. Logistics Engineering and Management. *Prentice Hall*, Englewood Cliffs (NJ), 1992.

[5] Bodek, N. TPM for Operators. *Productivity Press Inc.*, 1992.

[6] De Smet, R., Gelders, L., Pintelon, L. Case studies on disturbance registrations for continuous improvement. *Journal of Quality in Maintenance Engineering* **3** (1997) 91-108.

[7] Dekker, R., Wildeman, R. and Van der Duyn Schouten, F. A Review of Multi-Component Maintenance Models with Economic Dependence. *Mathematical Methods of Operations Research*, **45** (1997) 411-435.

[8] Geraerds, W.M.J., Towards a Theory of Maintenance. *The English University Press*, London, (1972) 297-329.

[9] Gits, C.W. On the Maintenance Concept for a Technical System: A Framework for Design. *Ph.D. Thesis*, TUEindhoven (The Netherlands), 1984.

[10] Golhar, D. Y. and Stamm, C. L., The Just-in-Time Philosophy: A Literature Review, International . *Journal of production Research*, **29** (1991) 657-676.

[11] Hartmann, E.H. Successfully Installing TPM in a Non-Japanese Plant: Total Productive Maintenance. *TPM Press Inc.*, 1992.

[12] Helgeson, .P., Birnie, D.P. Assembly Line Balancing Using the Ranked Positional Weight Technique. *Journal of Industrial Engineering*, **12** (1961) 394-398.

[13] Ho , W., Spearman, M. Factory Physics: Foundations of Manufacturing Management. *Irwin*, Chicago, 1996.

[14] Jonsson, P. The Impact of Maintenance on the Production Process - Achieving High Performance. *Lund University*, 1999.

[15] Kelly, A. Maintenance Organizations & Systems: Business-centred Maintenance. *Butterworth-Heinemann*, Oxford, 1997.

[16] Kelly, A. Maintenance Planning and Control. *Butterworths*, London, 1984.

[17] Law, A. M. and Kelton, W. D. Simulation Modeling & Analysis. *McGraw-Hill Inc.*, 1991.

[18] Lawrence, J. J. Use Mathematical Modelling to Give Your TPM Implementation Effort an Extra Boost. *Journal of Quality in Maintenance Engineering*, **5** (1999) 62-69.

[19] Lee, J., Wang, B. Computer-Aided Maintenance: Methodologies and practices. *Kluwer Academic Publishers*, Dordrecht, 1999

[20] Lyonnet, P. Maintenance Planning: Methods and Mathematics. *Chapman & Hall*, London, 1991.

[21] McCall, J. Maintenance Policies for Stochastically Failing Equipment: A Survey. *Management Science*, **11** (1965) 493-524.

[22] Moubray, J. Reliability Centered Maintenance. *Butterworth-Heineman*, Oxford, 1997.

[23] Nahmias, S., Production and Operations Analysis. *Irwin*, Chicago, 1997.

[24] Nakajima, S. TPM Development Program: Implementing Total Productive Maintenance. *Productivity Press Inc.*, 1989.

[25] Nakajima, S. TPM: Challenge to the Improvements of Productivity by Small Group Activities. *Maintenance Management International*, **6** (1986) 73-83.

[26] Nakajima, S. TPM: Introduction to Total Productive Maintenance. *Productivity Press Inc.*, 1988.

[27] Nowlan, F.S., Heap, H.F., Reliability Centred Maintenance. *United Airlines Publication*, San Fransisco, 1978.

[28] Pierskalla, W. and Voelker, J. A Survey of Maintenance Models: The Control and Surveillance of Deteriorating Systems. *Naval Research Logistics Quarterly*, **23** (1976) 353-388.

[29] Pintelon, L., Gelders, L, Van Puyvelde, F. Maintenance Management. *Acco Leuven/Amersfoort*, 1997.

[30] Pintelon, L., Waeyenbergh, G. A Practical Approach to Maintenance Modelling (in Flexible Automation and Intelligent Manufacturing, eds Ashayeri, J., Sullivan, W.G., Ahmad, M.M.). *Begell House*, inc., (1999) 1109-1119.

[31] Taylor, RW. A Linear Programming Model to Manage the Maintenance Backlog. *Omega*, **24** (1996) 217-227.

[32] Valdez-Flores, C. and Feldman, R. A Survey of Preventive Maintenance Models for Stochastically Deteriorating Single-Unit Systems. *Naval Research Logistics*, **36** (1989) 419-446.

[33] Van Dijkhuizen, G. Maintenance Meets Production - On the Ups and Downs of a Repairable System. *Institute for Business Engineering and Technology Application*, Print Partners Ipskamp, Enschede, 1998.

[34] Wireman, T., Total Productive Maintenance : An American Approach. *Industrial Press Inc.*, 1991.

Index